UNIVERSITY OF WISCONSIN-LA CROSSE
OFFICE OF INTERNATIONAL EDUCATION
308 NO. 16TH STREET
ROOM 1209 CENTENNIAL HALL
LA CROSSE, WI 54601

HOW TO FIND
CHEMICAL
INFORMATION

HOW TO FIND CHEMICAL INFORMATION

A GUIDE FOR PRACTICING CHEMISTS, EDUCATORS, AND STUDENTS

SECOND EDITION

Robert E. Maizell

Director, Technology Information Consultants
Science Park
New Haven, Connecticut

Formerly Manager
Business and Scientific Intelligence Centers
Olin Corporation
New Haven, Connecticut

A WILEY-INTERSCIENCE PUBLICATION
JOHN WILEY & SONS
New York • Chichester • Brisbane • Toronto • Singapore

Library of Congress Cataloging in Publication Data:

Maizell, Robert E. (Robert Edward), 1924–
  How to find chemical information.
  "A Wiley-Interscience publication."
  Includes bibliographical references and index.
  1. Chemical literature.   I. Title.
QD8.5.M34   1986   540′.7   86-15687
ISBN 0-471-86767-5

Printed in the United States of America

10 9 8 7 6 5 4

PREFACE

The author hopes that this book will be helpful for desktop reference and daily use and as a text (for self-study and in the classroom). The primary audiences intended are practicing chemists and engineers, educators, and students. Librarians should also find this volume helpful.

A great deal has happened in the field of chemical information since the first edition of this book was published in 1979. In many ways, these years have been the most exciting, productive, and creative in the history of the field.

This second edition achieves major improvements over the first. It is far more than just an updating. The material is treated in much greater depth, and there is a considerable amount of new material, including several new chapters. Whenever appropriate, the historical aspects are covered to provide the perspective for use today.

There are 19 chapters in this edition as compared to 15 in the first edition. New chapters include a separate chapter on the development of *Chemical Abstracts,* including a detailed year-by-year history of important policy and other changes; government information resources and services; analytical chemistry sources from the perspective of chemical structure; and a summary and evaluation of representative major trends in the field of chemical information.

Examples of changes include significantly improved chapters that cover *Chemical Abstracts* and the other essential classics such as *Beilstein* and *Gmelin.* In addition, much more space is devoted to the newer online techniques and sources.

The important chapters on patents, on physical properties, and on safety and environmental data have been significantly improved, as have the other chapters.

The reader should note that, whenever feasible, addresses, and frequently phone numbers, are given for key contacts. These lists of addresses appear at the ends of the appropriate chapters. They are designated in the text by an asterisk at the name or organization for which an address is given.

As in the first edition, this book emphasizes the more enduring of the classical tools of chemical information, the more significant newer tools,

v

and, most importantly, the fundamental methods and principles that the chemist and engineer need to cope with the constantly changing array of chemical information sources. The goal is to help the reader find and use the chemical information that is needed in the best way possible. Because this book emphasizes the more enduring principles, coverage of methods, sources, and tools, although very extensive, is judiciously selective. Because most chemists and engineers are employed in industry at one time or another in their careers, the emphasis is on approaches to meet more practical needs. However, the equally important needs of chemists and engineers in colleges and universities and in government or independent research and development are also addressed.

An emphasis of this book is evaluation and methods of evaluation. This is intended to help the reader evaluate both present and future sources and methods, as well as the actual data. Pros and cons are presented for many of the sources described.

Because strategies are important, these are discussed in detail in several chapters. Many types of sources are discussed in addition to printed and online tools.

As much as possible, an effort has been made to keep this book readable and understandable. Accordingly, unexplained jargon is kept to a minimum.

This book is based primarily on firsthand experience and study. The author knows the excitement of actively participating in some of the more important developments described. Whenever appropriate, field trips were made to the sources described, and there were extensive telephone discussions and correspondence to verify the accuracy of the data presented. The author expresses his thanks for the many courtesies extended to him.

A large number of friends were kind enough to read or discuss pages of the manuscript and to make useful suggestions. These included, in particular, Dr. W. V. Metanomski, manager of editorial development at Chemical Abstracts Service, and incoming (1986) chairman of the American Chemical Society Division of Chemical Information. His opinions were invaluable. Others who should be thanked for playing an important part in the review process are

Mr. Harry Alcock
Mr. Robert Bachman
Mr. Dale B. Baker
Mr. N. J. Barone
Dr. Carlos Bowman
Mr. W. M. Clarke
Dr. Michael Dixon
Dr. Alvin Fein
Dr. Sherman Fivozinsky
Prof. Dr. Ekkehard Fluck
Mr. Donald Hagen

Mr. Richard Hagstrom
Dr. Stephen Heller
Dr. Ernest Hyde
Mr. Michael Jones
Dr. Stuart Kaback
Dr. Henry Kissman
Dr. Leonard Krause
Mrs. Bonnie Lawlor
Dr. David Lide
Dr. Walter Lippert
Prof. Dr. Reiner Luckenbach

Ms. Laura Magistrate
Dr. Stuart Marson
Mr. Dan Meyer
Mrs. B. J. Molino
Mrs. Paula Moses
Mrs. Jane Myers
Dr. Philip Pollick
Dr. Wayne Pouchert
Dr. Peter Rusch

Mrs. Marie Scandone
Mrs. Carole A. Shores
Dr. Robert Stobaugh
Dr. Bruno Vasta
Dr. David Weisgerber
Mr. Jack Westbrook
Dr. Howard White
Ms. Joanne Witiak

The author may have inadvertently omitted some who contributed, and to these he extends his apologies.

Capable secretarial assistance was provided by Sheila Klein. Mr. James Smith, Senior Editor at John Wiley and Sons, was a strong supporter and advisor throughout. To Mona Maizell go sincere thanks for understanding, patience, and enthusiastic moral support and encouragement at every stage.

ROBERT E. MAIZELL

New Haven, Connecticut
December 1986

CONTENTS

HOW TO FIND
CHEMICAL
INFORMATION

1 BASIC CONCEPTS

Why use the literature and other chemical information sources? This question is asked with surprising frequency by both beginning and experienced chemists and engineers. A premise of this book is that use of the literature and other information sources is key to achieving success in research and development and other functions in the chemical profession.

Effective use of information helps avoid duplicating previously reported work. This achieves savings in time (1) and funds and avoids infringing on the proprietary rights of others. In addition, even if there is no directly related previous work, the chemist who makes effective use of information can plan and act on a solid foundation of background data. Further, as a source of ideas or for idea development, chemical information sources and tools are invaluable fountains of inspiration and serendipity. At lease one research study shows that creative chemists use the literature more than less creative chemists, although literature use does not necessarily make a person creative (2).

Published sources of chemical information are not always the first choice. It is often easier and more efficient to ask a colleague in the next laboratory—especially if that person is believed to be both knowledgeable and reliable and has the desired information readily at hand. Alternatively, the chemist may find it quicker to determine needed data in the laboratory than to consult the literature, particularly for common physical properties (such as melting or boiling points) that may be conveniently measured with readily available instruments.

Nevertheless, the chemist who knows how to use chemical information quickly and efficiently, and who has the required energy, imagination, and zeal, will usually have a clear advantage over the person who either lacks these skills and qualities or is too lazy to use them.

Although use of the literature and other chemical information sources is not easy, the often significant benefits make it well worth the additional effort required.

1.1. THE SURROGATE CONCEPT IN LITERATURE USE

Some personal familiarity with—and regular use of—journal and patent literature and other primary information sources is imperative. As these sources continue to grow, so do specialized services (such as computer-based online tools) intended to provide quicker, more effective access to primary literature and other information. These sources are now so important, numerous, and complex, and change so rapidly, that they are often most effectively used by chemists with specialized training in information science. These persons are designated by such job titles as chemical information specialists, information scientists, literature chemists, information chemists, or chemistry librarians.

Recognition is given to this field within the framework of the American Chemical Society by activities of the Division of Chemical Information to which these specialties are a prime interest and by publication of the *Journal of Chemical Information and Computer Sciences*. For laboratory chemists, these persons are surrogates who can be of considerable help in planning or performance of literature studies, as appropriate, and who, in addition, can sometimes advise and guide as to research project strategies.

Although laboratory chemists can find that surrogates are invaluable colleagues, there is no substitute for firsthand familiarity with the literature. It is a mistake to rely too heavily on a surrogate for information gathering needs because only by *firsthand* use of the literature can the chemist keep fully up to date and professionally stimulated.

Another consideration is capability of the surrogate in the field of chemistry under study. Some information chemists are highly skilled not only in computer or information sciences but also in specific fields of chemistry. Other information specialists, especially those who serve a diversified clientele in several disciplines, have only a generalist's knowledge of chemistry and may not have either time or opportunity to develop the highly specialized expertise that many laboratory chemists possess. In addition, the surrogate may concentrate on a specific topic as directed by the laboratory chemist. The result could be that the surrogate may not have the opportunity to recognize peripheral, but significant, data that could be identified by a person who specializes in the field and that could be crucial to success of this or related projects.

The best approach is a full partnership in which the laboratory chemist shares project goals, progress, and information about problems with the information scientist, and in which the two work closely together to achieve desired objectives in the most cost-effective manner. One way to achieve this is to require that qualified information scientists be full members of all major project teams.

In the ideal situation, laboratory chemist and information chemist develop search strategy (see Chapter 3 and 10) together. If computer-based, online searching (see Chapter 10) is to be done, the laboratory and information chemist preferably sit together, and both participate as search results begin to appear so that strategy can be modified if needed. If this is not possible, as may frequently be the case, the two should consult closely as soon as possible to review initial results and plan next steps. This kind of professional partnership can ordinarily yield results that are more than satisfactory.

1.2. THE ADMINISTRATOR AND INFORMATION

If a chemical organization or department is to flourish and prosper to the fullest, it is important that the administrators appreciate the value of literature and other information sources and of the surrogates (literature experts) mentioned in the previous section. Literature and information use and searching are probably the most economical forms of chemical activities and can be the most productive. There is significant infusion of pertinent new information and ideas into the organization, and other important benefits can be achieved, such as cost avoidance. Support by management of the chemical information center or library is vital to the fruitful work of chemists or engineers in any organization, be they professors, industrial or governmental employees, or students.

REFERENCES

1. "ACS Report Rates Information System Efficiency," *Chem. Eng. News* **47**(31), 45–46 (1969). This study by the American Chemical Society Corporation Associates is one of the better studies on the specific value of information tools.
2. R. E. Maizell, "Information Gathering Patterns and Creativity," *American Documentation*, **11**, 9–17 (1960).

2 INFORMATION FLOW AND COMMUNICATION PATTERNS IN CHEMISTRY

The chemist should be aware of the broad framework of information flow. For example, many chemists and other scientists obtain much of their information from colleagues by such informal means as face-to-face conversations, telephone calls, and correspondence. Informal contacts of this type have been called "invisible colleges" or networks. In addition, chemists acquire much information at national, international, regional, and local technical meetings sponsored by professional societies and trade associations.

Primary journals, patents, trade and review journals, trade literature, government reports, books, and printed abstracting and indexing services are all highly significant sources of information. The well-informed chemist is also aware of computerized online databases that permit retrospective searching, current alerting, and other capabilities. All of these modes are described in this book.

The sequence of communications between chemists often follows this approximate sequence of overlapping principal steps:

1. Conducting search of literature and other information sources, including contacts with colleagues, to identify previous work, to build on a foundation of facts, and to avoid replication of what others have already done.

2. Performing laboratory work.

3. Entering data in laboratory notebooks. These are internally used, proprietary documents. Notebooks are especially important in obtaining patent protection. Entries are best made in accordance with requirements of the organization with which the chemist is affiliated, hence exact mode of entry and type of data required is variable. Retention of notebooks is usually governed by special records retention procedures as determined by the chemist's organization.

4. Writing letters for in-house use. These are important to the inner workings of most organizations. Some locations find centralized correspondence centers useful. Others encourage personal filing systems. Letters are not a substitute for reports.

5. Writing research reports and related documents for in-house use. These are proprietary internal documents in most cases. An exception would be, for example, reports based on government sponsored research that, if unclassified, are usually available to the public.

6. Filing patent applications.

7. Informally exchanging results with colleagues in other organizations (face-to-face meetings, telephone calls, correspondence). Excluded from such an exchange would be information that is confidential, proprietary, or potentially injurious to a sound patent position.

8. Presenting results at professional society, trade association, and other technical meetings. Writing and submitting paper to journal editors for possible publication. In many organizations, these actions require clearance by research directors, public relations people, and legal staff to avoid premature release of information.

9. Publication of accepted papers in journals and/or issuing of patents. Some important journals and more recent U.S. patents are now available almost simultaneously in both printed and online forms.

10. Announcement of publications and patents by current awareness and alerting services.

11. Abstracting and indexing of published papers and issued patents by one or more of the major abstracting and indexing services. Concurrent input of abstracting and indexing data for both online and printed availability.

12. Printing of corresponding abstracts and indexes in full-size, microfilm, or microfiche form, and online availability.

13. Use of abstracts and indexes, both printed and online versions.

14. Summarization and evaluation in review articles, monographs, encyclopedias, and information analysis centers. Citations in other literature.

Some of the newer "office of the future" techniques can significantly telescope or otherwise shorten some of these steps by electronic capture of data in initial steps and subsequent reuse in later steps. For example, recording of data by use of a personal computer could permit not only initial capture of data, but also subsequent analysis and use in writing of letters, reports, patent applications, and papers, as well as exchange of these data with colleagues through electronic mail. This capability is readily available today to those chemists who have the required hardware and software.

It is important to emphasize also that communication frequently may not follow the neat series of steps described. Some of the steps may be omitted or bypassed, or may occur in a different order. Informal oral and written communication may occur unpredictably, for example.

3 SEARCH STRATEGY

One of the questions most frequently asked by chemists is how to conduct an effective information search—one that will yield optimum results quickly. Developing an answer to this question usually requires formulating answers to other, more specific questions. This process is helpful in developing an overall approach and philosophy to use of chemical literature.

The first step is to formulate the goal or objective—the information being sought—as precisely as possible to save time and avoid wasted effort. The search should be delimited within those parameters that reflect the chemist's precise interest.

Some questions that the chemist should ask in delimiting and conducting a search include the following examples:

1. Are my goals and objectives clearly defined? Do I know what I want, why I want it, and what I will do with it when I get it?

2. What information do I already have on hand? Have I looked thoroughly at this information to see what leads this might provide and also to avoid going over the same ground?

3. How soon do I need this information? How important is it to me or to my project? Answers to these questions will help determine how much time and effort to put into the search and will reflect priorities.

4. Before looking at the literature, have I talked with colleagues in my own organization or elsewhere who might have some information (or leads to information) at their fingertips?

5. What time period do I need to cover? Must I go all the way back to the beginning of modern chemical literature, or would a search of the last few years suffice? Can I limit the seach to the current year, at least at the outset? See also point **14** on time period as a factor.

6. Do I need to make a search international in scope? Or can I limit it to a specific country, language, organization, or geographic area where I believe most of the important work has been or is being done?

7. Can I limit the search to certain kinds of documents? For example, am I interested in patents only or in nonpatents only?

8. What specific aspect of the field, the chemical, or the chemistry am I interested in? (If I am interested in all aspects of a large field, I could be taking on more than is ordinarily possible.) An examination of the latest Collective Index to *Chemical Abstracts (CA)* will give some ideas of what and how much has been published on the topic.

9. What sources appear to be most fruitful in attacking the problem and obtaining the needed information? Before beginning a search it is important to list sources likely to be most productive. If completeness is an objective, all pertinent sources should be consulted; this also helps avoid bias that can come from consulting just one source. Use of multiple sources has many advantages.

10. How readily available are those sources to me, and do I know how to use them properly? How specific are they to the subject at hand? (I might wish to prefer a smaller, highly specialized source or service in an area of specific interest to a larger, more generalized source or service.)

11. In my overall search plan, have I worked out a strategy wherein those sources that I know will take a long time to respond (such as persons I need to correspond with) will have been contacted at the outset so that I will not be delayed?

12. Before beginning a detailed search of such sources as all of the complete indexes to *CA*, will I check selectively such potential sources of quick answers as, for example,

a. *Kirk–Othmer Encyclopedia of Chemical Technology* (see Section 12.1.A);
b. Desk handbooks (see Section 15.10);
c. *Beilstein* (see Section 12.4.A);
d. *Gmelin* (see Section 12.5.A);
e. Review sources (e.g., review articles) (see Chapter 11);
f. Those indexes of *CA* that are online (1967 to date); and
g. Monographs and treatises on the subject (see Chapter 12).

13. With regard to point 12, are there recently published reviews, encyclopedias, or monographs that contain much or all of the data I need and that may shorten (or even eliminate) need for additional searching at least for the immediate present?

14. Will I start with the latest available source? (This is usually the best procedure rather than use an archaic source.) If I use a source such as *CA*, will I work *backward* in time, that is, start with the latest index and work backward, if needed, to the earliest index? This is the preferred method. After all, why go all the way back to 1907 (the beginning of *CA*) if I have good reason to believe that everthing I need was published during the last five years, for example.

15. Have I developed a subject search plan and a flexible iterative subject search policy? This means, first, systematically developing an array of subjects or key words that I believe are most likely to cover my areas of interest. It then means—based on experience as the search progresses—modifying, as necessary, my original array by adding or deleting search terms and tools. Some of these changes may be a result of a need for greater breadth; others may reflect a need for being more specific. Such tools as the *Index Guide* to *Chemical Abstracts* (see Section 6.5) help in construction of an array.

16. Do I consider all significant sides of the question under investigation? For example, in looking for data on reaction of compound A with compound B, it is important to look under both compounds in the indexes if completeness is desired. Time can often be saved by first looking under the compound with the fewest number of total index entries and then, as necessary, under the other compound. Useful clues can often be obtained from studies that do not work or give opposite results from what is desired.

17. Should I consider attacking the problem through approaches other than subject such as author, originating organization, journal reference or patent number, or molecular formula?

18. Have I asked a surrogate, such as a chemical information scientist or similar person, to review my search strategy and plans before starting in order to get the benefit of that person's expert advice?

19. Do I keep a systematic record of my progress? (Sources looked at, information found, and index entries used should be meticulously recorded in a standard laboratory notebook rather than on loose scraps of paper.)

20. Am I alert to the possibility of serendipity—of unexpectedly finding information pertinent to another of my interests or otherwise important and stimulating but not directly related to my immediate question? Am I prepared to react to and use such information if I find it?

21. In implementing my strategy, do I take into account the personal qualities needed to achieve best results? Examples of such qualities include patience, orderliness, attention to detail, reasonable persistence (see point **22**), and imagination. Ability to cope with some initial frustration is also helpful.

22. Do I know when to stop? Can I recognize the point of diminishing returns? Do I know when I have all that I need? If I cannot find the information in the literature, might it be quicker or more economical to determine the needed information in the laboratory, or adopt some other approach? Is there someone reliable who I can write or telephone to get the data I need?

Several other points relating to strategy may be helpful. In point **22** and in Section 5.9, reference is made to the importance of personal contact. It is important to extend this concept to contacting of equipment and instrument manufacturers and other related vendors. These organizations can have a considerable amount of know-how about both products and processes, and they frequently have no special axe to grind. Accordingly, they may be willing to exchange appropriate information more freely than manufacturers of chemicals or other similar sources who have been, historically, much more secretive.

Another clue relates to use of negative results. One creative use of negative results is to recognize that when something does not work it could be very suitable for another purpose. For example, products that fail as candidates for one use could be proven ideal for the reverse or opposite of that use.

Additional hints on search strategy will be found in Chapter 10, especially Section 10.8. Although Chapter 10 relates to online computer-based searching, some of the same principles apply in other situations.

4 KEEPING UP TO DATE—CURRENT AWARENESS PROGRAMS

4.1. INTRODUCTION

With more than 450,000 articles, patents, books, reports, and other documents of chemical interest published annually, how can chemists and engineers keep up with new information in their fields? The task is difficult even for those who specialize in narrow fields. Fortunately, experience has shown that many persons can develop and implement manageable programs for keeping up to date.

At the outset, the chemist or engineer must accept the likelihood that 100% coverage, even in a relatively limited field of interest, probably is not achievable in any current awareness effort. There will be pertinent material that will be identified too late or not at all.

With this important limitation in mind, the first step is to clearly define current professional interests—areas in which to develop or maintain knowledge. It is better to state these interests in positive terms (to state what you are interested in) than in negative terms (to state what you are not interested in).

This statement is usually called a *profile*. It may include not only subjects but also authors and organizations whose activities the chemist wishes to follow because they are known to be active in specific fields, and citations to key references previously located.

The chemist or engineer planning a current awareness program needs to consider carefully the finite resources available. Other demands on resources as well as priorities also need to be taken into account. One of the most valuable assets any person has is time. Budgeting time spent in keeping up to date is essential to achieving career objectives. About five hours per week is a good rule of thumb for most chemists who have full-time jobs. That figure can easily be doubled or tripled for full-time students.

Chemists embarking on new projects about which they may know little or nothing will require considerably more time than the rule of thumb, especially in the initial phases. A full-fledged literature and patent search should precede all new laboratory projects to avoid duplicating previous efforts by others. Such a search could easily take several days or even weeks, depending on the project. Laboratory chemists and engineers might want to have this kind of study done by a chemical information specialist (or other surrogate.) This should save time and would be an appropriate allocation of skills. There are, however, important advantages to doing at least some of one's own searches (see Chapter 1).

In planning current awareness efforts, the chemist needs to consider available funds. Current awareness programs, particularly those based on computers, can become costly. Several hundred dollars per year is common for a computer-based awareness program for a chemist in industry.

The key, then, is to take stock of available resources, both personal and those of one's organization, and to develop a current awareness effort consistent with those resources and appropriate to needs. With a time-and-dollar budget clearly defined (but flexible enough to meet the unexpected), the chemist can plan with confidence, and will be better able to cope with continuing proliferation of chemistry journals and chemical information services, many of which are highly specialized and expensive. Each chemist needs to make careful decisions as to which publications are to be scanned or to be read cover to cover regularly, and for how many hours a week. The balance of the literature (except for special items identified through *Selective Dissemination of Information* programs as described later in this chapter) will simply need to be largely ignored, difficult though that may be from the perspective of personal discipline. Unless this kind of decision is made, the chemist can spend a lifetime doing nothing else but reading the literature.

Flexibility is, however, important too. Journals and services improve, become more expensive, deteriorate, or collapse, and new ones are introduced. Personal and organization interests, budgets, and policies change. Accordingly, reevaluation every few months is imperative.

4.2. READING CLUBS

Teams of chemists working on large projects, especially in industry, sometimes establish *reading clubs* or similar arrangements. Each person in the team agrees to accept responsibility for covering a given segment or type of literature. The results are then shared with colleagues. This is a delicate arrangement and does not function well unless there is a close working relationship and/or a clear, enforced directive from management.

4.3. PERSONAL PLAN OF ACTION

In developing a plan of action for keeping up to date, the chemist will find it helpful to visualize a triangle as shown in Figure 4.1.

The base of the triangle consists of general *newsy* (but important) material with which most chemists and engineers should maintain some awareness.

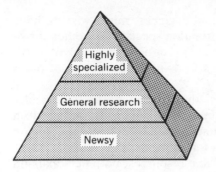

Figure 4.1. The Current Awareness Triangle.

The middle part of the triangle consists of relatively broad areas of chemistry within which there may be some specialization. The apex represents specific materials closely allied to unique individual interests and on which most current awareness time will be spent.

A.   Base of the Triangle: Chemical News Magazines

The general material of most widespread interest is reported in chemical *news* magazines. The chemist who wants to be well informed should systematically and promptly scan news magazines such as:

1. *Chemical & Engineering News* (weekly publication of the American Chemical Society).
2. *Chemical Week* (weekly publication of McGraw-Hill, Inc.).
3. *Chemical Marketing Reporter* (weekly publication of Schnell Publishing Co.).
4. *Chemical Engineering* (biweekly publication of McGraw-Hill, Inc.).
5. *Chemistry and Industry* (twice-monthly publication of the British Society of Chemistry and Industry which contains news, research, and other features).
6. *European Chemical News* (weekly publication of Industrial Press Ltd.).

It is wise for the chemist who wants to keep as current as possible to include more than one weekly chemical news magazine as part of a carefully thought-out reading program.

Basic though it is, *Chemical & Engineering News (C&EN)* alone is not enough for the American chemist or chemical engineer. Either *Chemical Week* or *Chemical Marketing Reporter* should be added. Both of these publications cover chemistry more from the industrial and business perspectives than does *C&EN*. This is important because almost all chemists are affected increasingly by costs, markets, government regulations, and other similar developments. For chemists in the United Kingdom and in other countries in Western Europe, examples of technoeconomic news publications of choice are *European Chemical News* and *Chemistry and Industry*. These important British publications are also useful to chemists in any country who want to keep up with worldwide chemical developments.

Although there is some overlap in coverage among these publications, each has unique features in addition to the differences in perspective noted. For example, contributions to *Chemistry and Industry* include *Highlights from Current Literature*, a monthly commentary on some of the most important and interesting recently published papers in scientific journals; there are also some original research and other papers of considerable value. *Chemical Week* and *Chemical Marketing Reporter* publish buyers' guides to chemical products. *Chemical Engineering* publishes semiannual inventories of new processes and plant construction or expansion plans. The McGraw-Hill publications and *European Chemical News* are distinguished by their speed and excellent coverage of fast-breaking, important stories and their willingness to evaluate the significance and implications of what they believe is important.

As mentioned, most of the chemical news publications are oriented primarily toward the industrial or business aspects of chemistry. But the boundary between this kind of chemistry and the more scientific aspects is blurred. Each kind spills over into and has implications for the other. Chemists who want to further their careers will develop a balanced current awareness program and will keep informed on both the scientific and business aspects, although they will usually concentrate on one or the other.

B. Middle Part of the Triangle: General Research Publications, Title Announcements, and Abstracts Groupings

A balanced reading program should contain two or three general research publications within which there may be some specialization and that can be skimmed selectively. Examples include:

1. *Angewandte Chemie—International Edition in English*—a top-quality monthly publication that includes important, fastbreaking developments, abstracts of important work published elsewhere, and some original research papers.

2. *Science*—almost all branches of science are covered in this weekly, although emphasis is on the biological aspects. Important breakthroughs are frequently announced here first. *Nature* is an equally good choice.

3. One or more of the American Chemical Society's primary journals (see page 196) and/or selected publications of the American Institute of Chemical Engineers such as *Chemical Engineering Progress* or *AIChE Journal*.

4. Other, including commercial and trade magazines. One good such magazine is *High Technology*.

In cases of uncertainty, the chemist or engineer should call on the local chemistry librarian or chemical information specialist for advice in making a selection. Teachers and more experienced colleagues can also help "fine-tune" choices to achieve a manageable portfolio.

The following examples also belong in the middle part of the triangle. Reading time and costs usually permit a selection of only one of these examples:

1. *Current Contents®, Physical, Chemical, and Earth Sciences Edition,* published by the *Institute for Scientific Information®*.* This weekly service reproduces tables of contents of approximately 1000 of the most important research journals (as well as many books) in chemistry and related sciences. The service permits the chemist to quickly scan contents of many journals of potential interest. If a pertinent article is noted, the chemist can consult the full original in a local chemistry library, or can order copies using *The Genuine Article*™ document delivery service. More details about this service are given in Section 5.2. Another edition, *Current Contents®, Engineering and Applied Sciences,* is of interest to engineers and technologists. Disadvantages of *Current Contents* include lack of immediate access to full abstracts and unreliability of many titles as *pointers* to what is pertinent. These same disadvantages apply to most services based primarily on titles. Cost for 1986 was $272 per year.

Chemical Titles is a similar publication issued every two weeks by Chemical Abstracts Service (CAS). Approximately 750 of the world's most important chemically oriented journals are covered. Each article is indexed by author and keyword with a reference to the journal in which it appears. Every article selected for coverage in *Chemical Titles* is subsequently reported in *Chemical Abstracts (CA).* Orientation is more strongly chemical than in any other comparable U.S. publication. Cost to American Chemical Society (ACS) members was $125 per year (1986).

2. *ACS Single Article Announcement (SAA),* published semimonthly, permits the chemist to skim through contents pages of ACS journals. Material of interest can be ordered directly from the ACS* on the form provided. Because of the narrow scope of this service, its utility is limited. Cost to ACS members was $60 per year (1986) (see Section 5.2).

3. One of the Section Groupings (see also Section 6.6) of *Chemical Abstracts* provides an excellent mechanism for keeping informed in broad or specialized areas by appropriate scanning. Available groupings include:

    a. Applied chemistry and chemical engineering.

    b. Biochemistry.

    c. Macromolecular chemistry.

    d. Organic chemistry.

    e. Physical, inorganic, and analytical chemistry.

*Throughout the text an asterisk will refer the reader to the list of addresses at the end of the chapter. Many will include phone numbers as well.

Because many sections are large, it is more feasible to skim quickly (to identify material of special interest) than to read exhaustively. The Section Groupings have unique advantages of offering broader coverage than any other comparable source, full abstracts, and keyword indexes—all at a reasonable cost. Cost to ACS members was $200 per year (1986).

4. Scanning of one or more sections of the Derwent *Chemical Patents Index* is often appropriate (see Section 13.11). This highly regarded tool offers comprehensive, rapid coverage of patents for most important industrial nations. Chemists and engineers working in industry find this tool of special value, since many important technological developments are first reported in patents. Cost is variable, depending on section and other factors, but it is more expensive than the other sources mentioned unless an organization already has a basic subscription and simply desires to order multiple copies.

C.   Apex of the Triangle: Specialized Journals, Selective Dissemination, and Other Approaches

The most vital part of any current awareness program is the apex of the triangle as depicted in Figure 4.1. This part represents the specific areas in which the chemist wants to keep as up to date as possible and that most closely correspond with the statement of interests or profile.

If there is a specialized journal or other publication that zeros in on this profile, this is one approach. Unfortunately, in many fields of chemistry, there are several journals in fields of specific interest, and some of these are often too thick, no matter how specialized. Special consideration should be given to so-called letters journals, such as *Tetrahedron Letters,* in which material is more likely to appear quickly than in the large, omnibus journals. A totally unique journal—the only one of its kind in the field—would be the prime candidate, although such journals are relatively rare.

Also falling within the apex are computer-based tools targeted specifically at current awareness. These tools are based on input of a statement of interests (profile) matched by computer versus the total volume of input of chemical interest received by a computerized information center (see Section 10.9). When a "match" or "hit" occurs, this is automatically recognized by the computer, printed, and mailed to the persons whose profiles are represented. This service is usually referred to as Selective Dissemination of Information (SDI).

The first known computer-based SDI effort in chemistry is the work of Maizell, Rice, and Freeman reported in 1964 (1). These investigators employed *Chemical Titles* as the basis for automatically keeping chemists informed in crucial areas of research.

One SDI approach uses so-called *standard interest profiles* or *macroprofiles*. These are profiles selected by an outside organization to meet the common interests of a broad group of chemists.

A good example is *ASCATOPICS®*, which is published and sold by the

Aerosols
Alkaloids—Isolation. Characterization, Synthesis
Catalysis
Chemical Reaction Mechanisms & Kinetics
Chemical Hazards—Health and Safety
Chromatography—Affinity
Chromatography—Gas, GLC, GSC
Chromatography—Gel Permeation & Ion Exchange
Chromatography—High Speed Liquid
Chromatography—Liquid
Chromatography—TLC & Paper
Colloid & Surface Chemistry
Computers in Chemistry
Detection & Identification of Narcotic Drugs & Poisons
Electrochemistry
Fluorescence, Luminescence, Phosphorescence
Food Additives
Liquid & Molecular Crystals
Mass Spectrometry
Molybdenum Chemistry
Optical Activity—Resolution, Racemization, ORD, CD
Organosilicon Compounds
Peptide Synthesis
Photochemistry
Polymers—Fibers & Films
Polymers—Plastics & Elastomers
Polymers—Preparation of New Monomers & Polymers
Polymers—Science & Technology
Selenium & Tellurium Chemistry
Specific Ion Electrodes
Spectroscopic Techniques—Atomic Absorption, Mossbauer,
   Gamma-ray, X-ray
Spectroscopic Techniques—IR, UV, Laser, Raman
Spectroscopic Techniques—EPR, ESR, NMR, PMR
Transition Metals Chemistry (Ir, Os, Pd, Pt, Re, Rh, Ru & Tc)

Figure 4.2. *ASCATOPICS®* in Chemistry. Reproduced with permission from *ASCATOPICS®*. Copyright owned by the Institute for Scientific Information®, Philadelphia, PA, U.S.A.

Institute for Scientific Information®, Philadelphia* (see also Section 8.3). Over 350 topics in the sciences and social sciences are available. Topics of special interest to chemists are shown in Figure 4.2. The list, which is revised and updated annually, is representative of what was available in 1986.

Entries in the report look like the example shown in Figure 4.3. Results are mailed to subscribers at a cost of $175 per year (1986 price for users in United States, Canada, and Mexico; cost is higher elsewhere). Users receive weekly computer-based reports containing complete bibliographic information for new items published on topics they select.

Principal advantages include speed of issue and availability of the document delivery service known as *The Genuine Article*™ [previously *OATS®*, (*Original Article Text Service*)] for convenience in getting copies of full original documents (see Section 5.2).

ASCATOPICS® is, however, limited to publications covered by the *Institute for Scientific Information®*. Patents, which are especially important to industrial research workers, are not included. The chemist is limited to information supplied in the computer printout; there are no abstracts in these reports.

An extremely important, valuable, and popular current awareness tool is *CA SELECTS*, which is published by CAS.* This is a further development and enhancement of the now discontinued macroprofile service initiated by the United Kingdom Chemical Information Service in 1971. *CA SELECTS* provides complete *CA* abstracts and bibliographic citations. One hundred sixty four fields are now (1986) covered in this service, which was introduced in 1976. Recent (1986) prices for each topic were $120 per year for 26 issues; rates are lower for *CA* subscribers and quantity orders. Advan-

```
ASCATOPIC                    (S0687)
PHOTOCHEMISTRY


                                            S0687


REPORT FOR   03 AUG 84                      PAGE   1
--------------------------------------------------------------
PHOTO//   PHOTODEGRADATION CF THE FUNGICIDE TOLCLOFOS-METHYL
DEGR//    IN WATER AND ON SOIL SURFACE
            MIKAMI N    IWANISHI K  YAMADA H    MIYAMOTO J
                                                  14 REFS
              J PESTIC S    9(2): 215-222,1984
            -----> CHECK TO ORDER TEAR SHEETS ----->( ) #SY313
          N MIKAMI, SUMITOMO CHEM CO LTD,TAKARAZUKA RES
          CTR,BIOCHEM & TOXICOL LAB,
          TAKARAZUKA, HYOGO 665, JAPAN
--------------------------------------------------------------
PHOTO//   PHOTOLYSIS AND HYDROLYSIS OF THE FUNGICIDE
LYS//     PROCYMIDONE IN WATER
            MIKAMI N    IWANISHI K  YAMADA H    MIYAMOTO J
                                                  10 REFS
              J PESTIC S    9(2): 223-228,1984
            -----> CHECK TO ORDER TEAR SHEETS ----->( ) #SY313
          N MIKAMI, SUMITOMO CHEM CO LTD,TAKARAZUKA RES
          CTR,BIOCHEM & TOXICOL LAB,
          TAKARAZUKA, HYOGO 665, JAPAN
--------------------------------------------------------------
LIGHT     THE LIGHT SIDE CF HORSERADISH-PEROXIDASE
CHEM//    HISTOCHEMISTRY (LETTER)
OXID//      MARTIN TL   MUFSON EJ    MESULAM MM       3 REFS
              J HIST CYTO  32(7): 793,JUL 1984
            -----> CHECK TO ORDER TEAR SHEETS ----->( ) #SY249
          TL MARTIN, HARVARD UNIV,DEPT NEUROL,
          BOSTON, MA 02115
--------------------------------------------------------------
          SURFACE CONDUCTIVITY OF CHLOROPHYLL-A IN VARIOUS
          AMBIENT GASES
            DESPAX B    PERES P     HURAUX C        19 REFS
              IEEE EL INS  19(3): 257-264,JUN 1984
            THESE ITEMS IN THIS PROFILE WERE CITED:
TRIBUTSCH H       PHOTOCHEM FHOTOBIOL   14    95 71
REUCROFT PJ       PHOTOCHEM PHOTOBIOL   10    79 69
SIMPSON WH        PHOTOCHEM PHCTOBIOL   11   319 70
            -----> CHECK TO ORDER TEAR SHEETS ----->( ) #SY286
          B DESPAX, UNIV TOULOUSE 3,CNRS,GENIE ELECT LAB,
          F-31062 TOULOUSE, FRANCE
--------------------------------------------------------------
LIGHT     (GE) LIGHT-ABSORPTION OF SILVER(I)-OXIDES
OXID//      FRIEBEL C    JANSEN M                    18 REFS
              Z NATURFO E  39(6): 739-743,JUN 1984
            -----> CHECK TO ORDER TEAR SHEETS ----->( ) #SY354
          M JANSEN, UNIV HANOVER,INST ANORGAN CHEM,
          D-3000 HANOVER, FED REP GER
```

Figure 4.3. Typical *ASCATOPICS®* Output. Reproduced with permission from *ASCATOPICS.®* Copyright owned by the Institute for Scientific Information®, Philadelphia, PA, U.S.A.

ACTIVATED NITRILES IN HETEROCYCLIC SYNTHESIS - THE
REACTION OF SUBSTITUTED CINNAMONITRILES WITH 2-
FUNCTIONALLY SUBSTITUTED METHYL-2-THIAZOLIN-4-ONE
DERIVATIVES
     SADEK KU   HAFEZ EAA   MOURAD AEE   ELNAGDI MH
                                           13 REFS
      Z NATURFO B   39(6): 824-828,JUN 1984
THESE ITEMS IN THIS PROFILE WERE CITED:

MCELROY WD   PHOTOCHEM PHOTOBIOL   10   153 69
MCELROY WD   PHOTOCHEM PHOTOBIOL   10   153 69
MCELROY WD   PHOTOCHEM PHOTOBIOL   10   153 69
-----> CHECK TO ORDER TEAR SHEETS ----->( ) #SY354
MH ELNAGDI, UNIV CAIRO,FAC SCI,DEPT CHEM,
CAIRO, EGYPT

INFRARED/   DIFFERENTIAL THERMAL-ANALYSIS AND INFRARED
LYS//   INVESTIGATIONS ON SOIL HYDROPHOBIC SUBSTANCES
     GIOVANNI.G LUCCHESI S   16 REFS
      SOIL SCI   137(6): 457-463,JUN 1984
-----> CHECK TO ORDER TEAR SHEETS ----->( ) #SY431
G GIOVANNINI, CNR,INST SOIL CHEM,
I-56100 PISA, ITALY

INHIBITORS OF DNA-SYNTHESIS (APHIDICOLIN AND
ARAC/HU) PREVENT THE RECOVERY OF RNA-SYNTHESIS
AFTER UV-IRRADIATION
     MAYNE LV   20 REFS
      MUTAT RES   131(5-6): 187-191,MAY-JUN 1984
THESE ITEMS IN THIS PROFILE WERE CITED:

KANTOR GJ   PHOTOCHEM PHOTOBIOL   25   483 77
KANTOR GJ   PHOTOCHEM PHOTOBIOL   25   483 77
KANTOR GJ   PHOTOCHEM PHOTOBIOL   25   483 77
-----> CHECK TO ORDER TEAR SHEETS ----->( ) #SY478
LV MAYNE, UNIV SUSSEX,SCH BIOL SCI,CELL & MOLEC
BIOL LAB, BRIGHTON BN1 9QG, E SUSSEX, ENGLAND

PHOTO//   CRITIQUE ON THE ANALYSIS OF PHOTOREPAIR IN CHICK-
LYS//   EMBRYO FIBROBLASTS
     BRONK BV   VANDEMER.WP   8 REFS
      MUTAT RES   131(5-6): 205-207,MAY-JUN 1984
THESE ITEMS IN THIS PROFILE WERE CITED:

CHIANG T   PHOTOCHEM PHOTOBIOL   30   525 79
CHIANG T   PHOTOCHEM PHOTOBIOL   30   525 79
CHIANG T   PHOTOCHEM PHOTOBIOL   30   525 79
-----> CHECK TO ORDER TEAR SHEETS ----->( ) #SY478
BV BRONK, CLEMSON UNIV,DEPT PHYS,
CLEMSON, SC 29631

SENSITIVITY OF BLOOM SYNDROME FIBROBLASTS TO
MITOMYCIN-C
     HOOK GJ   KWOK E   HEDDLE JA   37 REFS
      MUTAT RES   131(5-6): 223-230,MAY-JUN 1984
THESE ITEMS IN THIS PROFILE WERE CITED:

SMITH PJ   PHOTOCHEM PHOTOBIOL   36   333 82
SMITH PJ   PHOTOCHEM PHOTOBIOL   36   333 82
SMITH PJ   PHOTOCHEM PHOTOBIOL   36   333 82
-----> CHECK TO ORDER TEAR SHEETS ----->( ) #SY478
GJ HOOK, LUDWIG INST CANC RES,9 EARL ST,
TORONTO M4Y 1M4, ONTARIO, CANADA

Figure 4.3. Continued.

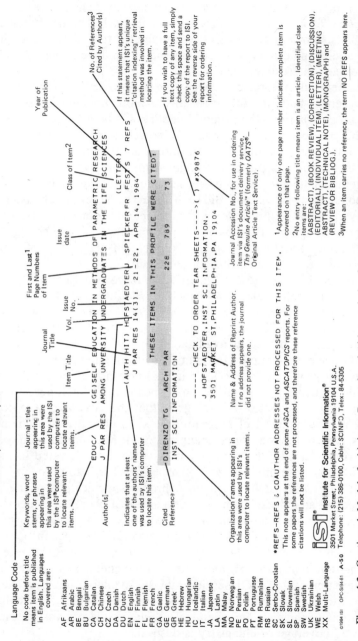

Figure 4.3. Continued. Model of a hit item. Reproduced with permission from *ASCATOPICS*®. Copyright owned by the Institute for Scientific Information®, Philadelphia, PA, U.S.A.

tages include excellent abstracts, comprehensiveness, and specific focus. The service is not as fast as *ASCATOPICS®*, although *CA SELECTS* is as current as the full printed *Chemical Abstracts*. The topics introduced by *CA SELECTS* in each year is shown in Table 4.1. The list of topics covered in 1986 is shown in Table 4.2. It is anticipated that new topics will continue to be added and others dropped, depending on developments in chemistry and changing interests of chemists.

Chemical Abstracts Service has announced a new series of current-awareness bulletins for those in the biotechnology industry. Called *CAS BioTech Updates,* the printed bulletins will be issued by subscription every other week starting in January, 1986. *CAS BioTech Updates* includes abstracts from *CA* and business information related to biotechnology processes, people in the industry, government activities, production, and pricing internationally. Each issue is divided into four sections: patents, papers, books and reviews, and BioTech Industry Notes. Each issue also contains keyword and author indexes. Four separate topics are offered in the series by subscription: Biosensors, Environmental Biotechnology, Genetic Engineering, and Pharmaceutical Applications. The cost of a one-year subscription to a topic is $195 for 26 issues.

A series related to *CA SELECTS* is *BIOSIS/CAS SELECTS*. Sources for this biologically related tool include *Biological Abstracts, Biological Abstracts/RRM (Reports, Reviews, Meetings),* and *CA*. Topics covered in 1986 are shown in Table 4.3. The price for each topic in 1986 was $105 per year, with quantity discounts available.

Other organizations, too numerous to mention here, also offer standard interest profile service. The principal advantages of the macroprofile concept are usually given as low cost, speed, and sometimes, breadth of coverage. The principal disadvantage is that the coverage may not focus sufficiently on material of specific interest to an individual chemist. This can result in too much nonpertinent material. Macroprofiles may also lack the flexibility needed by some chemists who are interested in more than one field, or whose interests vary sharply from time to time.

Customized SDI, based on a profile specifically designed to meet the needs of an individual chemist or a group of chemists, is usually considered more desirable than the macroprofile approach. Customizing costs more, but it is more likely to yield pertinent material, and it can be varied to meet changing interests. This service is available from CAS, the *Institute for Scientific Information®,* and others. If the chemist's organization has a large computer center, SDI service may be available locally. The chemist can elect to have a profile run against a variety of sources or so-called databases. For many chemists and chemical engineers, *CA* is a very attractive choice if broad coverage of chemistry is desired.

CAS approaches to customized SDI are known as "Customer Defined Services for SDI (Selective Dissemination of Information)." These include:

Corporate Updates, for which search profiles are prepared by CAS staff experts to suit the interests clients describe; Individual Search Service (ISS), for which clients write their own profiles; and Business Updates, for which business related output is drawn from CAS's *Chemical Industry Notes*.

Corporate Updates include *CA* abstracts and, where applicable, structure diagrams. Abstracts may be grouped under subject or by CAS Registry Number, whichever is preferred. Printouts from the ISS service include *CA* abstracts and keyword index entries. CAS Registry Numbers or Volume Index entries may be optionally included.

The *Institute for Scientific Information® (ISI®)* provides another form of weekly current alerting tailored to meet specific needs of individual chemists or groups of chemists. This important service is designated as *ASCA® (Automatic Subject Citation Alert®)* and its profiles employ a number of search terms and techniques to retrieve relevant information. These include: title words and word stems, names of source journals or cited authors, source journal or cited journal titles, frequently cited references or books, and author's organization. Citation searching, a special feature of *ASCA®*, often locates items that might otherwise be missed (see also Section 8.3). Output is based on approximately 5000 journals and includes authors, address of first author or reprint author, article title, complete journal citation, language codes, number of references, and the order number for *The Genuine Article™, ISI®'s* document delivery service (see Section 5.2). Items retrieved may include articles, book reviews, letters, editorials, meeting abstracts, and technical notes. Costs vary depending on scope as specified by the user. *ASCA®* is particularly distinguished by its speed of coverage and unique citation access. It is, however, not complete (e.g., patents are not included at this time), and abstracts are not given as a rule.

Another *ISI®* service is *Custom Contents®*. This permits users to specify the journals for which they would like to receive tables of contents on a current and regular basis. Minimum cost for this new custom service is $100 per year.

In addition to the types of programs and services just described, chemists can now initiate their own automatic SDI programs through most of the major online databases described in Chapter 10. In this approach, search strategies (profiles) are first created and then, after any testing, can be saved. Whenever predesignated files are updated, these are automatically run against the saved profiles, and results are mailed to the user-defined address. Success of this approach depends on the skill with which databases are selected and with which strategies are created, implemented, and updated. A less automatic option is to go online at regular intervals (as selected by the user) and then scan across any updated files with the strategy selected. This affords a greater degree of control and flexibility than the fully automatic approach in that changes in profile scope may usually be made more easily.

TABLE 4.1. CA SELECTS Introduction by Year Beginning in 1976

Year Started	Topic
1976	Forensic Chemistry
	High Speed Liquid Chromatography (Title changed to High Performance Liquid Chromatography in 1980)
	Mass Spectrometry
	Organosilicon Chemistry
	Photochemistry
	Psychobiochemistry
1977	Catalysis (Applied & Physical Aspects)
	Catalysis (Organic Reactions)
	Chemical Hazards (Title changed to Chemical Hazards, Health, & Safety in 1979)
	Coal Science & Process Chemistry
	Corrosion
	Electrochemical Reaction
	Electron & Auger Spectroscopy
	Electron Spin Resonance (Chemical Aspects)
	Gas Chromatography
	Gel Permeation Chromatography
	Ion Exchange
	Nuclear Magnetic Resonance (Chemical Aspects) (Divided into Carbon & Heteroatom NMR and Proton Magnetic Resonance in 1979)
	Paper & Thin-Layer Chromatography
	Radiation Chemistry
	Solvent Extraction
	Surface Chemistry (Physiocochemical Aspects)
1978	Analytical Electrochemistry
	Anti-Inflammatory Agents & Arthritis
	Antitumor Agents
	Atherosclerosis & Heart Disease
	Carcinogens, Mutagens, & Teratogens
	Colloids (Applied Aspects)
	Colloids (Physicochemical Aspects)
	Computers in Chemistry
	Engine Exhaust (Discontinued in 1981)
	Environmental Pollution
	Flavors & Fragrances
	Fungicides
	Gaseous Waste Treatment
	Herbicides
	Infrared Spectroscopy (Divided in 1979)
	Insecticides
	beta-Lactam Antibiotics
	Liquid Crystals
	Liquid Waste Treatment

TABLE 4.1. Continued

Year Started	Topic
	Metallo Enzymes & Metallo Coenzymes
	Moessbauer Spectroscopy (Discontinued in 1981)
	New Books in Chemistry
	Organofluorine Chemistry
	Organotransition Metal Complexes
	Photobiochemistry
	Pollution Monitoring
	Prostaglandins
	Raman Spectroscopy
	Recovery & Recycling of Wastes
	Silver Chemistry
	Solar Energy
	Solid & Radioactive Waste Treatment
	Steroids (Biochemical Aspects)
	Steroids (Chemical Aspects)
	Trace Element Analysis
1979	Adhesives
	Animal Longevity & Aging
	Batteries & Fuel Cells
	Biogenic Amines & the Nervous System
	Biological Information Transfer
	Blood Coagulation
	Carbon & Heteroatom NMR
	Colloids (Macromolecular Aspects)
	Crystal Growth
	Detergents, Soaps, & Surfactants
	Drugs & Cosmetic Toxicity
	Electrodeposition
	Energy Reviews & Books
	Flammability
	Food Toxicity
	Heat-Resistant & Ablative Polymers
	Infrared Spectroscopy (Organic Aspects)
	Infrared Spectroscopy (Physicochemical Aspects)
	Organic Reaction Mechanisms
	Organophosphorus Chemistry
	Proton Magnetic Resonance
	Substituent Effects and Linear Free Energy Relationships (Discontinued in 1981)
	Thermochemistry
	X-Ray Analysis & Spectroscopy
	Zeolites
1980	Atomic Spectroscopy
	Chemical Instrumentation

TABLE 4.1. Continued

Year Started	Topic
	Chemical Processing Apparatus
	Cosmochemistry (Discontinued in 1982)
	Electrophoresis
	Emulsion Polymerization
	Fluidized Solids Technology
	Fuel & Lubricant Additives
	Inorganic Fluorine Chemistry (Discontinued in 1982)
	Inorganic & Organometallic Reaction Mechanisms
	Ion-Containing Polymers
	Laser Applications
	Lasers & Masers (Discontinued in 1982)
	Macrocyclic Antibiotics (Discontinued in 1982)
	Nuclear Reactor Fuels (Discontinued in 1982)
	Optical & Photosensitive Materials
	Optimization of Organic Reactions
	Organic Stereochemistry
	Organoboron Chemistry & Boranes (Discontinued in 1982)
	Organotin Chemistry
	Plastic Films
	Polymer Morphology
	Porphyrins
	Solvent Effects (Discontinued in 1982)
	Surface Analysis
	Synfuels
	Synthetic Macrocyclic Compounds
	Thermal Analysis
	Ultrafiltration
	Ultraviolet & Visible Spectroscopy
1981	Amino Acids, Peptides, & Proteins
	Antioxidants
	Coatings, Inks, & Related Products
	Cosmetic Chemicals
	Distillation Technology
	Epoxy Resins
	Fats & Oils
	Inorganic Analytical Chemistry
	Inorganic Chemicals & Reactions
	Novel Sulfur Heterocycles
	Organic Analytical Chemistry
	Plastics Fabrication & Uses
	Plastics Manufacture & Processing
	Radiation Damage (Material Aspects) (Discontinued in 1982)
	Subatomic Particles (Discontinued in 1982)
	Synthetic High Polymers

24

TABLE 4.1. Continued

Year Started	Topic
1983	Block & Graft Polymers
	Catalyst Regeneration
	Chelating Agents
	Colorants & Dyes
	Electrochemical Organic Synthesis
	Emulsifiers & Demulsifiers
	Enhanced Petroleum Recovery
	Fiber-Reinforced Plastics
	Initiation of Polymerization
	Lubricants, Greases, & Lubrication
	New Plastics
	Novel Natural Products
	Novel Polymers from Patents
	Paint Additives
	Paper Additives
	Plastics Additives
	Polyesters
	Polymer Blends
	Polymer Degradation
1984	Automated Chemical Analysis
	Carbohydrates (Chemical Aspects)
	Controlled Release Technology
	Crosslinking Reactions
	Fermentation Chemicals
	Natural Products Synthesis
	Polyurethanes
	Water Treatment
1985	Corrosion-Inhibiting Coatings
	Drilling Muds
	Novel Pesticides & Herbicides
	Photosensitive Polymers
	Plasma & Reactive Ion Etching
	Radiation Curing
	Selenium & Tellurium Chemistry
	Water-Based Coatings

TABLE 4.2.   CA SELECTS[a] (1986)

Acid rain & acid air[b]
Adhesives
Amino acids, peptides, & proteins
Analytical electrochemistry
Animal longevity & aging
Anti-inflammatory agents & arthritis
Antioxidants
Antitumor agents
Atherosclerosis & heart disease
Atomic spectroscopy
Automated chemical analysis
Batteries & fuel cells
Beta-Lactam antibiotics
Biogenic amines & the nervous system
Biological information transfer
Block & graft polymers
Blood coagulation
Carbohydrates (chemical aspects)
Carbon & heteroatom NMR
Carcinogens, mutagens, & teratogens
Catalysis (applied & physical aspects)
Catalysis (organic reactions)
Catalyst regeneration
Ceramic materials (patents)[b]
Chelating agents
Chemical hazards, health, & safety
Chemical instrumentation
Chemical processing apparatus
Chemical vapor deposition[b] (CVD)
Coal science & process chemistry
Coatings, inks, & related products
Colloids (applied aspects)
Colloids (macromolecular aspects)
Colloids (physicochemical aspects)
Color science[b]
Colorants & dyes
Computers in chemistry
Conductive polymers[b]
Controlled release technology
Corrosion
Corrosion-inhibiting coatings
Cosmetic chemicals
Crosslinking reactions
Crystal growth
Detergents, soaps, & surfactants
Distillation technology
Drilling muds

Drug & cosmetic toxicology
Electrically conductive organics
Electrochemical organic synthesis
Electrochemical reactions
Electrodeposition
Electron & Auger spectroscopy
Electron spin resonance (chemical aspects)
Electronic chemicals & materials[b]
Electrophoresis
Emulsifiers & demulsifiers
Emulsion polymerization
Energy reviews & books
Enhanced petroleum recovery
Environmental pollution
Epoxy resins
Fats & oils
Fermentation chemicals
Fiber-reinforced plastics
Flammability
Flavors & fragrances
Fluidized solids technology
Food toxicity
Forensic chemistry
Free radicals[b]
Fuel & lubricant additives
Fungicides
Gas chromatography
Gaseous waste treatment
Gel permeation chromatography
Heat-resistant & ablative polymers
Herbicides
High performance liquid chromatography
Infrared spectroscopy (organic aspects)
Infrared spectroscopy (physicochemical aspects)
Initiation of polymerization
Inorganic analytical chemistry
Inorganic chemicals & reactions
Inorganic & organometallic reaction mechanisms
Insecticides
Ion-containing polymers
Ion exchange
Laser applications
Laser-induced chemical reactions[b]

TABLE 4.2. Continued

Liquid crystals	Polyesters
Liquid waste treatment	Polymer blends
Lubricants, greases, & lubrication	Polymer degradation
Mass spectrometry	Polymer morphology
Metallo enzymes & metallo coenzymes	Polymerization kinetics & process control
Natural product synthesis	
New books in chemistry	Polyurethanes
New plastics	Porphyrins
Novel natural products	Prostaglandins
Novel pesticides & herbicides	Proton magnetic resonance
Novel polymers from patents	Psychobiochemistry
Novel sulfur heterocycles	Radiation chemistry
Optical & photosensitive materials	Radiation curing
Optimization of organic reactions	Raman spectroscopy
Organic analytical chemistry	Recovery & recycling of wastes
Organic optical materials[b]	Selenium & tellurium chemistry
Organic reaction mechanisms	Silver chemistry
Organic stereochemistry	Solar energy
Organofluorine chemistry	Solid & radioactive waste treatment
Organophosphorus chemistry	Solvent extraction
Organosilicon chemistry	Spectrochemical analysis[b]
Organotin chemistry	Steroids (biochemical aspects)
Organo-transition metal complexes	Steroids (chemical aspects)
Paint additives	Surface analysis
Paper additives	Surface chemistry (physicochemical aspects)
Paper & thin-layer chromatography	
Phase transfer catalysis[b]	Synfuels
Photobiochemistry	Synthetic high polymers
Photochemical organic synthesis[b]	Synthetic macrocyclic compounds
Photochemistry	Thermal analysis
Photoresists[b]	Thermochemistry
Photosensitive polymers	Trace element analysis
Plasma & reactive ion etching	Ultrafiltration
Plastic films	Ultraviolet & visible spectroscopy
Plastic additives	Water-based coatings
Plastics fabrication & uses	Water treatment
Plastics manufacture & processing	X-Ray analysis & spectroscopy
Pollution monitoring	Zeolites

[a] *CA SELECTS* is a series of 164 current-awareness bulletins in 15 subject areas published every 2 weeks, 26 times per year. Each *CA SELECTS* topic is a separate printed publication, designed to bring researchers only the scientific and technical information they need. Each topic includes *CA* abstracts and associated bibliographic information, selected by computer from the CAS database according to a precise, special interest profile.

[b] New in 1986.

TABLE 4.3. Topics Covered by BIOSIS/CAS SELECTS

Allergy & Antiallergy	Immunochemical Methods
Bacterial & Viral Genetics	Interferon
Biochemistry of Fermented Foods	Leukotrienes and Slow-Reacting Sub-
Cancer & Nutrition	stances
Cancer Immunology	Lymphokines
DNA Replication	Mammalian Birth Defects
Endorphins	Metals in the Environment
Enzme Methods	Monoclonal Antibodies
Food & Drug Legislation	Neuroreceptors
Genetic Manipulation in Plants	Pediatric Pharmacology
Geriatric Pharmacology	Peptide & Protein Sequences
Hormones & Gene Expression	Plant Genetics
Hormones & Hormone Receptor In-	Vitamins
teractions	

4.4.   MEETINGS

As part of any current awareness program, the chemist needs to allocate time for attending professional society and trade association meetings. These meetings offer the opportunity to listen to and meet persons at the leading edge of research and development and obtain information long before it is published. Further, the opportunity to meet informally with colleagues in the same field of interest is invaluable. Many chemists and engineers prefer to attend more specialized smaller meetings such as Gordon Research Conferences or special symposia and seminars devoted to relatively narrow fields.

In contrast, huge *omnibus* national meetings, which attempt to cover all aspects of chemistry or chemical engineering, can be too broad in scope, the quantity of papers can be overwhelming, and the number of persons attending can dilute opportunities for personal contacts as well as for good question-and-answer sessions. Attending large national meetings can be useful only if there is careful preplanning, for example, by making plans to attend several symposia on specialized topics of clear interest to the individual.

How many meetings per year can or should a chemist or engineer attend?

Like reading journals, chemists could easily spend a lifetime attending meetings. Many organizations will send their key people to an average of one meeting each year, but this can vary widely, depending on meeting location, duration, and fees. The chemist should select the meetings to be attended as far in advance as feasible—at least several months.

A.   Getting Copies of Meeting Papers

Because no one can attend all meetings, what can chemists and engineers do when they see announcements of papers of interest, but when no one from their organization can attend?

Some actions to get copies of meeting papers can be attempted, but results are often uncertain. Reliance on ultimately seeing a published version of a meeting paper is risky at best. Only about half of all meeting papers are published according to some estimates. Those published often undergo extensive revision after oral presentation and may therefore vary significantly in content from the original version. The time lag between oral presentation and formal publication can be substantial—from several months to well over a year.

American Chemical Society (ACS) meeting papers and those of the American Institute of Chemical Engineers (AIChE) are best accessed initially through the collections of abstracts of meeting papers prior to national and some regional meetings. In the case of ACS, an index to the abstracts, based on words in the titles, speeds up identification of pertinent material.

Because meeting papers are not usually subject to rigorous review, and for other reasons, abstracts of them leave much to be desired. Chemists interested in learning more about the contents of a meeting paper can write the author and request a preprint. These are readily available, however, only for an estimated 50% or less of all meeting papers. Some suggestions on contacting authors for copies of their papers can be found in Section 5.3.

Obtaining meeting papers for several divisions of ACS is simplified because they publish preprints before or shortly after national meetings. At this writing, these divisions include:

1. Chemical Marketing and Economics.
2. Environmental Chemistry.
3. Fuel Chemistry.
4. Polymeric Materials: Science and Engineering, Inc. (formerly Organic Coatings and Plastics Chemistry).
5. Petroleum Chemistry, Inc.
6. Polymer Chemistry, Inc.

In addition, meeting paper copy services are maintained by the Agricultural and Food Chemistry, and Rubber Divisions. The Rubber Division, in fact, operates its own library service under the direction of Mrs. Ruth C. Murray.*

For AIChE papers, copies are often available directly from AIChE for several weeks after the meeting. Thereafter, photocopies are usually available from the Engineering Societies Library.*

As an aid to deciding which meetings to monitor and possibly attend, the chemical news magazines and professional society news publications publish lists on a regular basis. There are also publications that list future meetings of technical, scientific, medical, and management organizations and universities, for example, *Scientific Meetings*.* Another source is the *World Meetings* series compiled and marketed by Macmillan.*

Conference Papers Index is an index of over 500,000 papers presented at scientific and technical meetings worldwide from 1973 to the present. It is available in both printed form and online. The publisher is Cambridge Scientific Abstracts.*

Meeting paper preprints that are not published elsewhere are among the many kinds of documents covered by *Chemical Abstracts*.

Index to Scientific and Technical Proceedings and Books, published by the *Institute for Scientific Information®*,* can help chemists and others locate conference papers published in book or journal form. This is available online (1978 to date), and there is a printed version published monthly.

The British Library Lending Division publishes *Index to Conference Proceedings Received*, which is available online through *BLAISE–LINE*.*

During recent years the ACS has conducted experiments on alternative ways of providing chemists with access to meeting papers. The principal approach considered is known as *teleconferencing* or *audiographics*. In the ACS approach, chemists at sites distant from national meetings are provided with copies of slides in advance. At the time of delivery of the paper, chemists at these remote sites can hear the speakers and can ask questions, employing modified telephone lines. Video access is not provided, except for the previously mentioned slides, which can be displayed by local projectionists at the remote sites. The ACS has tried this on a limited basis with a few selected symposia. The experiment is of interest because it offers the potential of important savings in travel time and costs.

The tools just mentioned, and others that either exist now or will almost inevitably be developed in the future, are an encouraging indication that chemists should be able to get a much better "handle" on meeting and conference papers than ever before.

4.5. RESEARCH IN PROGRESS; DISSERTATIONS

Identification of research in progress can be used by chemists and engineers to:

- Avoid unwarranted duplication of research effort and expenditure.
- Locate possible sources of support for research on a specific topic.
- Identify leads to the published literature or participants for symposia.
- Obtain information to support grant or contract proposals.
- Stimulate new ideas for research planning or innovations in experimental techniques.
- Acquire source data for technological forecasting and development.
- Survey broad areas of research to identify trends and patterns or reveal gaps in overall efforts.
- Learn about current work of a specific research investigator, organization, or organizational unit.

Information about research underway is frequently difficult to obtain, since most investigators are understandably reluctant to allow premature disclosure of results. There are, however, sources that can help identify the scope of some of the work in progress.

Until a few years ago, the leading source of this kind in the United States had been the Smithsonian Science Information Exchange, Inc. (SSIE), which covered primarily federally sponsored research in progress. However, after years of highly successful operation under the direction of David Hersey, SSIE was discontinued in 1981 as part of budget-cutting efforts at that time.

Fortunately, part of SSIE's functions were assumed by, and are still available through, the National Technical Information Service (NTIS) in the form of an online file known as *Federal Research in Progress (FRIP)*, which may be accessed through such systems as *DIALOG® Information Services* (see Section 10.12A). In addition, the previously developed SSIE file is still online, at least for the present, and historical information therein complements *FRIP*.

Use of these two files is a convenient way to determine what research some federal agencies may have underway in areas of interest. *FRIP* is, however, far from complete. Agencies contributing are Department of Agriculture, National Aeronautics and Space Administration (NASA), National Bureau of Standards, National Institutes of Health, National Institute of Occupational Safety and Health, National Transportation Research Board, U.S. Geological Survey, and Veterans Administration. An attempt is to be made to obtain input from other agencies. Addition of all other major federal departments is needed to make this database more fully useful.

Information included in *FRIP* consists of the following elements (if the submitting agency is able to provide all the data):

Sponsoring agency.	Abstract.
Starting date.	Progress.
Completion date.	Subject index terms.
Performing organizaton (doing the work).	References.
Investigators.	Funding.
Title.	

Unfortunately, abstracts in *FRIP* lack the important editorial review for uniformity and completeness of detail that had been provided by SSIE staff, many of whom had graduate degrees in chemistry and other sciences. In addition, deep indexing and highly skilled retrieval assistance previously supplied by SSIE is missing. The demise of SSIE has left a crucial void that desperately needs filling. See also Chapter 9.

Doctoral dissertations offer the chemist another way of obtaining information as soon as possible and can contain some information that may never be published in any other form. Although chemists can sometimes borrow dissertations through local librarians if the university where the work was done is known, this procedure is often time consuming, and no loan copies are available in many cases.

A better approach is to start with the *Comprehensive Dissertation Index (CDI)*. This permits searching for pertinent dissertations by subject, author, title, and university. This service, which is computer based and available online through *BRS®* Information Technologies and *DIALOG®* Information Services (see Section 10.4), is a product of UMI (University Microfilms International).* *CDI* helps locate the abstracts and/or the dissertations, most of which are available from UMI for a fee. The abstracts are published by UMI in *American Doctoral Dissertations (ADD)* or *Dissertation Abstracts International (DAI)*. Online versions of these include abstracts for more recent years only.

The program is based on agreements between individual degree-granting institutions and UMI. Most American universities now participate in this file, which goes back to 1861. In addition to U.S. dissertations, many from Canada, and an increasing number from other countries are included.

Chemical engineers and some chemists should also be interested in the listing of recent Ph.D. dissertations that appears annually in the January issue of *Chemical Engineering Progress*. This list is categorized by subject.

CA coverage of dissertations is discussed in Section 7.15.

The American Chemical Society's *Directory of Graduate Research* (2) is another source that can provide clues to ongoing research. Research interests and recent publications of faculty members in chemistry and chemical engineering are specified. All major colleges and universities in the United States and Canada are included. The compilation is available in both printed and online forms.

4.6. SUMMARY

This chapter makes the point that a workable personal plan of action for current awareness can be developed by any chemist or engineer who is eager to keep up to date. Persons who are able to develop and implement such plans can expect to find that timely information, of the right kind and in manageable amounts, can help provide an important competitive edge. Choice of which current awareness tools and services to use depends in large measure on career goals and on present or future employment. A chemist who does research and teaches in a university, and plans to stay there, would have a different personal plan of action for keeping up to date than a person working in industrial research. Another very important factor is the field of chemistry. Every person should, however, strive to have a well rounded, balanced plan. This should typically include a combination of firsthand browsing and reading of selected sources, meaningful personal contacts at meetings and elsewhere, and perusal of automated (computer-based) output based on a carefully selected individual profile or an appropriate group or macroprofile. Writing the plan in a notebook, revising it as needed, and keeping of a log of current awareness strategies, efforts, and results can help provide structure, incentive, and a sense of accomplishment.

REFERENCES

1. R. R. Freeman, J. T. Godfrey, R. E. Maizell, C. N. Rice, and W. H. Shepard, "Automatic Preparation of Selected Title Lists for Current Awareness Services and as Annual Summaries," *J. Chem. Doc.,* **4,** 107–112 (1964).
2. American Chemical Society, *Directory of Graduate Research,* Washington, D.C., 1985. Updated at regular intervals.

LIST OF ADDRESSES

American Chemical Society
1155 16th St., N.W.
Washington, D.C. 20036
(202-872-4600)

American Chemical Society
Rubber Division
University of Akron
Akron, OH 44325

BLAISE–LINE
2 Sheraton Street
London, England W1V 4BH

Cambridge Scientific Abstracts
5161 River Road
Bethesda, MD 20816

Chemical Abstracts Service
P.O. Box 3012
Columbus, OH 43210
(800-848-6538)

Engineering Societies Library
345 East 47th Street
New York, N.Y. 10017

Institute for Scientific Information
3501 Market Street
Philadelphia, PA 19104
(800-523-1850)

Macmillan Publishing Co, Inc.
866 Third Avenue
New York, N.Y. 10022

Scientific Meetings
P.O. Box 81622
San Diego, CA 92138

University Microfilms International
300 North Zeeb Road
Ann Arbor, MI 48106
(800-521-0600, except Michigan)

5 HOW TO GET ACCESS TO ARTICLES, PATENTS, TRANSLATIONS, SPECIFICATIONS, AND OTHER DOCUMENTS QUICKLY AND EFFICIENTLY

5.1. INTRODUCTION

Many chemists report that one of the most frustrating experiences they encounter in their professional work is gaining quick access to needed articles, patents, books, and other documents not available in a local chemistry library or bookstore. The problems of obtaining access to or copies of documents quickly and efficiently have been brought home to thousands of chemists and engineers by their use of new automatic and other alerting services (see Chapter 4). Some of these services are so fast that they list many materials before they are received by the local chemistry library. Some materials listed are so "exotic" that obtaining copies becomes a challenging, sometimes formidable task.

Difficulty in document access is most likely to be experienced by chemists who lack ready access to strong chemistry libraries. It is also more likely to occur when the document sought is from less frequently used sources, such as journals from remote countries, material written in unfamiliar languages, or from material that is older or out of print.

Another complication in locating papers is use of very brief abbreviations in some sources for identifying journals and other references. For example:

B.			*Berichte der Deutschen Chemischen Gesellschaft* or *Chemische Berichte*
C.r.	or	*C.R.*	*Comptes Rendus Hebdomadaires des Séances de l'Académie des Sciences*
C.	or	*C.Z.*	*Chemisches Zentralblatt*

Fortunately, there are steps the informed chemist or engineer can take to get much of what is needed. What are some of these options and solutions?

5.2. DOCUMENT DELIVERY SERVICES

One of the most important and reliable sources for any chemist seeking copies of more recent chemical documents not available in a local library or information center is the Document Delivery Service (DDS) of Chemical Abstracts Service (CAS), which was initiated under the leadership of James L. Wood in 1980. The start-up of DDS marked the successful filling of a major void that had previously existed in obtaining of hard-to-get documents cited by CAS in all of its services and publications. DDS files are comprehensive, and the service is exceptionally efficient, knowledgeable, and courteous.

Documents covered include:

- Soviet documents for the last 15 years
- Patents

     All U.S. patents since 1971

     Most foreign patents for the last 5 years
- All other documents cited by CAS in the last 10 years

Cost (in 1986) of service from DDS was $16 per document regardless of length, although if the *CA* or *Chemical Industry Notes* abstract number is known, the charge was reduced to $14. Most orders are filled and shipped within 24 hours after receipt. There can be a delay if the document requested is to be sent on loan and there are others previously interested. Orders can be placed by letter, telephone, Telex®, or online. As could be expected, DDS does not provide translations, multiple copies, or copies of complete monographs, proceedings, and the like. Documents for *CA* abstracts based on abstracts from another service are not provided.

Another excellent choice for the chemist is *The Genuine Article*™ service, formerly *Original Article Text Service (OATS®)*, of the *Institute for Scientific Information® (ISI®)* (see Figure 5.1). This important service covers over 6800 journals. Tearsheets or high-quality photocopies are provided for nearly any article covered by *ISI®*. Orders can be placed by mail, phone, Telex, facsimile, and even *online* (see Chapter 10). Especially responsive service is available for orders placed by telephone "hot line" (toll free). *The Genuine Article*™ approach has many advantages and apparently avoids potential copyright problems (see Section 5.4). This service is best for those relatively recent publications covered by *ISI®*.

Any article published from 1982 to the present, and consisting of 10 or fewer pages, costs $8.00 including first class delivery in the United States, Canada, or Mexico, or $9.00, including air mail, to all other locations (1986 costs). For every additional 10 pages or fraction thereof, there is a surcharge, and there is a royalty surcharge for a small number of journals. An additional charge of $1.00 applies to "hot line" orders.

THE Genuine Article™ — ISI®'s Document Delivery Service

Fill in oval with ISI Accession Number

1986 Price Structure* —

Per article from 1982 to present, 10 pages or less, ordered with ISI® Accession Number:

$8.00 - U.S.A., Canada, and Mexico.

$9.00 - All other locations (price includes airmail delivery).

For every additional 10 pages or fraction, add $2 per article.

For orders placed without ISI Accession Numbers, add $.25 per article.

For some articles a royalty surcharge is applied to cover royalties set by publishers. This surcharge varies from item to item.

Contact ISI for full information on other ordering and payment arrangements.

*Rates are subject to change without notice.

Payment must accompany orders placed by mail.

© 1985 ISI OPC-S/85-283

A-83B

File # ___ 0

Name

Organization

Address

City   State/Province   Country

ZIP/Postal Code   Telephone

Journal

Volume No.   Issue No.

Pages   Date

Author

I need ISI Stamps. Please send me:

___ $9.00's ___ $8.00's ___ $2.00's ___ $1.00's

___ $ .50's ___ $ .25's ($50 minimum order)

Please send ___ The Genuine Article Order Cards.

Please send ___ The Genuine Article Multi-Order Forms.

(Both order cards and forms are provided for your convenience . . . it's not necessary to use them to order articles.)

Pay for The Genuine Article orders the convenient way — with ISI Stamps. Place them here.

Residents of NJ, NY, PA & CA must add sales tax.

Figure 5.1. The *Genuine Article*® form. For more recent prices, consult text and contact ISI®. Copyright owned by Institute for Scientific Information,® Philadelphia, PA.

Another service is provided by *ACS Single Article Announcement* published by the American Chemical Society. This service covers only ACS journals. A related service available from ACS is its Timely Tearsheets Service (see also Section 4.3.B). There are also a large number of other document delivery services. Many of these will accept orders placed online in addition to more conventional routes.

5.3. REPRINTS

One traditional and inexpensive way of obtaining copies is writing authors for reprints. The first step is to be certain the reference and address are correct; this is simply a matter of proofreading in many cases. Beyond this, it is helpful to supply as much incentive as possible to encourage authors to fulfill requests for reprints quickly.

A politely worded, typed letter is helpful, preferably with indication of why the reprint desired is of interest. Even better is reciprocity, as for example, enclosing a reprint of one's own work in the same or related fields. This can help establish an ongoing rapport that may result in access to information prior to publication, depending on how close a professional bond can be established. It is also desirable to enclose a self-addressed, stamped reply envelope and/or to use a tool such as *Request-A-Print*® (see Figure 5.2).

Some authors have an antipathy to answering correspondence, especially requests for reprints. In this event, a telegram or phone call may be necessary to get what is wanted.

Although the reprint approach is by far the most economical way to get copies of documents, the often lengthy time delays could seriously interfere with ambitious work plans and schedules. The uncertainty of whether or not a sought-after reprint will *ever* be received can be extremely frustrating.

Another alternative, especially for less recent material, is for the chemist to locate one or more abstracts of the document. This can be done by a search of the abstracting services. Abstracts can help confirm whether the document is pertinent and worth ordering. See Chapters 6–8.

Even if the document sought is not available in local chemistry libraries, *equivalents* may be (see also Section 5.7.3 on equivalents). It is common for an author to write about the same or similar work in several publications and languages.

Another option is to see if the author has written about the work as part of a series. In that case, an earlier or later paper may be readily available and may suffice.

Also, review articles or books on the subject may contain sufficient information or the work may be adequately described in publications of other authors writing on the same subject. See Chapter 11 on Reviews.

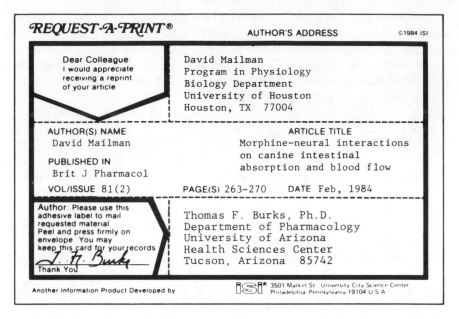

Figure 5.2. *Request-A-Print®* form.

5.4.   HELP AVAILABLE FROM CHEMISTRY LIBRARIANS

Chemistry librarians have a wide variety of tools for locating documents, for example, *CASSI* (1). This tool shows which major libraries in the United States and abroad receive publications cited by *CAS* since 1907. *CASSI* also includes references to journal literature cited by *Chemisches Zentralblatt* from 1830 to 1969 and several hundred titles covered in *Beilstein's Handbuch der Organischen Chemie* (see Section 12.4.A). Other information useful in locating and obtaining document copies is included. Librarians can be of great help in identifying abbreviated or otherwise cryptic journal references using tools such as that just mentioned.

A common library practice has been to obtain a photocopy of a needed document from local or other libraries. The revised U.S. copyright law went into effect in 1978, and its provisions must be observed in obtaining and providing photocopies. A paper by Weil and Polansky (2) reviews the subject and helps explain why document delivery services provide copies of some documents but loan the originals of others.

Other actions that librarians can take include borrowing from another library (using the familiar procedure known as interlibrary loan), purchasing copies of documents, and helping chemists determine whether equivalents or translations are available (see Section 5.7.2.b).

5.5. ONLINE SERVICES

A wide variety of services permit and encourage online ordering of documents (see also Chapter 10). This can be a convenient next step after completing a literature or patent search online in services such as *DIALOG*® and others described in Chapter 10. In addition, the full texts of an increasing number of documents are available online as discussed in Section 10.13 and elsewhere in this book. These materials include, for example: all U.S. patents since 1975 (see Section 13.14); ACS journals; and the *Kirk–Othmer Encyclopedia of Chemical Technology* (see Section 12.2).

5.6. PATENTS

Patent copies are not subject to copyright laws, but they may be more difficult to obtain than articles and other documents, since many chemists (and the librarians who serve them) are usually less familiar with all details of patent systems, especially non-United States patent systems.

Writing to inventors for patent copies is not a favored approach. Instead, copies of patents can be obtained from a variety of other sources.

The chemist should first consider, however, that an article may have been written about the subject matter of the patent by the inventor or someone else in the same organization. The same article could supplant the need for the patent and may contain even more useful information. Articles are, however, not a 1:1 substitute for patents; they lack the legal content and significance of patents and are frequently not as detailed. In addition, if the patent of interest is not readily available, an alternative is to determine the availability of equivalents (see Section 13.17).

Sources of patents include selected federal, state, municipal, and university libraries that maintain numerical sets of U.S. patents. See Table 5.1 for a list of these libraries.

These libraries, designated as Patent Depository Libraries, receive current issues of U.S. Patents and maintain collections of earlier issued patents. The scope of these collections varies from library to library, ranging from patents of only recent years to all or most of the patents issued since 1790.

These patent collections are open to public use and each of the Patent Depository Libraries, in addition, offers the publications of the U.S. Patent Classification System (e.g., The Manual of Classification, Index to the U.S. Patent Classification, Classification Definitions, etc.) and provides technical staff assistance in their use to aid the public in gaining effective access to information contained in patents. With one exception, as noted in Table 5.1, the collections are organized in patent number sequence.

Facilities for making paper copies from either microfilm in reader printers or from the bound volumes in paper-to-paper copies are generally provided for a fee.

TABLE 5.1. Locations of Libraries That Have Copies of U.S. Patents Arranged in Numerical Order in Addition to Files at the U.S. Patent Office, Arlington, VA[a,b,c]

1. **Christine A. Bain**
   Science/Health Science/Technology
     Reference Services
   The New York State Library
   Empire State Plaza
   Albany, New York 12230
   (518) 474-7040

2. **Eulalie Brown, Head**
   Government Publications and Maps
     Department
   General Library
   The University of New Mexico
   Albuquerque, New Mexico 87131
   (505) 277-5441

3. **Eric Esau, Head**
   Physical Sciences Library
   Graduate Research Center
   University of Massachusetts
   Amherst, Massachusetts 01003
   (413) 545-1370

4. **Colleen Sue McDonald**
   Anchorage Municipal Libraries
   Z.J. Loussac Public Library
   524 West 6th
   Anchorage, Alaska 99501
   (907) 264-4481

5. **Margaret Bean, Associate Librarian**
   Engineering Transportation Library
   312 UGL
   The University of Michigan
   Ann Arbor, Michigan 48109-1185
   (313) 764-7494

6. **Jean Kirkland, Head**
   Department of Microforms
   Price Gilbert Memorial Library
   Georgia Institute of Technology
   Atlanta, Georgia 30332
   (404) 894-4508

7. **John F. Vandermolen, Head**
   Science and Technology Department
   Auburn University Libraries
   Auburn University, Alabama 36849
   (205) 826-4500, Ext. 21

8. **Susan Ardis, Head**
   McKinney Engineering Library
   Room 1.3 ECJ
   The University of Texas at Austin
   Austin, Texas 78712
   (512) 471-1610

9. **Laurel Blewett, Documents Librarian**
   Business Administration/Government Documents Department
   Troy H. Middleton Library
   Louisiana State University
   Baton Rouge, Louisiana 70803
   (504) 388-2570

10. **Rebecca Scarborough, Head**
    Government Documents Department
    Birmingham Public Library
    2100 Park Place
    Birmingham, Alabama 35203
    (205) 226-3680

11. **Marilyn McLean, Curator of Science**
    Boston Public Library
    Copley Square
    Boston, Massachusetts 02117
    (617) 536-5400, Ext. 265

12. **Dorothy Solomon, Patent Depository Library Representative**
    Buffalo & Erie County Public Library
    Lafayette Square
    Buffalo, New York 14203
    (716) 856-7525, Ext. 267

13. **Michaele Lee Huygen**
    Library
    Montana College of Mineral Science and Technology
    Butte, Montana 59701
    (406) 496-4284

14. **Karen Spak, Head Librarian**
    Alliance College Library
    Cambridge Springs, Pennsylvania 16403
    (814) 398-2098

TABLE 5.1. Continued

15. **Robert Poyer, Coordinator of Public Services**
    Medical University of South
    Carolina
    171 Ashley Avenue
    Charleston, South Carolina 29425
    (803) 792-2371

16. **Diane A. Richmond, Head**
    Science and Technology Informa-
    tion Center
    Business/Science/Technology Divi-
    sion
    Chicago Public Library
    425 N. Michigan Avenue
    Chicago, Illinois 60611
    (312) 269-2865

17. **Rosemary Dahmann, Head**
    Science and Technology Depart-
    ment
    Public Library of Cincinnati and
    Hamilton County
    800 Vine Street
    Cincinnati, Ohio 45202-2071
    (513) 369-6936

18. **Siegfried Weinhold, Head**
    Documents Collection
    Cleveland Public Library
    325 Superior Avenue
    Cleveland, Ohio 44114-1271
    (216) 623-2870

19. **Judy Erickson, Head**
    Reference Services
    Engineering and Physical Sciences
    Library
    University of Maryland
    College Park, Maryland 20742
    (301) 454-3037

20. **W. David Gay, Documents Librar-
    ian**
    Evans Library–Documents Divi-
    sion
    Texas A&M University
    College Station, Texas 77843-5000
    (409) 845-2551

21. **Lawrence J. Perk, Head**
    Special Materials
    Ohio State University Libraries
    1858 Neil Avenue Mall
    Columbus, Ohio 43210
    (614) 422-6286

22. **Johanna Johnson, Patent Librarian
    & Assistant Manager**
    Dallas Public Library
    1515 Young Street
    Dallas, Texas 75201
    (214) 749-4176

23. **Robert Jackson, Senior Librarian**
    Business, Science and Technology
    Denver Public Library
    1357 Broadway
    Denver, Colorado 80203
    (303) 571-2122

24. **Barbara Klont, First Assistant**
    Technology & Science Department
    Detroit Public Library
    5201 Woodward Avenue
    Detroit, Michigan 48202
    (313) 833-1450

25. **Frank Adamovich, Documents Li-
    brarian**
    Patent Collection
    The University Library
    University of New Hampshire
    Durham, New Hampshire 03824
    (603) 862-1777

26. **Christine Kitchens, Assistant Head**
    Government Documents Depart-
    ment
    Broward County Main Library
    100 S. Andrews Avenue
    Fort Lauderdale, Florida 33301
    (305) 357-7444

27. **Barbara Kile, Director**
    Division of Government Publica-
    tions and Special Resources
    The Fondren Library
    Rice University
    Houston, Texas 77251-1892
    (713) 527-8101, Ext. 2587

TABLE 5.1. Continued

28. **Mark Leggett, Head**
    Business, Science and Technology
    Division
    Indianapolis–Marion County Public
    Library
    P.O. Box 211
    Indianapolis, Indiana 46206
    (317) 269-1741

29. **Bruce B. Cox, Documents Librarian**
    Linda Hall Library
    5109 Cherry Street
    Kansas City, Missouri 64110
    (816) 363-4600

30. **Alan Gould, Assistant Professor, Libraries**
    Engineering Library
    Nebraska Hall, 2nd Floor West
    University of Nebraska—Lincoln
    Lincoln, Nebraska 68588-0410
    (402) 472-3411

31. **Mary Honeycutt, Coordinator of State Library Services**
    Arkansas State Library
    One Capital Mall
    Little Rock, Arkansas, 72201-1081
    (501) 371-2090

32. **Billie Connor, Department Manager**
    Science and Technology
    Los Angeles Public Library
    630 West Fifth Street
    Los Angeles, California 90071-2097
    (213) 489-4658 (temporary)

33. **Margaret Hayes, Technical Reports and Patents Librarian**
    Kurt F. Wendt Engineering Library
    University of Wisconsin
    215 North Randall Avenue
    Madison, Wisconsin 53706
    (608) 262-6845

34. **Barbara C. Shultz, Head**
    Business/Science Department
    Memphis and Shelby County Public
    Library and Information Center
    1850 Peabody Avenue
    Memphis, Tennessee 38104
    (901) 725-8876

35. **Edward Oswald, Business Librarian**
    Business and Science Department
    Miami–Dade Public Library
    101 West Flagler Street
    Miami, Florida 33130-2585
    (305) 375-5230

36. **Theodore R. Cebula, Science and Business Coordinator**
    Milwaukee Public Library
    814 West Wisconsin Avenue
    Milwaukee, Wisconsin 53233
    (414) 278-3247

37. **Edythe Abrahamson, Head**
    Technology and Science Department
    Minneapolis Public Library and Information Center
    300 Nicollet Mall
    Minneapolis, Minnesota 55401
    (612) 372-6570

38. **Donna M. Hanson, Science Librarian**
    University of Idaho Library
    Moscow, Idaho 83843
    (208) 885-6235

39. **Paul H. Murphy, Science Librarian**
    Vanderbilt University
    Science Library
    419-21st Avenue South
    Nashville, Tennessee 37240-0007
    (615) 322-2775

40. **Amanda Putnam**
    Reference Department
    University of Delaware Library
    Newark, Delaware 19717-5267
    (302) 451-2238

41. **Nicholas Patton, Principal Librarian**
    Science & Technology Division
    Newark Public Library
    P.O. Box 630
    Newark, New Jersey 07101
    (201) 733-7815

42. **Richard Hill, Head**
    New York Public Library Annex
    521 W. 43rd Street
    New York, New York 10036-4396
    (212) 714-8529

TABLE 5.1. Continued

43. **Charles Wilt, Librarian**
    The Franklin Institute Library
    20th Street & Parkway
    Philadelphia, Pennsylvania 19103
    (215) 448-1227

44. **Catherine M. Brosky, Head**
    Science & Technology Department
    Carnegie Library of Pittsburgh
    4400 Forbes Avenue
    Pittsburgh, Pennsylvania 15213
    (412) 622-3138

45. **Cheryl Hunt, Patent Reference Librarian**
    Business/Industry/Science/Patent
       Department
    Providence Public Library
    150 Empire Street
    Providence, Rhode Island 02903
    (401) 521-8726

46. **Jean M. Porter, Head**
    Documents Department
    D. H. Hill Library
    North Carolina State University
    Box 7111
    Raleigh, North Carolina 27695-7111
    (919) 737-3280

47. **Carolyn Baber, Head**
    Government Publications Department
    University Library
    University of Nevada—Reno
    Reno, Nevada 89557-0044
    (702) 784-6579

48. **Claire E. Hoffman, Librarian**
    Government Publications Department
    University Library Services
    Virginia Commonwealth University
    901 Park Avenue
    Richmond, Virginia 23284-0001
    (804) 257-1104

49. **Thomas K. Andersen, Head**
    Government Publications Section
    California State Library
    Library–Courts Building
    P.O. Box 2037

Sacramento, California 95809
(916) 322-4572

50. **Stephanie Wunsh**
    Applied Science Department
    St. Louis Public Library
    1301 Olive Street
    St. Louis, Missouri 63103
    (314) 241-2288, Ext. 390

51. **Craig Smith**
    Reference & Circulation
    Documents Section
    Oregon State Library
    Salem, Oregon 97310
    (503) 378-4239

52. **Julianne Hinz, Head**
    Documents Division
    Marriott Library
    University of Utah
    Salt Lake City, Utah 84112
    (801) 581-8394

53. **Joanne Anderson, Senior Librarian**
    Science and Industry Section
    San Diego Public Library
    820 E Street
    San Diego, California 92101
    (619) 236-5813

54. **Harold N. Wiren, Engineering Librarian**
    Engineering Library, FH-15
    University of Washington
    Seattle, Washington 98195
    (206) 543-0740

55. **Jeanne Oliver**
    Reference Department
    Illinois State Library
    Centennial Building
    Springfield, Illinois 62756
    (217) 782-5430

56. **Vicki W. Phillips, Head Document Librarian**
    Oklahoma State University Library
    Stillwater, Oklahoma 74078
    (405) 624-6546

TABLE 5.1. Continued

57. **Mary-Jo DiMuccio, Program Manager**
Patent Information Clearinghouse[c]
1500 Partridge Avenue
Building No. 7
Sunnyvale, California 94087
(408) 730-7290

58. **Timothy Lee Wherry**
Engineering/Computer Science
    Subject Specialist
Noble Science and Engineering Library
Arizona State University
Tempe, Arizona 85287
(602) 965-7609

59. **Mary Hubbard, Head**
Science/Technology Department
Toledo/Lucas County Public Library

325 Michigan Street
Toledo, Ohio 43624
(419) 255-7055, Ext. 212

60. **Diane H. Smith, Head**
Documents Section
C207 Pattee Library
Pennsylvania State University Libraries
University Park, Pennsylvania 16802
(814) 865-4861

61. **William Felker**
Free Library of Philadelphia
Logan Square
Philadelphia, Pennsylvania 15213
215-686-5330

[a]All of the libraries listed offer CASSIS (Classification And Search Support Information System), which provides direct, online access to Patent Office data.
[b]This list is updated regularly in the Official Gazette of the U.S. Patent Office.
[c]The patents in this facility are arranged by Patent Office Class and Subclass, beginning with 1836.

Owing to variations in the scope of patent collections among the Patent Depository Libraries and in their hours of service to the public, anyone contemplating use of the patents at a particular library is advised to contact that library, in advance, about its collection and hours, so as to avert possible inconvenience.

The most complete patent library in the United States is at the U.S. Patent Office, Arlington, VA (see Section 13.10). Many libraries in industrial and other chemical research laboratories have extensive files of chemical patents in microfilm or printed form. The chemist can also obtain copies of patents from government patent offices or from private organizations that specialize in procurement of patents. United States patents can be obtained for $1.50 each (1986 price) from the Commissioner of Patents and Trademarks.*

Addresses of major patent offices around the world, along with prices for copies of patents, are listed in the Introduction to the Semiannual Volumes of Chemical Abstracts (found in the first issue of each volume). Most government patent offices, although they charge relatively low prices, cannot match the delivery speed and more personalized service of commercial procurement organizations. Examples of such organizations include:

Air Mail Patent Service
Box 2232
Arlington, VA 22202

Derwent Publications, Ltd.
128 Theobalds Road
London WCIX 8RP, England

(Note that Derwent can provide copies of patents via facsimile
transmission.) The U.S. address is:

Derwent, Inc.
6845 Elm Street
McLean, VA 22101

IFI Plenum/Data Co.
2001 Jefferson Davis Highway
Arlington, VA 22202
(800-368-3093)

Rapid Patent Service
2221 Jefferson Davis Highway
Arlington, VA 22202
(800-336-5010)

As previously mentioned, another major source for copies of recent patents of chemical interest is the CAS Document Delivery Service. As noted in Section 13.14, the full texts of all U.S. patents since 1975 are available through the online file known as *LEXPAT*, which is a product of Mead Corp.

5.7. TRANSLATIONS

One of the most vexing information problems facing chemists who are natives of the United States, Canada, the United Kingdom, and other English speaking nations is that of coping with interesting material published in a foreign language. It is a particularly troublesome problem in the United States, where relatively few chemists are multilingual.

This is still important, despite the fact that about 69% of the world's chemical literature (papers) is in English. Over the last 50 years or so, the use of English in this form of chemical literature has made steady and continuous progress. Statistics accumulated by Chemical Abstracts Service (CAS) with reference to abstracted papers (patents and books are not included) are shown in Table 5.2.

Some publications appear totally or partially in English, even though this

TABLE 5.2. Language of Publication of Journal Literature Abstracted in Chemical Abstracts (As Percentage of Total Journal Literature Abstracted).

Language	1961	1966	1972	1978	1984
English	43.3	54.9	58.0	62.8	69.1
Russian	18.4	21.0	22.4	20.4	15.7
Japanese	6.3	3.1	3.9	4.7	4.0
German	12.3	7.1	5.5	5.0	3.4
Chinese	a	0.5	a	0.3	2.2
French	5.2	5.2	3.9	2.4	1.6
Polish	1.9	1.8	1.2	1.1	0.7
Spanish	0.6	0.5	0.6	0.7	0.6
Bulgarian	0.4	0.3	0.5	0.5	0.4
Italian	2.4	2.1	0.8	0.6	0.4
Czech	1.9	0.9	0.6	0.5	0.3
Others	7.3	2.6	2.6	1.0	1.6

^{a}Included in "Others" for year.

is not the native language of that country. Journals published in Czechoslovakia, Japan, Sweden, and West Germany are among countries where this is sometimes the case. Among the most outstanding examples are *Chemische Berichte* and *Liebigs Annalen der Chemie,* which are published in German or English, with summaries in both languages and English tables of contents and indexes. Other examples include:

Acta Chemica Scandinavica

Angewandte Chemie, International Edition in English

Bulletin of the Chemical Society of Japan

Collection of Czechoslovak Chemical Communications

Denki Kagaku (The Electrochemical Society of Japan)

Japan Chemical Week

Kunststoffe–German Plastics

Propellants, Explosives, Pyrotechnics

Svensk Papperstidning

Although English is the dominant language in papers, the situation is quite different with regard to patents. As Table 6.9 shows, Japanese patents constitute (1984) 54.3% of the total abstracted by CAS. This represents a sharp increase over the percentage of just five years earlier (1979). With this type of document (patents), the need for translations is much more acute. As explained in section 5.7.5, availability of English-language equivalents in some cases can help considerably.

What are some options open to the chemist or engineer who learns about intriguing research published in an unfamiliar foreign language?

There are several alternative courses of action that can often help resolve the apparent dilemma relatively quickly and at minimal cost. Some of these involve working closely with a chemical information specialist or research librarian. The following paragraphs list these actions.

1. Locate and Use a Suitable Abstract. This may be found in an abstracting service or sometimes in the original journal. (Many foreign-language journals publish abstracts in several languages, one of which is usually English.) Sometimes the abstract indicates that the material is not worth pursuing further—that it is not as pertinent as the title or other preliminary information indicated. Conversely, the abstract may suggest that the publication *is* of interest and worth delving into further. The abstract may contain sufficient data to satisfy immediate interest and need, but it is not wise to rely on the abstract alone. The preferred route, if the abstract indicates that the publication is of interest, is to try to look at the material in a language the chemist can understand. How to achieve this is described in some of the following paragraphs.

2. Locate and Use an Already Available Translation

a. Many journals are translated into English cover to cover, especially Russian-language publications, for example:

- *Biochemistry (Biokhimiya)*
- *Bulletin of the Academy of Sciences of the USSR: Division of Chemical Science (Izvestiya Akademii Nauk SSSR, Seriya Khimicheskaya)*
- *Chemical and Petroleum Engineering (Khimicheskoe i Neftyanoe Mashinostroenie)*
- *Chemistry and Technology of Fuels and Oils (Khimiya i Tekhnologiyu Topliv i Masel)*
- *Colloid Journal of the USSR (Kolloidnyi Zhurnal)*
- *Doklady Biochemistry (Doklady Akademii Nauk SSSR)*
- *Doklady Chemical Technology (Doklady Akademii Nauk SSSR)*
- *Doklady Chemistry (Doklady Akademii Nauk SSSR)*
- *Doklady Physical Chemistry (Doklady Akademii Nauk SSSR)*
- *Fibre Chemistry (Khimicheskie Volokna)*
- *High Energy Chemistry (Khimiya Vysokikh Energii)*
- *Inorganic Materials (Izvestiya Akademii Nauk SSSR, Neorganicheskie Materialy)*
- *Instruments and Experimental Techniques (Pribory i Tekhnika Eksperimenta)*
- *Journal of Analytical Chemistry of the USSR (Zhurnal Analiticheskoi Khimii)*

- *Journal of Applied Chemistry of the USSR (Zhurnal Prikladnoi Khimii)*
- *Journal of General Chemistry of the USSR (Zhurnal Obshchei Khimii)*
- *Journal of Organic Chemistry of the USSR (Zhurnal Organicheskoi Khimii)*
- *Journal of Structural Chemistry (Zhurnal Strukturnoi Khimii)*
- *Kinetics and Catalysis (Kinetika i Kataliz)*
- *Metallurgist (Metallurg)*
- *Protection of Metals (Zashchita Metallov)*
- *Refractories (Ogneupory)*
- *Russian Chemical Reviews (Uspekhi Khimii)*
- *Russian Journal of Inorganic Chemistry (Zhurnal Neorganicheskoi Khimii)*
- *Russian Journal of Physical Chemistry (Zhurnal Fizicheskoi Khimii)*
- *Soviet Chemical Industry (Khimicheskaya Promyshlennost)*
- *Soviet Electrochemistry (Elektrokhimiya)*
- *Soviet Journal of Bioorganic Chemistry (Bioorganicheskay Khimiya)*
- *Soviet Journal of Coordination Chemistry (Koordinatsionnaya Khimiya)*
- *Soviet Journal of Glass Physics and Chemistry (Fizika i Khimiya Stekla)*
- *Soviet Radiochemistry (Radiokhimiya)*

CASSI (1) is a convenient source for identifying and locating journals that are translated cover to cover. Another good source is *Journals in Translation,* published by the British Library Lending Division.

To determine whether a translation may already be available, it is desirable to check with one of the better known publishers of cover-to-cover translations. These include:

Plenum Publishing Corp. (Consultants Bureau)
233 Spring Street
New York, N.Y. 10013
This is the largest commercial publisher of cover-to-cover Russian to
   English technical translations. Consultants Bureau translations are said
   by the publisher to appear about six months after appearance of the
   original.

The Royal Society of Chemistry
Burlington House
London WIV OBN, U.K.

Pergamon Press (U.S. office of a U.K. firm)
Maxwell House, Fairview Park
Elmsford, N.Y. 10523

Verlag Chemie (U.S. office)—now known as VCH Publishers
1020 N.W. 6th Street
Deerfield Beach, FL 33441

American Institute of Chemical Engineers
345 East 47th Street
New York, N.Y. 10017

b. Ask a chemistry librarian to check with one of the so-called translation clearing houses to see if a translation already exists. This may be the case with material of widespread interest or importance. In the United States, one major source is the National Translation Center, John Crerar Library, which in 1984 became part of the University of Chicago science library system.* The center charges a small fee for searching its files, and it publishes a monthly *Translations Register Index,* which lists some 18,000 translations each year.

3. Identify Equivalent Journal Articles. As previously noted, the same author may publish on the same research in journals issued in several different countries and languages. Even authors whose native tongue is English sometimes publish first in foreign-language publications and subsequently publish the same or similar material in English. The best professional society and other publications will not accept outright double publication, but it occurs in journals with less rigid standards and in some trade publications on the applied aspects of chemistry. Identification of such duplicates is straightforward; the chemist merely needs to check the author indexes to *Chemical Abstracts* and related publications.

4. Locate Review Articles or Books That Adequately Describe the Work of Interest. See Chapter 11 on Reviews.

5. Locate Equivalent or Corresponding Patents Published in English. Identification of corresponding patents can be done with concordances or patent indexes such as those of the International Patent Documentation Center, Derwent, *Chemical Abstracts,* and IFI (see Chapter 13 on patents). These concordances provide information about "families" of closely related patents applied for (or issued in) different countries.

6. Forego Translations of the Full Document. Instead work with those parts of the publication that can be handled with a limited amount of dictionary look-up. Examples of well-regarded dictionaries are cited in References (3)– (5). This approach involves scanning for chemical equations (which require no translation) and for such parts as tables, graphs, and abstracts that may require use of a dictionary for only a few key words. (In some cases, journals in other languages will have already translated synopses or abstracts and captions into English.) It is surprising how often this simple approach will provide results at minimum cost and with acceptable time delay.

7. *Identify People*. Identify a person within the chemist's own organization who can provide an on-the-spot oral translation. The advantage of this approach is that questions and answers between chemist and translator can readily be exchanged on points that require amplification. In addition, an oral translation session can be limited to parts of the publication most pertinent—key sections within the text that the chemist can identify to the translator in the course of their session. Such in-house translations are usually the only way to handle confidential or proprietary materials.

8. *Obtain a Custom-Made Translation*. These can be done by commercial translation houses or by free-lance translators. Quality and accuracy of such translations vary widely. The work should be done by a translator who has a proven reputation for top quality work. The translator must have knowledge of the languages to be translated "to" and "from," and must also have adequate training as a chemist or engineer. (Those organizations that translate cover-to-cover versions of chemical journals are good sources for custom-made translations or can make recommendations.) Translation rates vary, depending on such factors as language, type of document, and speed of delivery required. In 1985, commercial rates for a top quality translation from languages such as German or French were approximately $50–80 per 1000 words translated into English. These rates include perfect typed copy. The comparable rate for Japanese was $90–120 and for Russian $80–100. One way to control the cost of translations is to request a *rough draft* rather than a finished product. By saving polishing and editing, this should help minimize costs. It is particularly appropriate for a field with which the chemist is already familiar. One way of identifying translators is to contact the American Translator's Association,* which publishes a translation services directory that may be helpful. Another obvious source is the local telephone book's "yellow pages." Some of the firms engaged in computer-based or computer-assisted translations may also be helpful (see also Section 5.10).

5.8. SPECIFICATIONS AND STANDARDS

The American Society for Testing and Materials, now known simply as ASTM*, is the principal developer and publisher of U.S. standards, including analytical and other methods.

The American National Standards Institute (ANSI),* is another good source, and can help identify both U.S. and foreign standards.

A major supplier of industry and government specifications, standards, and associated documents, including codes and manuals, is Information Handling Services® (IHS®).* An IHS® division, Global Engineering* helps identify the documents and can provide copies quickly. Global Engineering can be reached through a toll free telephone number. IHS® in addition offers an online database known as *Tech Data®*, which can be used to help locate specifications and standards, as well as to search several bibliographic and full-text files of engineering interest. Examples of the files searchable on *Tech Data*™ (see Section 10.4), which is made available in cooperation with BRS® (a sister company of IHS®) include: *Engineering Index, Kirk–Othmer Encyclopedia of Chemical Technology, National Technical Information Service, Patdata®* (an index to U.S. patents from 1975), and *Superindex®* (contains indexes from approximately 600 reference books). Online systems and databases are fully discussed in Chapter 10.

5.9.   PERSONAL CONTACTS

Chemists and engineers should always consider personal contacts with authors of papers and inventors of patents as a major source of important information. Requesting reprints (Section 5.3) is just a first step. A logical second or concurrent step is to talk with (or write to) authors or inventors of papers and patents of interest for specific detail about, for example, what was actually done, what worked best or worst and why, and what future plans are.

The reasons why this type of personal contact is so important are that: publications often contain only relatively limited detail on what is of interest; they are edited to conform to space limitations, legal and other requirements; and they are often out of date. In addition, chemists are often willing to say over the phone or in other informal contacts what they may not be willing to put in a formal publication or any other written form. Face-to-face contact is always best, but this is not always feasible.

The recommendation of personal contact is made regardless of whether authors or inventors being contacted are in colleges and universities or in industry, government, or other chemical facilities. Chances for success with this type of contact are best when authors are in colleges and universities. Nevertheless, it is an approach always worth trying. The worst that can usually happen is a polite "no," but the best can mean an important insight into a research effort or other project. A prerequisite is discretion on the part of the chemist making the inquiry and clear indication of some knowledge of, and interest in, the field of chemistry being discussed.

5.10.   OTHER OPTIONS, RECENT DEVELOPMENTS, AND FUTURE OUTLOOK

It may be necessary for the chemist to personally visit a chemistry library in another city to obtain needed information as completely and quickly as possible. Prior arrangements should be made through the local chemistry librarian to increase chances of success on such visits. A few examples of some of the stronger chemistry libraries in the United States include:

Chemists Club Library
52 East 41st Street
New York, N.Y. 10017

Linda Hall Library
5109 Cherry Street
Kansas City, MO 64110

Engineering Societies Library
345 East 47th Street
New York, N.Y. 10017

New York Public Library, Science and Technology Division
Fifth Avenue and 42nd Street
New York, N.Y. 10018

Indiana University Chemistry Library
Bloomington, IN 47405

Purdue University
West Lafayette, IN 47907

University of California
Berkeley, CA 94720

John Crerar Library of the University of Chicago
5730 S. Ellis Avenue
Chicago, IL 60637

University of Houston
4800 Calhoun Road
Houston, TX 77004

Library of Congress
Washington, D.C. 20541

University of Rochester
Rochester, N.Y. 14627

Libraries of comparable strength are located in Canada, major European nations, and Japan. In the United Kingdom, an outstanding source of scientific documents is the British Library Lending Division.* Services are provided on a worldwide basis. Another good United Kingdom source is the Library of the Royal Society of Chemistry;* John Kennedy is the librarian.

No discussion of document sources would be complete without mention of the U.S. National Technical Information Service (NTIS).* Unclassified federally funded research reports and other documents are a prime emphasis of this large and important facility, which is part of the U.S. Department of Commerce. In addition, the U.S. Joint Publications Research Service* translates and abstracts selected foreign technical publications for federal agencies. Many of these reports and translations are available to the public through NTIS (see also Section 9.2). For hard-to-get articles and other documents published in other countries, especially in the East Bloc, the U.S. State Department in Washington may be able to help.

Many chemists belong to organizations that have laboratories, plants, or other representatives and contacts abroad. Persons at these overseas locations are resources who can be called on for aid in obtaining copies of documents.

An interesting speculation is that the present journal publication system will eventually be augmented, or perhaps even replaced by selective, computer-based distribution of copies of articles and other documents on a large scale. Copyright payments could be built into the cost of such a system. The concept seems unlikely, although continued major increases can be expected in the number of full-text documents available online.

Research on automatic (computer-based) translation has sputtered along for several decades now. Recently, however, there has been a revival of interest, and a major breakthrough could potentially take place in the 1980s if the economics (costs) fit well. This is still not a widely used technique in the United States, except probably by the defense branches of the federal government.

However, computer-aided translation has been offered by the International Telephone and Telegraph Corporation's Engineering Support Center, Harlow, England, for such languages as German, French, Spanish and English; initial emphasis is on electronics technology.

Weidner Communications Inc. of Northbrook, Illinois, has offered a translation system aimed at owners of personal computers, and Challenge Systems of Richardson, TX, and Logos Corp. of Waltham, MA, have also been active in this area. Computer-assisted translation is now offered by Berlitz/Agnew Tech Division.* This includes a foreign-language translation network (Berlitz Linguanet) with centers in San Francisco, Los Angeles, Houston, Chicago, Washington, Philadelphia, and New York.

REFERENCES

1. *CASSI,* Chemical Abstracts Service, Columbus, Ohio, 1985. (Covers 1907–1984.)
2. B. H. Weil and B. F. Polansky, "Copyright Basics and Consequences," *J Chem. Inf. Comput. Sci.* **24**(2), 43–50.
3. A. M. Patterson, *German–English Dictionary for Chemists,* 3rd ed., Wiley, New York, 1950.
4. A. M. Patterson, *French–English Dictionary for Chemists,* 2nd ed., Wiley, New York, 1952.
5. L. Callaham, *Russian–English Chemical and Polytechnical Dictionary,* 3rd ed., Wiley, New York, 1975.

LIST OF ADDRESSES

American Chemical Society
1155 Sixteenth St., N.W.
Washington, D.C. 20036

ANSI
1430 Broadway
New York, N.Y. 10018

American Translators Association
109 Croton Avenue
Ossining, N.Y. 10562

ASTM
1916 Race Street
Philadelphia, PA 19103

Berlitz/Agnew Tech Division
6415 Independence Ave.
Woodland Hills, CA
91367 (213-340-5147)

British Library Lending Division
Boston Spa, Wetherley
West Yorkshire LS 23 7BQ
United Kingdom

**Commissioner of Patents and
   Trademarks**
Washington, D.C. 20231

Global Engineering
2625 Hickory Street
Santa Ana, CA 92707
   (800-854-7179)

Information Handling Services
15 Inverness Way East
Englewood, CO 80150

Institute for Scientific Information
3501 Market Street
Philadelphia, PA 19104

**Library of the Royal Society of
   Chemistry**
Burlington House
London WIV OBN
United Kingdom

**University of Chicago Science
   Library**
5730 S. Ellis Avenue
Chicago, IL 60637

**U.S. Joint Publications Research
   Service**
1000 N. Globe Road
Arlington, VA 22201

**U.S. National Technical Information
   Service**
5285 Port Royal Road
Springfield, VA 22161

6 CHEMICAL ABSTRACTS SERVICE— HISTORY AND DEVELOPMENT

6.1. INTRODUCTION

The history of Chemical Abstracts Service (CAS) is the story of the development of the world's most outstanding tool of scientific and technical information. CAS development, from its founding in 1907 to the present, reflects accurately the history of chemistry as a science, and of other related sciences and technologies, extending into the biological, engineering, and computer sciences. The editors and directors of CAS (see list in Table 6.1), perhaps because of their unique positions, or more likely because of their unique abilities, are outstanding individuals in the history of scientific and technical information and in the history of chemistry.

Careful study of a recent historical perspective such as this should help chemists use CAS tools more effectively because it summarizes development of these tools over time and indicates when the most important policy changes took place. This perspective also provides an appreciation for the complexity and magnitude of the changes and improvements introduced, and the amount of research and other investigations involved. For example, the efforts that went into the hallmark transformation of CAS into its present computer-based mode rank near the top of management and research achievements in recent American chemical history. Figures 6.1 and 6.2 will help provide some perspective on related abstracting and indexing efforts worldwide. Chapter 7 details some specifics on the current use of *CA*.

6.2. PHYSICAL PLANT AND STAFF

Much of the growth of chemical information can be traced both in the number of pages and abstracts in *Chemical Abstracts (CA)* (Table 6.2A), and in the buildings occupied by CAS staff.

During its first year of operation (1907), *CA* was edited at the National Bureau of Standards in Washington, D.C. In 1908, offices were moved to the University of Illinois at Urbana. Beginning in 1909, and for many years

TABLE 6.1. Editors and Directors of CAS

Name	Years in Office
Directors	
E. J. Crane	1956–1958
Dale B. Baker	1958–1986
Ronald L. Wigington	1986–
Editors	
William A. Noyes, Sr.	1907–1908
Austin M. Patterson	1909–1913
John J. Miller	1914 (3 months)
E. J. Crane[a]	1914 (9 months)
E. J. Crane	1915–1958
Charles L. Bernier	1958–1961
Fred A. Tate[a]	1961–1967
Russell J. Rowlett, Jr.	1967–1982
David W. Weisgerber	1982–

[a] Acting editor.

thereafter, quarters for CAS staff was a crowded (desk-to-desk) floor in the chemistry building of The Ohio State University (OSU) in Columbus. A new building on the campus was occupied in 1955, but that soon became inadequate despite addition of a floor in 1961. Finally, in 1965, CAS moved to its present American Chemical Society owned Olentangy River Road site adjacent to the OSU campus, and in 1973 a second CAS building was constructed on that site.

These two buildings now house a total staff of over 1200 persons, as compared to employment levels of just 3 part-time staff in 1907, and 58 full-

1830 Chemisches Zentralblatt and predecessors (discontinued 1969)

1907 Chemical Abstracts

1926 British Abstracts (discontinued 1953)

1927 Nippon Kagaku Soran (Second Series) (changed 1974)

1940 Bulletin Signalétique (discontinued 1983)

1953 Referativnyi Zhurnal, Khimiya

Figure 6.1. First abstract journals in chemistry and dates started. Copyright 1984 by the American Chemical Society, used with permission.

Chemical Abstracts (1907)

Derwent Central Patent Index and predecessors (1951)

Referativnyi Zhurnal, Khimiya (1953)

Current Abstracts of Chemistry and Index Chemicus® (1960)

Chemischer Informationsdienst (1970)

Kagaku Gijutsu Bunken Sokuho, Kagaku, Kagakukogyo Hen (Current Bibliography on Science and Technology, Chemistry and Chemical Engineering) (1974)

Figure 6.2. Major abstract journals in chemistry today and dates started. Copyright 1984 by the American Chemical Society, used with permission.

time and 32 part-time staff as late as 1950 (see Table 6.2B). This employment increase is easily explained by the explosive growth of chemical literature, the many functions and services that CAS now offers, and the fact that most abstracting is now done in-house.

The change to in-house abstracting took place gradually in the 1960s and 1970s and has largely replaced a system of using primarily volunteer abstractors and section editors from around the world, individuals who are paid a token fee for their efforts. This change was brought about by successful development of computer-aided production of *CA* that facilitated abstracting and indexing of documents in a single unified and highly efficient step.

The volunteer abstractor system, now almost entirely outmoded, (there are now about 650 such volunteers), served an important function in its day. There are no accurate statistics prior to 1963. In 1963, 77% of all abstracts were prepared by volunteer abstractors. From then on, the number of abstracts prepared by volunteers has steadily declined as shown in the following list.

Year	Percentage	Year	Percentage	Year	Percentage
1963	77.0	1970	42.0	1977	16.4
1964	67.0	1971	42.7	1978	12.7
1965	68.1	1972	37.1	1979	8.6
1966	64.6	1973	36.6	1980	6.4
1967	50.6	1974	32.9	1981	5.6
1968	52.0	1975	25.1	1982	4.5
1969	49.9	1976	17.9	1983	4.0
				1984	3.7

TABLE 6.2A. Chemical Abstracts Publication Record 1907–1984

Year	Vol.	Number of Abstracts Papers	Number of Abstracts Patents	Number of Abstracts Books	Total Abstracts	Total Abstracts To Date	Patent Equiva-lents	Total Documents Cited	Total Documents Cited To Date	Pages of Abstracts	Issue Index Pages	Vol. Index Pages	Total Pages Published[a]
1907	1	7,994	3,853		11,847	11,847				3,074		363	3,437
1908	2	11,414	3,658	97	15,169	27,016				3,416		473	3,889
1909	3	11,455	3,806	198	15,459	42,475				3,020		341	3,361
1910	4	13,006	3,754	785	17,545	60,020				3,314		727	4,041
1911	5	15,892	5,014	776	21,682	81,702				3,926		845	4,771
1912	6	15,740	6,919	535	23,194	104,896				3,544		799	4,343
1913	7	19,025	6,946	659	26,630	131,526				4,096		834	4,930
1914	8	16,468	7,920	727	25,115	156,641				3,872		725	4,597
1915	9	12,200	6,159	622	18,981	175,622				3,379		580	3,959
1916	10	10,519	5,265	324	16,108	191,730				3,180		492	3,672
1917	11	10,921	4,680	344	15,945	207,675				3,470		524	3,994
1918	12	9,283	4,074	524	13,881	221,556				2,712		503	3,215
1919	13	10,957	3,741	542	15,240	236,796				3,338		589	3,927
1920	14	13,619	4,432	1,275	19,326	256,122				3,826		846	4,672
1921	15	15,211	4,265	975	20,451	276,573				4,059		783	4,842
1922	16	18,070	5,142	886	24,098	300,671				4,365		1,156	5,521
1923b	17	19,507	4,749	1,059	25,315	325,986				3,924		1,008	4,932
1924	18	20,523	5,084	1,036	26,643	352,629				3,740		1,135	4,875
1925	19	20,951	5,475	671	27,097	379,726				3,618		1,155	4,773
1926	20	23,103	6,099	1,036	30,238	409,964				3,842		1,406	5,248
1927	21	25,037	7,872	582	33,491	443,455				4,098		1,413	5,511
1928	22	28,153	9,936	1,046	39,135	482,590				4,878		1,727	6,605
1929	23	29,082	17,867	1,344	48,293	530,883				5,614		1,821	7,435
1930	24	32,731	21,246	1,169	55,146	586,029				6,066		2,142	8,208
1931	25	32,278	18,904	1,546	52,728	638,757				6,161		2,282	8,443
1932	26	37,403	20,678	1,380	59,461	698,218				6,184		2,270	8,454
1933	27	36,139	28,051	1,963	66,153	764,371				6,024		2,172	8,196
1934c,d	28	38,371	21,824	1,375	61,570	825,941				3,798		1,157	4,955
1935	29	42,593	19,241	1,579	63,413	889,354				4,204		1,330	5,534
1936	30	41,927	20,836	1,809	64,572	953,926				4,346		1,528	5,874
1937	31	44,032	19,006	1,697	64,735	1,018,661				4,498		1,485	5,983
1938	32	45,917	19,515	1,496	66,928	1,085,589				4,782		1,542	6,324
1939	33	45,414	19,893	1,801	67,108	1,152,697				4,860		1,607	6,267
1940	34	40,624	11,635	1,421	53,680	1,206,377				4,170		1,384	5,554
1941	35	35,588	17,176	1,330	54,094	1,260,471				4,184		1,464	5,648
1942	36	30,479	14,334	833	45,646	1,306,117				3,684		1,232	4,916
1943	37	30,523	11,473	1,673	43,669	1,349,786				3,470		1,200	4,670
1944	38	30,440	11,494	1,766	43,700	1,393,486				3,306		1,175	4,481
1945	39	22,824	9,357	1,491	33,672	1,427,158				2,782		1,137	3,919
1946	40	29,943	8,810	825	39,578	1,466,736				3,853	144	1,591	5,588
1947	41	30,461	7,925	902	39,288	1,506,024				3,909	142	1,496	5,547
1948	42	35,867	7,002	1,127	43,996	1,550,020				4,623	168	1,740	6,531
1949	43	40,612	11,390	1,439	53,441	1,603,461				4,769	183	1,991	6,943
1950	44	47,496	10,063	1,539	59,098	1,662,559				5,592	210	2,350	8,152
1951c	45	50,657	10,417	1,959	63,033	1,725,592				5,340	242	2,534	8,116
1952	46	56,419	12,185	1,543	70,147	1,795,739				5,890	265	2,574	8,729
1953	47	61,273	11,906	1,912	75,091	1,870,830				6,444	294	2,821	9,559
1954	48	67,606	11,083	1,926	80,615	1,951,445				7,151	297	3,237	10,685
1955	49	74,664	9,926	1,732	86,322	2,037,767				8,264	324	3,721	12,309
1956	50	78,009	12,350	2,037	92,396	2,130,163				8,768	355	4,522	13,645
1957	51	84,205	16,822	1,498	102,525	2,232,688				9,353	392	4,397	14,142
1958	52	95,736	21,920	1,274	118,930	2,351,618				10,628	486	4,873	15,987
1959	53	98,680	26,760	1,756	127,196	2,478,814				11,557	525	5,471	17,553
1960	54	104,484	27,675	2,096	134,255	2,613,069				13,014	552	6,686	20,252
1961	55	118,337	26,249	2,307	146,893	2,759,962	7,609	154,502	2,767,571	13,999	658	8,322	22,979
1962	56,57	140,168	26,467	2,716	169,351	2,929,313	5,787	175,138	2,942,709	16,725	758	9,373	26,856
1963c	58,59	141,016	26,240	4,148	171,404	3,100,717	8,400	179,804	3,122,513	15,298	1245	9,136	25,679
1964	60,61	161,489	26,422	2,082	189,993	3,290,710	13,375	203,368	3,325,881	16,608	1498	10,176	28,282
1965	62,63	165,770	29,225	2,088	197,083	3,487,793	19,312	216,395	3,542,276	17,963	1820	11,074	30,857
1966	64,65	181,715	35,031	3,557	220,303	3,708,096	28,940	249,243	3,791,519	20,700	2104	12,660	35,464
1967	66,67	202,684	36,797	3,046	242,527	3,950,623	26,766	269,293	4,060,812	22,815	2683	14,023	39,521
1968	68,69	198,035	31,720	2,753	232,508	4,183,131	19,180	251,688	4,312,500	22,103	2621	15,703	40,427
1969	70,71	210,344	39,424	2,552	252,320	4,435,451	33,026	285,346	4,597,846	23,533	3220	18,071	44,824
1970	72,73	230,902	43,044	2,728	276,674	4,712,125	33,068	309,742	4,907,588	23,792	3777	21,144	48,713
1971	74,75	262,127	43,405	3,444	308,976	5,021,101	41,129	350,105	5,257,693	24,690	5744	21,151	51,585
1972	76,77	280,143	51,179	3,104	334,426	5,355,527	44,622	379,048	5,636,741	27,386	5877	20,791	54,054
1973	78,79	269,711	48,683	2,611	321,005	5,676,532	35,544	356,549	5,993,290	25,865	5708	20,123	51,696
1974	80,81	272,235	58,436	2,953	333,624	6,010,156	42,039	375,663	6,368,953	26,282	5522	21,180	52,984
1975	82,83	317,472	68,471	6,291	392,234	6,402,390	62,011	454,245	6,823,198	31,100	6711	25,158	62,969
1976	84,85	317,985	67,176	5,744	390,905	6,793,295	67,603	458,508	7,281,706	31,390	6891	27,310	65,591
1977	86,87	348,059	55,441	6,637	410,137	7,203,432	68,088	478,225	7,759,931	32,253	7263	28,980	68,496
1978	88,89	363,195	57,343	7,804	428,342	7,631,774	70,217	498,559	8,258,490	33,368	7593	27,677	68,638
1979	90,91	370,771	58,738	7,378	436,887	8,068,661	78,854	515,741	8,774,231	34,878	8010	30,002	72,890
1980	92,93	407,342	61,998	6,399	475,739e	8,544,400	72,937	548,676	9,322,907	38,188	8697	32,197	79,064
1981	94,95	373,973	71,180	5,434	450,587	8,994,987	98,739	549,326	9,872,233	37,074	10,098	35,650	82,822
1982	96,97	381,257	70,774	5,758	457,789	9,452,776	99,658	557,447	10,429,680	37,970	10,595	36,072	84,637
1983b	98,99	371,389	74,948	5,416	451,753	9,904,529	95,811	547,564	10,977,244	34,394	10,498	35,368	80,260
1984	100,101	380,692	73,907	5,970	460,569	10,365,098	111,239	571,808	11,549,052	34,876	11,216	39,135	85,227

[a]Total includes "List of Periodicals" pages, which are not shown separately, for those years in which that list was published in *CA*.

[b]Type size decreased.

[c]Page size increased.

[d]Two-column format adopted.

[e]About 17,000 of the abstracts published in 1980 resulted from a 14-day shortening of processing time.

TABLE 6.2B. Number of CAS Employees

Year	Part Time	Full Time	Year	Part Time	Full Time
1907	3	0	1946	29	42
1908	5	0	1947	33	63
1909	1	4	1948	40	59
1910	0	4	1949	34	48
1911	2	4	1950	32	58
1912	13	5	1951	40	63
1913	10	5	1952	59	75
1914	13	5	1953	55	86
1915	3	6	1954	29	78
1916	13	14	1955	49	95
1917	15	6	1956	36	97
1918	11	6	1957	46	126
1919	8	5	1958	42	126
1920	8	5	1959	44	204
1921	12	6	1960	51	240
1922	7	8	1961	67	297
1923	8	7	1962	60	375
1924	11	8	1963	36	431
1925	8	8	1964	34	544
1926	11	11	1965	29	691
1927	6	11	1966	37	927
1928	15	13	1967	—	945
1929	· 10	16	1968	—	935
1930	16	20	1969	—	954
1931	9	19	1970	—	978
1932	10	17	1971	—	984
1933	9	18	1972	—	1006
1934	13	18	1973	—	1080
1935	7	21	1974	—	1162
1936	9	34	1975	—	1148
1937	13	35	1976	—	1130
1938	13	29	1977	—	1142
1939	11	31	1978	—	1157
1940	6	25	1979	—	1159
1941	14	24	1980	—	1181
1942	23	24	1981	—	1189
1943	23	25	1982	—	1200
1944	23	25	1983	—	1223
1945	31	31	1984	—	1271

The peak in the number of volunteer abstractors was reached in 1966 with 3292 as compared to 648 in 1984. No attempt should be made to correlate the number of volunteer abstractors with the percentage of abstracts prepared by them. Other factors enter into play: growth of literature, productivity of individual abstractors, total journal coverage versus individual document assignment, and so on. Table 6.3 shows how the system has evolved over the years.

This arrangement (volunteer abstractors) made possible economical (although slow) abstracting of documents in many chemical specialties and in a variety of languages. To the volunteers, their function was a badge of honor and a matter of genuine pride. It was a way to perform a needed service for the chemical profession, and it was an excellent way to keep up with new developments in fields of interest.

TABLE 6.3. Transition to In-House Abstracting: Number of Volunteer Abstractors

Year	Total	Year	Total	Year	Total
1907	129	1933	338	1959	1536
1908	171	1934	442	1960	1668
1909	205	1935	339	1961	2250
1910	252	1936	428	1962	2870
1911	219	1937	449	1963	3137
1912	237	1938	438	1964	3030
1913	219	1939	448	1965	3232
1914	248	1940	449	1966[a]	3292
1915	203	1941	474	1967	3245
1916	187	1942	474	1968	3107
1917	172	1943	456	1969	2921
1918	159	1944	457	1970	2819
1919	155	1945	444	1971	2745
1920	221	1946	496	1972	2549
1921	230	1947	486	1973	2236
1922	253	1948	590	1974	2084
1923	276	1949	641	1975	1831
1924	260	1950	624	1976	1542
1925	231	1951	771	1977	1406
1926	255	1952	835	1978	1255
1927	261	1953	943	1979	1029
1928	272	1954	1029	1980	949
1929	338	1955	1200	1981	830
1930	390	1956	1406	1982[b]	699
1931	415	1957	1283	1983	666
1932	438	1958	1396	1984	648

[a] In 1966 there were 1811 domestic and 1481 foreign volunteer abstractors.
[b] In 1982 there were 102 domestic and 597 foreign volunteer abstractors.

In addition to producing *CA,* the offices at Columbus now handle keyboarding for composition of all CAS publications and American Chemical Society (ACS) primary journals. This work was previously done at Mack Printing Company in Easton, PA. The change was accomplished in a gradual transition from 1965 to 1980. Furthermore, starting in 1974, CAS took over computer handling of all ACS membership records, subscription fulfillment, accounting, financial, personnel, and other business records for ACS headquarters in Washington, D.C., as well as in Columbus. Other functions and services of CAS are described throughout this chapter. In 1984, the budget for CAS was approximately $70 million. In 1985, CAS's budget grew to $75 million and its staff increased to over 1300 full-time persons. The budget for 1986 was $85 million.

6.3. *CA* ABSTRACTS

For many years, *CA* provided exceptionally full and complete abstracts, often running a full column or even more. Emphasis on briefer, more *findings-oriented* abstracts began to evolve beginning in about 1950. Unlike earlier abstracts, no attempt was made to record in the abstract all new data reported in original documents. In a few areas of synthetic organic chemistry and analytical chemistry, however, abstracts continued to provide lengthy, detailed procedures and extensive lists of numerical data. Finally, in 1970, instructions (reflected in the 1971 edition of the CAS *Directions to Abstractors*) were issued to have essentially all abstracts prepared in a concise, *findings-oriented* form. This was gradually implemented over the 1970–1971 period.

Abstracts are now concise and compact, although, fortunately, indexing remains as full as ever. The *CA* abstract today is intended to provide access to the original literature, but not to serve as a substitute for it, a very important distinction. Use of Registry Numbers and of newer type fonts and page arrangements helped achieve some of the compaction, although the primary means is simply use of fewer words. The average abstract text is now about 100 words long.

Few chemists would quarrel with this policy of briefer abstracts, particularly now that the new and efficient CAS Document Delivery Service is available (see Section 5.2). Problems could, however, potentially arise with (1) documents in less familiar foreign languages when there are no English language "equivalents," or (2) possibly in cases where large numbers of pages of the full text of documents may need to be obtained. In these cases, the older forms of more lengthy abstracts might still be useful.

Since 1967, abstracts have been numbered sequentially in each volume. Before that time, the number linked to index entries represented a column, and the small letter, or still earlier a numerical superscript, described the fraction of the column in which the entry of interest could be found.

CA issues have been published weekly since 1967, with about half of the sections appearing one week and the other half the next week. Before that time, all sections were published together on a biweekly basis, and prior to 1961, on a twice-a-month basis.

6.4. *CHEMICAL ABSTRACTS* INDEXES

Chemists have always depended heavily on *CA*'s comprehensive, reliable, in-depth indexing based on the original documents. The expertise with which CAS indexes are constructed is the single most important reason why *CA* is such an indispensable tool for chemists and for scientists in other disciplines around the world. Use of these indexes in an online mode (with computers) further enhances their utility. See Figure 6.3 on the development of indexes to *CA*.

CA indexes had been published annually until 1962, when they began to be published semiannually (every six months). *CA* Collective Indexes combine into single, organized listings the content of 10 individual Volume Indexes. Ten-year (decennial) Collective Indexes were published for *CA* from 1907 to 1956; five-year (quinquennial) Indexes have been produced since 1956 (see Table 6.4 and 6.5). These also show the date of the start of each of the major types of *CA* Volume Indexes and of the associated *Index Guides*.

Use of Collective Indexes reduces the repetitive work that would otherwise be needed to search individual Volume Indexes. Thus, a search of *CA* index content from 1907 through 1981 requires separate searches in 10 Collective Indexes rather than in 95 Volume Indexes. Table 6.4 indicates the individual indexes that comprise each of the Collective Indexes.

Volume Indexes to *CA* are based on a controlled vocabulary. To provide chemists with more rapid indexing of the contents of individual *CA* issues, a form of quick indexing (designated as the Keyword Index) has appeared with each issue beginning in 1963. Because the Keyword Index is based on uncontrolled vocabulary, it is relatively superficial. It is not intended to provide the precise and reliable access of the Volume Indexes. However, this distinction is now less important for chemists who can search *CA* online since rapid access to both types of indexes is possible online.

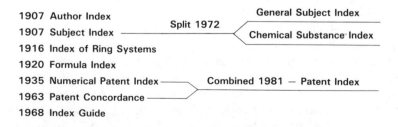

Figure 6.3. Development of *CA* Indexes.

TABLE 6.4. CA Collective Indexes: Historical Development

	Collective Index										
	1st	2nd	3rd	4th	5th	6th	7th	8th	9th	10th	11th
Years	1907–1916	1917–1926	1927–1936	1937–1946	1947–1956	1957–1961	1962–1966	1967–1971	1972–1976	1977–1981	1982–1986
CA volumes	1–10	11–20	21–30	31–40	41–50	51–55	56–65	66–75	76–85	86–95	96–105
Author Index	X	X	X	X	X	X	X	X	X	X	X
Subject Index	X	X	X	X	X	X	X	X			
Index Guide								X	X	X	X
Chemical Substance Index									X	X	X
General Subject Index									X	X	X
Formula Index		b	b	b	X	X	X	X	X	X	X
Index of Ring Systems	a	a	a	a	a	a	X	X	X	X	X
Numerical Patent Index				b	X	X	X	X	X		
Patent Concordance							X	X	X		
Patent Index										X	X

[a] For the 1st through the 6th Collective Indexes, Ring System Information was included in the Introduction to the *Subject Index*; for the 7th through the 11th Collective Indexes, the *Index of Ring Systems* was bound with the *Formula Index*.

[b] Two special Collective Indexes are available for searches of patents and formulas: The *10-Year Numerical Patent Index to Chemical Abstracts* (1937–1946) and the *27-Year Collective Formula Index to Chemical Abstracts* (1920–1946).

TABLE 6.5. Collective Indexes to CA: Vital Statistics (data for 1982–1986 are estimates)

	1st 1907–1916	2nd 1917–1926	3rd 1927–1936	4th 1937–1946	5th 1947–1956	6th 1957–1962	7th 1962–1966	8th 1967–1971	9th 1972–1976	10th 1977–1981	11th 1982–1986
Documents cited											
Nonpatent abstracts	138,436	170,493	357,507	370,117	559,180	510,373	804,749	1,118,615	1,478,249	1,896,992	(1,886,000)
Patent abstracts	53,294	47,741	186,455	142,693	104,247	119,426	143,385	194,390	293,945	304,700	(413,000)
Total abstracts	191,730	218,234	543,962	512,810	663,427	629,799	948,134	1,313,005	1,772,194	2,201,692	(2,299,000)
Patent equivalents	—	—	—	—	—	7,609	75,814	153,169	251,819	400,081	(530,000)
Total documents cited	191,730	218,234	543,962	512,810	663,427	637,408	1,023,948	1,466,174	2,024,013	2,601,773	(2,829,000)
Author Index											
No. of author entries	230,000	284,000	653,000	615,000	795,000	820,000	1,800,000	3,020,000	4,668,000	5,847,000	(6,133,000)
Pages	1,980	2,452	3,095	3,541	5,074	6,164	9,607	13,845	19,128	24,684	(25,800)
Books	2	2	2	2	4	4	6	8	11	14	(15)
Patent Index											
Number of abstracted and equivalent patents	—	—	—	142,693	104,247	127,035	219,199	347,559	545,764	704,534	(943,000)
Pages	—	—	—	182	144	172	352	761	1,374	3,643	(5,000)
Books	—	—	—	1	1	1	1	[a]	1	2	(3)
Subject Index											
Index entries	307,000	620,000	1,378,000	1,609,000	3,241,000	3,289,000	4,735,000	6,876,000	—[d]	—[d]	—[d]
Pages	2,843	4,139	4,885	6,386	13,740	12,659	24,632	46,384[c]	—	—	—
Books	2	3	3	4	11	7	13	18	—	—	—

General Subject Index											
Index entries	—	—	—	—	—	—	—	—	3,746,000	5,345,000	(6,042,000)
Pages	—	—	—	—	—	—	—	—	16,797	25,508	(28,800)
Books	—	—	—	—	—	—	—	—	10	15	(17)
Chemical Substance Index											
Index entries	—	—	—	—	—	—	—	—	7,351,000	9,264,000	(10,158,000)
Pages	—	—	—	—	—	—	—	—	40,891	56,112	(61,000)
Books	—	—	—	—	—	—	—	—	25	32	(36)
Formula Index											
Index entries	—	—	—	560,000[b]	714,000	867,000	1,869,000	2,556,000	3,189,000	3,493,000	(3,744,000)
Pages	—	—	—	2,077[b]	2,968	3,869	7,007	11,769	16,251	19,904	(21,300)
Books	—	—	—	2[b]	3	3	4	7	9	11	(13)
Index Guide											
Index entries	—	—	—	—	—	—	—	—	148,000	149,000	149,000
Pages	—	—	—	—	—	—	—	2,280	1,440	1,605	(3,500)
Books	—	—	—	—	—	—	—	2	1	1	(2)
Summary											
Total books	4	5	5	9	19	15	24	35	57	75	(86)
Total pages	4,823	6,591	7,980	12,186	21,926	22,864	41,598	62,203	95,881	131,456	(146,000)
Distribution time following completion of collective period	—	—	—	—	65 months	40 months	44 months	24 months	20 months	14 months	—

[a] Bound with author index.
[b] This Collective Formula Index covered the 27-year period from 1920–1946.
[c] Includes Registry Handbook–Number Section, which was issued as part of the Eighth Collective Index.
[d] The Subject Index was subdivided into General Subject and Chemical Substance Indexes beginning with the ninth collective period.

6.5. *CHEMICAL ABSTRACTS* NOMENCLATURE

For years (since 1945), CAS has published guidelines to, and summaries of, its approach to naming of chemical compounds. But it was not until 1972 that CAS extended systematic nomenclature in its indexes to most substances and decided to use *trivial* or nonsystematic names only rarely. In almost all cases, nomenclature currently used by CAS for chemical substance index headings is that of IUPAC (International Union of Pure and Applied Chemistry).

CAS staff believe that this has resulted in substance index names generally easier to interpret and easier to derive, and in improved groupings of structural families together in the alphabetically ordered index. Some chemists believe that this policy change makes use of *CA* unnecessarily complex, although public debates that followed original announcement of the change have almost completely subsided. Fortunately, however, the *CA Index Guide,* first introduced in 1968, provides a guide to trivial names and cross-references these to systematic names. Some cross-references were reintroduced into the Volume Indexes during the ninth collective period (1972—1976) for several hundred substances, but CAS could not afford to do this for all substances covered.

In addition to the *Index Guide,* other major aids to identification of correct nomenclature are the *Registry Handbook—Number Section,* first published in 1974, and the *Registry Handbook—Common Names* first published in 1978. By 1976, use of Registry Numbers (see Section 6.10) in abstracts had been extended to 40 sections. Use of Registry Numbers ensures unambiguous identification regardless of the potential vagaries of nomenclature. (Extension of use of Registry Numbers to the abstract text of the remaining 40 sections is not anticipated by CAS staff. Their studies are said to show that this would interfere with readability of these sections.)

6.6. *CHEMICAL ABSTRACTS* SECTIONS

The evolution of *CA* sections is a particularly helpful way to trace the growth of *CA* and of chemistry as a science. The existence of sections, initiated with *CA* at the outset in 1907, is intended to make it possible for chemists and other users of *CA* to limit scanning, reading, or searching of *CA* to specific sections, with a reasonable degree of assurance that most of what is pertinent will probably be found, providing the suggested cross-references are also pursued. However, no studies on the effectiveness of sections in regard to their intended function have been published to date.

Existence of sections also makes it possible to publish the Section Groupings. These provide a relatively affordable tool for chemists to keep up to date in broad, related areas of chemistry without the necessity of reading, or paying for, the full *CA*. The first Section Grouping (Biochemistry) was introduced in 1962, and the four additional Groupings (Organic Chemistry; Macromolecular; Applied Chemistry and Chemical Engineering; and Physical, Inorganic, and Analytical Chemistry) were introduced in 1963.

In 1945, the number of sections in *CA* was increased from the original 30 to 31 with the addition of the section *Synthetic Resins and Plastics*. In 1962 abstracts were arranged by subsection, and permanent section cross-references were provided. The number of sections was increased to 73 in 1962, to 74 in 1963, and to the present total of 80 in 1967.

The first published guide on the scope of each of the sections was made available to the chemical community in 1970. This guide became increasingly important as the development of online searching made it possible to more easily and efficiently limit searches to specific sections.

National concern with oil supply and prices and with other energy sources was reflected in reorganization of the *CA* sections on energy (51 and 52) in 1974. A separate section on safety, in response to a resurgence of interest in the same year, was, however, not provided. Instead, 1974 saw the emergence of emphasis on indexing of the safety aspects. At about the same time, CAS increased coverage of toxicology, although pinpointing of a specific year is not possible because this change took place gradually.

In 1982, *CA* sections were reorganized to more accurately represent additional changes in the development of chemistry. For example, the new sections *Biochemical Genetics* and *Biomolecules and Their Synthetic Analogs* were introduced. The complete list of *CA* sections at present is shown in Table 6.6.

6.7. PATENTS

Improved and expanded coverage of patents by CAS reflects the growing recognition by chemists of the importance of patents as unique and timely sources of chemical information (see also Chapter 13). This improvement may also reflect the stimulus provided by the competition of Derwent Publications Ltd., generally acknowledged as the worldwide leader in coverage of patent information.

CAS now covers all patents of chemical or chemical engineering interest from 18 countries and two international organizations—an especially sharp increase over earlier years. In addition, for patents issued by another 8 countries, all patents of chemical or chemical engineering interest are covered for those patents issued to individuals or organizations residing in the granted country (see Table 6.7).

Table 6.7 shows the development of CAS patent coverage since 1907. The

TABLE 6.6. List of *CA* Sections

Abstract Sections[a]

Biochemistry Sections

1. Pharmacology
2. Mammalian Hormones
3. Biochemical Genetics
4. Toxicology
5. Agrochemical Bioregulators
6. General Biochemistry
7. Enzymes
8. Radiation Biochemistry
9. Biochemical Methods
10. Microbial Biochemistry
11. Plant Biochemistry
12. Nonmammalian Biochemistry
13. Mammalian Biochemistry
14. Mammalian Pathological Biochemistry
15. Immunochemistry
16. Fermentation and Bioindustrial Chemistry
17. Food and Feed Chemistry
18. Animal Nutrition
19. Fertilizers, Soils, and Plant Nutrition
20. History, Education, and Documentation

Organic Chemistry Sections

21. General Organic Chemistry
22. Physical Organic Chemistry
23. Aliphatic Compounds
24. Alicyclic Compounds
25. Benzene, Its Derivatives, and Condensed Benzenoid Compounds
26. Biomolecules and Their Synthetic Analogs
27. Heterocyclic Compounds (One Hetero Atom)
28. Heterocyclic Compounds (More Than One Hetero Atom)
29. Organometallic and Organometalloidal Compounds
30. Terpenes and Terpenoids
31. Alkaloids
32. Steroids

33. Carbohydrates
34. Amino Acids, Peptides, and Proteins

Macromolecular Chemistry Sections

35. Chemistry of Synthetic High Polymers
36. Physical Properties of Synthetic High Polymers
37. Plastics Manufacture and Processing
38. Plastics Fabrication and Uses
39. Synthetic Elastomers and Natural Rubber
40. Textiles
41. Dyes, Organic Pigments, Fluorescent Brighteners, and Photographic Sensitizers
42. Coatings, Inks, and Related Products
43. Cellulose, Lignin, Paper, and Other Wood Products
44. Industrial Carbohydrates
45. Industrial Organic Chemicals, Leather, Fats, and Waxes
46. Surface-Active Agents and Detergents

Applied Chemistry and Chemical Engineering Sections

47. Apparatus and Plant Equipment
48. Unit Operations and Processes
49. Industrial Inorganic Chemicals
50. Propellants and Explosives
51. Fossil Fuels, Derivatives, and Related Products
52. Electrochemical, Radiational, and Thermal Energy Technology
53. Mineralogical and Geological Chemistry
54. Extractive Metallurgy
55. Ferrous Metals and Alloys
56. Nonferrous Metals and Alloys
57. Ceramics

58. Cement, Concrete, and Related Building Materials
59. Air Pollution and Industrial Hygiene
60. Waste Treatment and Disposal
61. Water
62. Essential Oils and Cosmetics
63. Pharmaceuticals
64. Pharmaceutical Analysis

Physical, Inorganic, and Analytical Chemistry Sections
65. General Physical Chemistry
66. Surface Chemistry and Colloids
67. Catalysis, Reaction Kinetics, and Inorganic Reaction Mechanisms
68. Phase Equilibriums, Chemical Equilibriums, and Solutions
69. Thermodynamics, Thermochemistry, and Thermal Properties
70. Nuclear Phenomena
71. Nuclear Technology
72. Electrochemistry
73. Optical, Electron, and Mass Spectroscopy and Other Related Properties
74. Radiation Chemistry, Photochemistry, and Photographic and Other Reprographic Processes
75. Crystallography and Liquid Crystals
76. Electric Phenomena
77. Magnetic Phenomena
78. Inorganic Chemicals and Reactions
79. Inorganic Analytical Chemistry
80. Organic Analytical Chemistry

[a]Guidelines used to assign abstracts to *CA* Sections according to their subject content are summarized in the publication, "Subject Coverage and Arrangement of Abstracts by Sections in Chemical Abstracts" (Subject Coverage Manual), 1982 ed.

following lists detail the improvements that have been made in the last 30 years. In 1953, chemical and chemical engineering patents issued to resident assignees and/or inventors in 18 countries were covered in *CA:*

Australia	Germany (West)	Netherlands
Austria	Great Britain	Norway
Belgium	Hungary	Spain
Canada	India	Sweden
Denmark	Italy	Switzerland
France	Japan	United States

In addition, among the 18 countries listed, were four countries whose chemical and chemical engineering patents issued to both resident and nonresident assignees and/or inventors were covered in *CA:*

France	Great Britain
Germany (West)	United States

TABLE 6.7. Development of CAS Patent Coverage by Country[a-d]

Country	1907	10	13	18	21	29	34	48	51	53	56	58	59	60	63	64	65	66	67	68	69	70	71	72	74	75	76	77	78	79	82[d]
Australian						x	x	x	x	x	x	x	x	x	x	x	x	x	x	x	x	x	x	x	x	x	x	x	x	x	+
Australian (Petty Patents)																															+
Austrian			x	x	x	x	x	x	x	x	x	x	x	x	x	x	x	x	x	x	x	x	x	x	+	+	+	+	+	+	+
Belgian				x	x	x	x	x	x	x	x	x	x	x	x	+	+	+	+	+	+	+	+	+	+	+	+	+	+	+	+
Brazilian Patente											+				+											+	+	+	+	+	+
Brazilian Pedido																											+	+	+	+	+
British		+	+	+	+	+	+	+	+	+	+	+	+	+	+	+	+	+	+	+	+	+	+	+	+	+	+	+	+	+	+
British Amended													+	+	+	+	+	+	+	+	+	+	+	+	+	+	+	+	+	+	+
British U.K. Patent Appl.																					+	+	+	+	+	+	+	+	+	+	+
Canadian		x	x	x	x	x	x	x	x	x	x	x	x	x	x	x	x	x	x	x	x	x	x	x	x	x	x	x	x	x	x
Czechoslovakian										x	x	x	x	x	x	x	x	x	x	x	x	x	x	x	x	x	x	x	x	x	x
Danish			x	x	x	x	x	x	x	x	x	x	x	x	x	x	x	x	x	x	x	x	x	x	x	x	x	x	x	x	x
Finnish													x	x	x	x	x	x	x	x	x	x	x	x	x	x	x	x	x	x	+
French	+	+	+	+	+	+	+	+	+	+	+	+	+	+	+	+	+	+	+	+	+	+	+	+	+	+	+	+	+	+	+
French Addition																					+		+	+			+	+			
French Demande																					+	+	+	+	+	+	+	+	+	+	+
French Medicinal															+	+	+	+	+	+	+	+	+	+	+						
French Addition to Medicinal																					+			+							
German (East)													x	x	x	x	x	x	x	x	x	x	x	x	x	x	x	x	x	x	+
German Patentschrift	+	+	+	+	+	+	+	+	+	+	+	+	+	+	+	+	+	+	+	+	+	+	+	+	+	+	+	+	+	+	+
German Auslegeschrift																					+	+	+	+	+	+	+	+	+	+	+
German Offenlegungsschrift																						+	+	+	+	+	+	+	+	+	+
Hungarian				x	x	x	x	x	x	x	x	x	x	x	x	x	x	x	x	x	x	x	x	x	x	x	x	x	x	x	x
Hungarian Halasztott																								x	x	x	x	x	x	x	x

Hungarian Teljes
Indian
Israeli
Italian
Japanese Tokyo Koho
Japanese Kokai Tokkyo Koho
Netherlands
Netherlands Application
Norwegian
Polish
Romanian
South African
Spanish
Swedish
Swiss
Swiss Examined Applications
United States
U.S. Patent Application
U.S. Published Patent Application
U.S. Reissue
U.S. Defensive Publication
U.S.S.R.
European Patent Appl.
PCT International Appl.

[a] + = all patents of chemical or chemical engineering interest, that is, patents issued to nationals or nonnationals.

[b] x = chemical and chemical engineering patents issued to individuals or organizations resident in the granting country (i.e., nationals) or resident in countries not listed above.

[c] Starting in 1980 all patent equivalents, both nationals and nonnationals issued by 26 countries and two industrial property organizations, are included in the *Patent Index*.

[d] No changes of coverage have occurred in 1983 and 1984. Beginning in 1986, *CA* has covered all patents from the People's Republic of China; these started to issue in 1985.

Since 1953, coverage has been extended as follows for resident assignees and/or inventors:

1956	Czechoslovakia	1967	Romania
1958	Israel	1968	South Africa
1958	USSR	1976	Brazil
1959	Germany (East)	1976	Italy (dropped)
1960	Finland	1979	European Patent applications
1960	Poland	1979	Patent Cooperation Treaty (PCT) international applications

Nonresident coverage has been extended as follows:

1963–1966, 1975 Belgium
1965–1967, 1976 Netherlands
1968 South Africa
1975 Austria, Canada, and Japanese Kokai Tokkyo Koho
1976 Brazil and Japanese Tokkyo Koho
1978 Israel, Romania, Switzerland, USSR
1982 Australia, Germany (East), India

In 1984, chemical and chemical engineering patent documents from 18 countries are covered for both resident and nonresident assignees and/or inventors plus the two international industrial property organizations. Patent documents are covered from the other 8 countries for resident assignees and/or inventors. Patent equivalent coverage of nonresidents was extended to all 26 countries beginning in 1980. Coverage of all patents from the People's Republic of China began in 1986 starting with the first such patents issued (1985).

Patents have come to be recognized by chemists as a source of much data that never appears elsewhere, although how much has never been definitively demonstrated. Patents present new information quickly, often for the first time anywhere, especially in the form of so-called *quick issue* or unexamined patent applications from Europe and Japan. The popularity of the unexamined patent application has grown as national patent offices increasingly began to employ this approach as a way to cope with growing backlogs of applications. CAS recognized this trend by beginning to cover Netherlands applications in 1964, West German applications in 1968, and Japanese applications in 1972, to name the most outstanding examples.

In the 1970s, two new multinational patent systems (the European Patent Office and the Patent Cooperation Treaty) were organized to reduce the cost and effort involved in filing in individual countries. CAS began to cover these as soon as the first patent applications appeared in 1979.

Multiple filing (in many countries) shows the commercial importance of patents, and knowledge of the existence of applications in English can point the way to obtaining copies of English-language equivalents. To reflect this, CAS has provided a patent concordance since 1963. Prior to that, CAS provided *title-only* abstracts to equivalents.

Derwent is now the only English-language service that provides the convenience of English-language abstracts for equivalent patents previously abstracted on a worldwide basis.

In 1981, CAS introduced a new *Patent Index,* which combined the content of the *Numerical Patent Index* and *Patent Concordance* and considerably expanded the identifying information presented for patents. Competition in the concordance approach to chemical patents is provided by Derwent and by INPADOC (International Patent Documentation Center).

Another evidence of the growing importance attached to patents is the 1979 decision by CAS to additionally index chemical substances from patent claims, even though these may be unsupported by examples, data, and so on. Prior to that, CAS had not provided indexing for this type of "paper chemistry." Even today, CAS is more selective in its coverage of patent information than is Derwent. For searches that focus on worldwide patents, Derwent has become the preferred source (see Section 13.11).

Another advantage of patents is that they are regarded by many as a good indication of the innovativeness of not only individuals and organizations, but also of nations. The recent sharp increase in the number of chemical patents from Japan is said to be one indicator of the additional challenge that country poses to U.S. industrial leadership as well as of opportunity for cooperation. Tables 6.8 and 6.9 show trends in patent coverage by CAS classified by national origin and country of publication.

TABLE 6.8. *CA* **Abstracts of Patents by Country of Origin (where the reported work was done)**

Country of Origin[a]	1979		1984	
	Number	Percentage	Number	Percentage
Japan	22,489	38.3	42,138	57.0
United States of America	12,728	21.6	11,078	15.0
Union of Soviet Socialist Republics	5,461	9.3	5,170	7.0
West Germany	4,875	8.3	4,229	5.7
East Germany	994	1.7	1,706	2.3
Czechoslovakia	931	1.6	1,514	2.1
United Kingdom	1,677	2.9	1,278	1.7
France	1,827	3.1	1,208	1.6
Romania	283	0.5	777	1.1
Switzerland	1,187	2.0	755	1.0
All others	6,286	10.7	4,054	5.5
	58,738	100.0	73,907	100.0

[a]Ten top countries in 1984.

TABLE 6.9. *CA* **Abstracts of Patents by Country of Publication**[a]

Country of Publication[a]	1979		1984	
	Number	Percentage	Number	Percentage
Japan	21,276	36.3	40,096	54.3
United States of America	9,991	17.0	7,781	10.5
West Germany	9,263	15.8	5,760	7.8
European Patent Org.	322	0.5	5,178	7.0
Union of Soviet Socialist Republics	5,108	8.7	4,996	6.8
East Germany	933	1.6	1,748	2.4
Czechoslovakia	881	1.5	1,476	2.0
France	1,725	2.9	933	1.3
United Kingdom	1,957	3.3	895	1.2
Romania	292	0.5	796	1.1
Patent Cooper. Treaty	18	0.03	777	1.1
Poland	2,333	4.0	673	0.9
All others	4,639	7.9	2,798	3.6
	58,738	100.0	73,907	100.0

[a]Ten top countries in 1984.

6.8. CHEMICAL ABSTRACTS SERVICE IN TRANSITION

The Baker–Tate–Rowlett era, which spanned more than two decades beginning in 1958, was among the most meaningful and successful in the history of CAS. This era saw transition to:

a. Start of CAS's first computer-based publication, *Chemical Titles* (1961)

b. Start of computer-based current awareness services (1962–1963)

c. Initiation of the CAS Registry System (1965)

d. Beginning of computerized production (1967)

e. Largely in-house abstracting (1970s)

f. Start of *CA SELECTS* (1976)

g. Beginning of *CAS ONLINE* and Document Delivery Service (1980)

With (1) the inclusion of abstracts in *CAS ONLINE* beginning in April, 1983; (2) the start of full-scale subject searching of *CA* in December, 1983 through *CAS ONLINE;* and (3) the start of an international network for scientific and technical information with West Germany in September, 1983, CAS is well along in its transition from a relatively passive organization to one of great vigor. CAS is now a full service organization. It offers a more complete range of chemical information services than any other organization.

An extensive chronology of key events in the history of CAS, including major policy changes, appears in Table 6.10.

TABLE 6.10. Milestones in the History of Chemical Abstracts Service (CAS)[a]

1907
Chemical Abstracts, Volume 1, Number 1, January 1, 1907
W. A. Noyes, Sr., Editor
Office in National Bureau of Standards, Washington, D.C.; later (in September) at
  University of Illinois, Urbana
Covered the world's publicly available chemical and chemical engineering litera-
  ture and grouped the abstracts into 30 sections by subject matter
Covered all granted patents from Great Britain, France, Germany, and the United
  States. [After 1949, the British documents covered were the examined applica-
  tions; after 1979, unexamined applications were also covered. French unexamined applications began to be covered in 1969. German (West) unexamined
  applications began to be covered in 1970. United States unexamined applications available through NTIS began to be covered in 1974.]
Volume *Author* and *Subject Indexes* introduced (Vol. 1) with uninverted chemical
  compound names as index headings
1908
List of Journals Abstracted published in *CA* Volume 2, Issue 1
1909
Austin M. Patterson, *CA* Editor (2nd)
CA office moved to The Ohio State University (OSU), Columbus, Ohio
1910
Patent coverage extended
  Canada, granted
1912
Volume *Numerical Patent Index* introduced (Vol. 6) (discontinued in 1915)
1913
Patent coverage extended
  Austria, granted
  Denmark, Norway, Sweden, granted (after 1968, examined applications)
1914
John J. Miller, *CA* Editor (3rd) (3 months); E. J. Crane, Acting Editor (9 months)
1915
E. J. Crane, *CA* Editor (4th)
Year added as part of journal reference in abstract heading
1916
Volume 10 *Subject Index* Introduction provided with list of organic radicals and
  Index of Ring Systems
Volume 10 *Subject Index*—format changed so that each entry began on a separate
  line; systematic index nomenclature introduced for chemical compounds (in-
  verted names); page fractions added to *CA* reference
1917
First Decennial Collective Index (Vols. 1–10) (1907–1916) included *Subject* and
  Author Indexes with Ring Index in *Subject Index* Introduction
1918

TABLE 6.10 Continued

Patent coverage extended

  Japan, examined applications (beginning in 1972, unexamined applications were also covered)

1920

Used *title only* abstracts for patent documents that were equivalent to previously abstracted documents. Cross-references to the abstracts were given (discontinued in 1961)

Volume *Formula Index* introduced (Vol. 14)

1921

Patent coverage extended

  Belgium, granted

  Netherlands, granted (in 1964 unexamined applications began to be covered)

  Switzerland, granted (in 1982 examined applications were also covered)

1922

List of Periodicals Abstracted by Chemical Abstracts included list of library holdings

1928

CA office moved to expanded facilities at new McPherson Chemistry Building on OSU campus

1929

Chemical Abstracts, Volume 23, Number 1 given new cover with *CA* Logo

Patent coverage extended

  Australia, granted (after 1949 the examined applications were covered)

  Hungary, granted (in 1971 examined applications were added and in 1972 unexamined applications)

  Italy, granted (discontinued 1976)

  Union of Soviet Socialist Republics, granted

1930

The Little CA house organ first issue (discontinued after issue 110 in 1966)

1934

CA Issue page in two columns; abstracts indexed by column number and numerical page fraction

1935

Volume *Numerical Patent Index* reintroduced (Vol. 29)

1937

1,000,000 th abstract published

1940

The Ring Index published

1942

CA established an office in the Library of Congress to facilitate the acquisition of hard-to-get publications

1944

30-Year Collective Numerical Patent Index (1907–1936) published by SLA

TABLE 6.10. Continued

1945

Volume 39 *Subject Index* Introduction included Discussion of the Naming of
   Chemical Compounds for Indexing; published separately as *The Naming and Indexing of Chemical Compounds by Chemical Abstracts*

CA Sections expanded from 30 to 31

ACS Photocopy Service introduced via U.S. Department of Agriculture (discontinued in 1956)

1947

CA Issues—letters substituted for column fraction numbers

1948

Patent coverage extended
   India, examined applications

1949

10-Year Collective Numerical Patent Index to CA (Vols. 31–40) (1937–1946) published

1951

27-Year Formula Index (Vols. 14–40) (1920–1946) published

1953

Patent coverage extended
   Spain, granted

1955

CA located in building on OSU campus exclusively for *CA*

R&D Department established

2,000,000 th abstract published

1956

Chemical Abstracts Service (CAS) established

ACS Member personal use pledge on *CA* subscription started (Member rate
   dropped in 1965)

E. J. Crane, first Director of CAS

CAS on self-supporting basis

Patent coverage extended
   Czechoslovakia, granted

1957

Fifth Decennial Index (Vols. 41–50) (1947–1956) contained for the first time *Numerical Patent Index* and *Formula Index*

1958

Dale B. Baker, CAS Director (2nd)

Charles L. Bernier, *CA* Editor (5th)

Patent coverage extended
   Israel, unexamined applications

CA Issue Patent Index introduced (Vol. 52)

Bibliography of Chemical Reviews (name changed to *Bibliography of Reviews in
   Chemistry* in 1962 and discontinued that year)

TABLE 6.10. Continued

1959

Patent coverage extended
  East Germany, granted
First Index produced by Varitype–Fotolist
  (Vol. 53 *Formula Index*)

1960

Fourth floor added to *CA* building on OSU campus
Patent coverage extended
  Finland, granted (in 1968 the examined applications were covered)
  Poland, granted
The Ring Index, 2nd ed. published

1961

Fred A. Tate, Assistant Director and Acting Editor (6th)
CA Issues published every two weeks
See references for patent duplicates included in *Numerical Patent Index* (discontinued in 1963)
Diacritical marks dropped and special characters converted into English equivalents
Chemical Titles introduced; computer produced with Keyword-in-Context (KWIC) Index
CAS Russian Photocopy Service introduced (since 1980, under umbrella of CAS Document Delivery Service)
NewsCASter issued for informed staff
ACS purchase of 60 acres property along Olentangy River Road

1962

CA Sections expanded from 31 to 73
CA Issues have abstracts arranged by subsection; permanent section cross-references included
Biochemistry Sections published as separate *CA* Section Grouping
Volume *Subject Index*
  assumption notes introduced; nomenclature and indexing formalized for coordination, inorganic, and organometallic compounds, addition compounds, charge-transfer complexes, salts of organic bases, ferrocene and its analogs; multiple entries (e.g., esters) provided
Formula Index included separate entries for each positional isomer
Sixth Collective Index (Vols. 51–55) (1957–1961); first five-year Collective Index

1963

CA Sections expanded from 73 to 74
Four additional *CA* Section Groupings offered
  Organic Chemistry
  Macromolecular
  Applied Chemistry and Chemical Engineering
  Physical and Analytical Chemistry [in 1982 changed to Physical, Inorganic, and Analytical Chemistry]
CA Issue Keyword Index introduced (Vol. 58)

TABLE 6.10. Continued

Patent Concordance introduced (Vol. 58); elimination of *see* references in *Numer-ical Patent Index*

Issue number added as part of journal reference in abstract heading

CAS Open Forum presented at ACS National Meeting

As they became available, began including national or international patent classifications in the bibliographic information in patent document abstracts with the inclusion of U.S. Patent Classification Numbers in U.S. patent abstracts

3,000,000 th abstract published

1964

CAS Advisory Board held first meeting (discontinued in 1982)

1965

CAS building located on Olentangy River Road

First Marketing Plan developed

Language designation added to abstract heading for journal articles

Chemical Titles available in computer readable form

Chemical Abstracts available on microfilm

CAS Registry System installed

Chemical-Biological Activities (CBAC) printed and computer-readable file in-troduced; this was the first service to include Registry Numbers (printed service discontinued in 1971)

1966

Desk-Top Analysis Tools produced

Board of Directors Committee on CAS started

Steroid Conjugates Bibliography published

1967

Russell J. Rowlett, Jr., *CA* Editor (7th)

CA Issues published weekly

CA Sections expanded from 74 to 80

CA Abstracts numbered sequentially with check digits

Patent coverage extended

   Romania, granted

Polymer indexing begun for specific homopolymers and copolymers; Polymer Structural Repeating Unit (SRU) indexing

Nomenclature included *(E)–(Z)* and *(R)–(S)* systems

Subject Index included subdivision of large headings

B and R used in Volume *Subject Index* for references to books and reviews

First Index to be computer produced (Vol. 66 *Formula Index*)

Hetero-Atom-in-Context (HAIC) Index introduced (Vol. 66) (discontinued in 1971)

Polymer Science and Technology (POST) printed and computer-readable file in-troduced (printed service discontinued in 1971)

CAStings first issue (discontinued in 1972)

Selective Dissemination of Information began experimentally; (name changed to *Basic Journal Abstracts* in 1968; discontinued at the end of 1972)

Polymer and Plastic Business Abstracts introduced (name changed to *Plastics In-dustry Notes* in 1968)

TABLE 6.10. Continued

Seventh Collective Index (Vols. 56–65) (1962–1966) contained for the first time *Patent Concordance*

1968

CA Issue Indexes formatted in upper and lower case type font (Vol. 69)

Abstract heading changes
   Author's name inverted (last name first)
   Year of publication follows journal title

Patent coverage extended
   South Africa, unexamined applications

Index Guide introduced (Vol. 69)

Registry II version of CAS Registry System installed

CA Condensates introduced (incorporated into *CA SEARCH* in 1978)

4,000,000 th abstract published

1969

B used as prefix to references for book titles in Volume *Author Index*

Volume 71 *Subject* and *Formula Indexes* include Registry Numbers for the first time

Registry Number Index introduced (Vol. 71) (discontinued in 1971)

ACCESS introduced [name changes to *Chemical Abstracts Service Source Index (CASSI)* in 1970]

Naming and Indexing of Chemical Compounds (8CI *Blue Book*) published

Provided illustrative *Introduction* to *CA* Issues

Patent Concordance available in computer readable form

1,000,000 th Registry Number assigned

1970

Findings-oriented abstracts introduced

Unified Document Analysis begun

CA Collective Indexes (1–7 Collective) available on microfilm

Chemical Abstracts Service Source Index (CASSI) available in computer-readable form

Subject Coverage and Arrangement of Abstracts by Section in CA available

1971

Expanded coverage of *Plastics Industry Notes* and changed name to *Chemical Industry Notes*

5,000,000 th abstract published

1972

CA Sections 1–5, 35–46 were the first computer-produced abstracts with Registry Numbers and highlighting (italics) of chemical substance names

Indexing of chemical reactants in synthetic and preparative studies extended

Systematic nomenclature introduced for most substances

Structural diagrams included in *CA* Issues and Volume Indexes according to new guidelines

Volume *Subject Index* divided into *Chemical Substance Index* and *General Subject Index*

Eighth Collective Index (Vols. 66–76) (1967–1971) computer produced

CA on microfiche available experimentally (offered as a service in 1977)

80

TABLE 6.10. Continued

Naming and Indexing of Chemical Substances for *Chemical Abstracts* during the Ninth Collective Period (1972–1976) (Section IV, Vol. 76 Index Guide)
CAS Report introduced
CA Integrated Subject File (CAISF) produced (discontinued at the end of 1978)
2,000,000 th Registry Number assigned
1973
Second CAS building on Olentangy River Road site constructed
First Workbook *Using Chemical Abstracts Service Printed Information Services* available
First workshop offered on how to use *CA*
CA Subject Index Alert (CASIA) produced (incorporated into *CA SEARCH* in 1978)
Chemical Industry Notes (CIN) available in computer-readable form
1974
CA Sections on energy (Sections 51 and 52) reorganized
Safety indexing emphasized
Bibliographic Guide for Editors and Authors (BIOSIS/CAS/Ei) published
Registry III version of CAS Registry System installed
Registry Handbook—Number Section published
First foreign CAS workshop presented in West Germany
6,000,000 th abstract published
1975
Special Bibliographies in *Chemical Titles* published (precursors of *CA SELECTS*)
CA Issues completely computer produced
Registry Numbers in abstracts extended to 34 Sections
CAS Editorial Advisory Board established (discontinued in 1982)
Computer-readable Abstract Text Services offered (discontinued in 1981)
   Ecology and Environment
   Food and Agricultural Chemistry
   Energy
   Materials
CBAC expanded with three additional *CA* Sections
Volume Index entries added to *CBAC* and *POST*
CAS assumed responsibility for ASTM CODEN assignment
CA Collective Indexes (1–8 Collectives) available on microfiche
First stand-alone Workbook *CAS Printed Access Tools* published
3,000,000 th Registry Number assigned
1976
Volume Number included as prefix to *CA* abstract number in Issues
CA SELECTS introduced
Hierarchies of *General Subject Index* headings included in *Index Guide*
Patent coverage extended
   Brazil, unexamined applications
Registry Numbers in abstracts extended to 40 *CA* Sections

TABLE 6.10. Continued

1977
Volume *Formula Index* coverage extended to unspecific derivatives
International CODEN Directory published
Parent Compound Handbook published (discontinued in 1983)
7,000,000 th abstract published
4,000,000 th Registry Number assigned

1978
CA Issue Keyword Index format includes indentures
REG/CAN FILE available (discontinued in 1981)
CA BIBLIO FILE available (discontinued in 1981)
CA SEARCH introduced
P for patent in Issue Keyword and Author Indexes
Registry Handbook—Common Names available in microfilm and microfiche
CAS Document Copy Service available (July–December); now part of Document
 Delivery Service

1979
Indexing of chemical substances from patent claims (unsupported by examples,
 data, etc.) initiated
Patent coverage extended
 European Patent Applications
 Patent Cooperation Treaty International Applications
Individual Search Services (ISS) bulletins introduced
CAS Updates introduced
8,000,000 th abstract published
CAS Library Document Copy Service (noncopyrighted documents) available; now
 part of Document Delivery Service

1980
CAS ONLINE, a chemical structure search and display system, introduced
Search Assistance Desk established
5,000,000 th Registry Number assigned
CAS Document Delivery Service introduced

1981
Volume and Issue *Patent Index* introduced (combined *Numerical Patent Index*
 and *Patent Concordance)*
BIOSIS/CAS SELECTS published

1982
David W. Weisgerber, *CA* Editor (8th)
CA Sections reorganized
Tenth Collective Index published (Vols. 86–95) (1977–1982); includes *Patent Index*
 for the first time
IUPAC names given for new and hypothetical elements
9,000,000 th abstract published

1983
Abstracts available on *CAS ONLINE*
CAS User Councils formed
6,000,000 th Registry Number assigned

TABLE 6.10. Continued

CA File component (bibliographic and indexing information) of *CAS ONLINE* introduced; STN[SM] International, the Scientific and Technical Information Network, established
1984
10,000,000 th abstract published by *CA*
CAOLD File component (pre-1967 references to chemical substances) of *CAS ONLINE* introduced
Ring Systems Handbook published
1985
Keyword-Out-of-Context (KWOC) Index to *Chemical Abstracts Service Source Index (CASSI)* published, available on microfiche
7,000,000 th Registry Number assigned
Patent coverage extended
    People's Republic of China (first abstracts appear in 1986)
Abstract texts searchable experimentally via *CAS ONLINE*
Markush searching for specific chemicals made possible in *CAS ONLINE*
1986
Ronald L. Wigington succeeds Dale B. Baker as Director

[a] Some of the coverage changes (changes in policy), such as extension of coverage to conference proceedings, government reports, and the like, cannot be pinpointed to date implemented because they occurred gradually, and without much fanfare, as did expansion of coverage of areas such as toxicology. Many indexing and nomenclature changes also took place gradually. In many cases, the type and tone of literature covered molded indexing policies rather than specific decisions on specific dates.

6.9. COMPUTERIZATION

The most important change in the history of CAS is its adoption of computer-based production and, associated with that, development of computer-based tools for searching and current alerting. This trend began in the early 1960s, aided initially by a series of extensive grants from the National Science Foundation for CAS to act as a prototype for automation of secondary information services in the United States. It is still underway, although many of the major objectives were achieved in the 1970s, and a landmark was attained with initiation of *CAS ONLINE* in 1980.

When CAS officials first started to share their computerization goals with other members of the ACS (primarily through papers and open forums at ACS National Meetings, as well as in the pages of *Chemical & Engineering News*), there were many who listened with considerable skepticism, perhaps based on concern that these efforts represented an undue departure from established goals, or perhaps based on a lack of understanding. Feelings pro and con ran high, and there was spirited, sometimes emotional, discussion at almost every ACS National Meeting.

In any event, the magnitude of the success ultimately achieved in this complex and difficult task cannot be overestimated. It required outstanding management, first-class research, and development of computer technology that has made CAS a leader far beyond the chemical community.

As evidence of this, in 1977 the EPA (Environmental Protection Agency) selected CAS as the organization best qualified to compile EPA's inventory of toxic substances. In 1983, the U.S. Patent Office stated its interest in employing the CAS approach for the automatic composition (input) and retrieval of information found in all U.S. patents, and CAS was subsequently chosen as part of the team to develop and install an automated system for processing patent applications.

In retrospect, it appears that the CAS thrust toward computerization started with formation of a research and development office in 1955 under the leadership of Karl F. Heumann, who is now with the Federation of American Societies for Experimental Biology. Subsequent directors included the British chemist, G. Malcolm Dyson, Kenneth H. Zabriskie, Jr., and Ronald L. Wigington. The present R & D Director is Nick A. Farmer who was appointed to that position in 1984. Table 6.11 summarizes these changes.

G. Malcolm Dyson, along with Hans Peter Luhn of International Business Machines Corporation, conceived of the idea of *Chemical Titles (CT)* in April, 1959. This was to be the first computer-readable service produced by CAS. The initial sample copy of *CT* was produced in 1960, and full-scale production was started in 1961. A novel computer-produced keyword-in-context (KWIC) index appeared in each issue.

In 1965, the second computer-based CAS service began to appear. This was *Chemical–Biological Activities (CBAC),* which contained abstracts, index entries, and full bibliographic information.

In the drive toward CAS computerization, the appointment of Fred A. Tate as Assistant Director and Acting Editor in 1961 provided aggressive leadership and, in addition, much of the necessary technical interface with the outside world.

TABLE 6.11. CAS R & D Directors and When They Held Office

Year	Name
1955–1959	Karl F. Heumann
1959–1963	G. Malcolm Dyson
1963–1967	Kenneth H. Zabriskie, Jr.
1967–1968	Robert W. White[a]
1968–1984	Ronald L. Wigington
1984–	Nick A. Farmer

[a] Acting director.

The first successful demonstration of use of CAS tapes outside CAS was achieved by Robert E. Maizell (Olin Corporation) and Charles N. Rice (Eli Lilly and Company) in 1962–1963 (1). They invented, developed, and implemented a method to use *CT* tapes for alerting of individual chemists or groups of chemists. This was the precursor for all computer-based alerting and searching in chemistry as practiced today. At about the same time, Purdue University ran sample *CT* tapes for students and faculty.

In 1965, CAS offered its first batch-search service based on the *CT* file. In 1967, the first experimental computer information center using CAS files was established at the University of Nottingham by The Chemical Society (London). Starting in 1965 and continuing today, CAS provided online searching of its structure files for the National Cancer Institute in Washington, D.C. Finally, in 1973, the first public online services based on CAS files were offered by System Development Corporation, Informatics, Inc., and Science Information Associates. Today, these CAS services are conveniently available through online vendors around the world as listed in Table 10.3. Furthermore, in 1980, CAS started its own online service *CAS ONLINE*.

One of the many benefits of computerization to chemists and other users of *CA* has been significant improvement in the promptness and appearance of the printed indexes. For example, the Collective Index for 1962–1966 appeared 44 months after completion of the collective period. For 1977–1981, this was reduced to just 14 months, an improvement of an astonishing 30 months, even though the size of the Collective index had more than tripled. Because of availability of CAS computerized input to online search services, *CA* index entries are now available simultaneously with, or earlier than, publication of the printed abstract issues. Evolution of CAS computer files and of associated equipment is summarized in Tables 6.12 and 6.13.

TABLE 6.12. CAS Computer File Availability—Batch and Online: Selected Key Dates

1962–1963	First experiments by Maizell and Rice using batch searches of *CT* for current awareness.
1965	CAS offered first batch-search service based on *CT* file.
1967	First experimental computer information center using CAS files established at the University of Nottingham by The Chemical Society (London)
1973	First online services based on CAS files offered by System Development Corporation (*CA Condensates* file), Informatics Inc., (NLM *TOXLINE* file, which includes *Chemical-Biological Activities*), and Science Information Associates (both *CA Condensates* and *TOXLINE*). Lockheed *DIALOG*® was licensed to offer online searches in 1974, and *Bibliographic Retrieval Services (BRS*®) in 1976.
1980	Initiation of *CAS ONLINE*.

TABLE 6.13. Acquisition of Key Computer Devices at CAS

Year	Device
1961	IBM 1401 (removed in 1965)
1964	IBM 1410 (upgraded to 7010 in 1966, removed in 1967)
1965	IBM 360/40 (removed in 1968)
1967	IBM 360/50 (removed in 1970)
1967	IBM 2280 Film Recorder (Functioned as a photocomposer. Replaced in 1974—donated to the Smithsonian Institute.)
1968	IBM 360/50 (second 360/50, removed in 1970)
1970	IBM 360/65 (removed in 1972)
1972	IBM 370/165 (Upgraded to 370/168 virtual system in 1974. Upgraded to 370/168, Model 3 in 1979. Attached Processor added in 1981. Still in use.)
1973	Autologic APS-4 photocomposition system (replaced in 1981—used only as a backup)
1980	XEROX 9700 Laser Printer installed
1981	Autologic APS-5 photocomposition system (still in use)
1982	IBM 3081 D16 (upgraded to 3081K48 by 1984)
1983	7 DEC VAX 780 Minicomputer systems operating UNIX installed for internal support
1984	6 additional DEC VAX 780 Minicomputer systems operating UNIX for internal support
1984	Second XEROX 9700 Laser Printer added
1980–1984	15 DEC PDP-11/45 and 14 DEC PDP-11/44 minicomputers added to support private and public *CAS ONLINE* processing
1985	529 terminals/personal computers and 55 printers operating from the 13 DEC VAX 780 UNIX systems. An additional number of terminals/ personal computers are functioning as follows:
	258 Distributed online edit with 2 DEC PDP-11/44 minicomputers
	114 Work at Home
	55 TSO (IBM's time sharing option)
	10 Name address (plus 31 IBM PCs also used for UNIX included in the 529 terminals/PCs listed)
	25 *CAS ONLINE*—internal CAS use
	56 Composition Services with 2 IV Phase minicomputers
	16 Registry Services with 1 DEC PDP-11/34 minicomputer
	22 H5-5 Heading systems with one IV Phase minicomputer
	10 Operator consoles
	4 Online Graphics with 2 DEC PDP-11/34 minicomputers
	12 Miscellaneous
	IBM 3090-M-200 to replace IBM 370-168

Online and offline computer-based searching of *CA* is in the process of totally transforming the information seeking patterns and habits of thousands of scientists and engineers throughout the world. This capability gives chemists the potential of outstanding searching power and speed unheard of just 10–15 years ago.

6.10. CHEMICAL ABSTRACTS SERVICE CHEMICAL REGISTRY SYSTEM

G. Malcolm Dyson, Director of Research at CAS during 1959–1963, originally conceived the idea of the Registry. He brought his structurally encoded files of boron and fluorine compounds to CAS, and this was eventually greatly modified and expanded into the present Registry. In 1965, Harry L. Morgan of the CAS staff, building upon brilliant work supplied by D. J. Gluck of the du Pont Company (2), perfected an algorithm for generating a unique and unambiguous computer representation of a chemical substance, that is, a unique connection table (3). This algorithm is the foundation of the CAS Chemical Registry System.

Initiation of the CAS Chemical Registry System in 1965 marked a quantum leap forward not only in the development of CAS but also in the development of all chemical information and related tools. CAS Registry Numbers are now widely used in a number of other publications and are recognized as the most uniquely accurate and unambiguous manner of identifying a chemical structure, other than by drawing and communicating the full chemical structure.

The principal component of the Registry structure record is a topological description in the form of a connection table listing of atoms and bonds that make up a substance's two-dimensional structure. Other portions of the record define known stereochemical characteristics of the molecule, identify any labeled atoms or atoms with abnormal valences, and indicate presence of charges in salts and complexes. Table 6.14 shows the growth of the Registry System over the years.

Efforts to develop a substructure search system started in the late 1960s. The interest and imagination of chemists had been stimulated by presentations on substructure searching at the open forums sponsored by CAS at ACS National Meetings. Substructure searching via nomenclature was first demonstrated as soon as computer-readable tapes containing indexed and named chemical substances became available. A full-fledged structure and substructure search system based on the CAS Registry file, *CAS ONLINE,* was introduced in 1980. This is more fully discussed in Section 10.11, but a brief description is given here.

CAS involvement with substructure search systems began in the late 1960s, with the development of an experimental system to search Registry II structure files. This system had screen and atom-by-atom search capabilities comparable to those of the present *CAS ONLINE* system but was restricted by the computer technology of the time; it was a tape-based batch mode system, with screen and atom-by-atom search manually encoded by the searcher and with answers limited to Registry Numbers. While the system did show the feasibility of substructure search procedures built around the CAS Registry structure files, it was hampered by lack of sufficiently powerful computer hardware and by the inability to provide structure diagrams for answers. As a consequence, the system was never promoted as a CAS service, although it was released on an experimental basis to several outside organizations. It was also used by the National Cancer Institute in a chemical information system to support their biological screening programs.

A few years later, several Swiss chemical firms—Ciba Ltd., J. R. Geigy Ltd. (now Ciba–Geigy Ltd.), F. Hoffmann–LaRoche & Co., Ltd., and Sandoz Ltd.—decided to jointly develop a computer system for chemical information based on the CAS Registry System and formed the Basel Information Center for Chemistry (BASIC). Their system used the experimental CAS substructure search system to search a CAS Registry structure file licensed from CAS as well as private structure files built and maintained by a system based on CAS Registry II programs. The BASIC group made a number of improvements to the original CAS search system, developing several new screen types and revising the screen dictionary; these improvements reduced the costs of search screen generation and (by providing better screen-out) atom-by-atom searching, the two time-consuming aspects of substructure search. At the same time, BASIC upgraded the private structure file aspects of their system with the CAS Registry III programs.

In the late 1970s, CAS reexamined the possibilities of substructure search as a CAS service. While the earlier work discussed above had demonstrated the feasibility of using an initial screen search followed by atom-by-atom search, it was clear that the large size of the file (then already over 5 million substances) would present problems. The inverted file organization usually used for online searching could be expected to lead to unacceptably slow response times, due to the very large file size. Some of the screens which would normally be generated from a query would be assigned to very large numbers of substances, leading to the necessity to intersect very large lists of potential retrievals.

The result of the investigation was a new approach to large-file searching. [Reprinted with permission from the American Chemical Society (P. G. Dittmar et al., "The CAS ONLINE Search System") published in the *Journal of Chemical Information and Computer Sciences* **23,** 93–102 (1983).]

Substances within the file can be searched to identify those that share structural characteristics and that therefore may share properties or activities of interest. Each search answer provides a structure diagram, a CAS Registry Number, a *CA* Index name and up to 50 synonyms (including common and trade names), and, at the searcher's option, *CA* abstract numbers and bibliographic references for the 10 most recent references to the

TABLE 6.14. Growth of CAS Chemical Registry Files

Year	Number of Substances Registered	Substances on File at Year End
1965	211,934	211,934
1966	313,763	525,697
1967	270,782	796,479
1968	230,321	1,026,800
1969	287,048	1,313,848
1970	288,085	1,601,933
1971	351,514	1,953,447
1972	277,563	2,231,010
1973	437,202[a]	2,668,212
1974	319,808	2,988,020
1975	372,492	3,360,512
1976	347,515	3,708,027
1977	369,676	4,077,703
1978	364,226	4,441,929
1979	346,062	4,787,991
1980	353,881	5,141,872
1981	424,230	5,566,102
1982	361,706	5,927,808
1983	418,905	6,346,713
1984	563,390[b]	6,910,103
1985	544,618[c]	7,454,721

[a] Registry III implementation added 77,650 substances.
[b] Input of substances indexed prior to 1965 added 140,760 substances.
[c] Input of substances indexed prior to 1965 added 151,431 substances.

substance in the literature. In 1983, CAS added the capabilities to display abstracts for some 3.5 million documents, and to search CAS bibliographic and index data back to 1967. Abstracts may now also be searched.

Sufficient funds are now on hand for CAS to begin registering substances cited prior to 1965 in *CA Collective Formula Indexes*. As a first step, each substance in the *Seventh Collective Formula Index* (1962–1966) is being compared with the CAS Registry File. If the substance is not already in the Registry File, it will receive a new Registry Number. Results of pre-1965 registration are available in *CAS ONLINE*. More than 140,000 substances reported between 1962 and 1965, and their *CA* references, were added to the CAS computer files during 1984 and made available for searching through *CAS ONLINE*. During 1985, over 150,000 more pre-1965 substances were registered.

The following describes what CAS considers to be chemical substances, which are indexed and registered in the CAS Registry System as separate entities:

> Chemical elements and their isotopes, including hypothetical elements; chemical compounds and their stereoisomers, including fully defined and incompletely defined derivatives; ions; free radicals; specific antibiotics, enzymes, hormones, polypeptides, polysaccharides, and polynucleotides; polymers of specific compounds; mixtures resulting from an intentional admixing (prior to use) of individual components; elementary particles, including certain class designations; alloys of specific metals; specific minerals (as distinct from rocks); and substances possessing alphanumeric and trade-name designations (except a very few which have come to be regarded as generic).

Note the provision for inclusion of hypothetical elements, ions, and incompletely defined derivatives, all of which can be the subject of research reported in the literature. This is a composite statement based on paragraphs in the *CA Chemical Substance Index* Introduction, the *CA Index Guide,* and several published papers on the CAS Registry System. A further discussion of the CAS Registry System is given in Chapter 10.

6.11.   *CA SELECTS*

In 1976 CAS began publication of what has proved to be the extremely popular *CA SELECTS* series. This has now grown to a total of 164 different topics available in 1986 (see Table 4.2).

Anyone walking through the facilities of almost any modern chemical R & D organization will almost surely see copies of *CA SELECTS* in both laboratories and office areas. Not even in its most recent heydays did the full *CA* enjoy this kind of personal use and identification with individual chemists.

Reasons for this popularity are many. The major reason is probably the handy grouping in individual, rather modest-sized, easy-to-read booklets of recent abstracts pertaining to specific, limited areas of chemistry that are the subjects of the most intense current research of commercial interest. CAS staff have worked hard at keeping the list of *CA SELECTS* topics flexible and consistent with the changing interest of chemists, and they both add and delete topics accordingly (see Table 4.1).

The novelty of *CA SELECTS* is that it transcends *CA* Section boundaries; an abstract may be included in more than one topic, whether the emphasis of the document is primary or secondary to the topic. *CA SELECTS* focuses on very specialized areas. On the other hand, *CA* sections reflect subdisciplines of chemistry, and an abstract is placed in only one section, based on primary emphasis of the original document, with cross-reference to related sections. For a further discussion of *CA SELECTS,* see Chapter 4.

6.12.   COOPERATIVE ACTIVITIES

CAS cooperation with other abstracting and indexing services is important because this reflects the increasingly interdisciplinary trends in science and technology and because of the potential of improvements in coverage of some selected areas. Cooperation frequently takes place within the framework of activities of such organizations as the International Council of Scientific Unions—Abstracting Board (ICSU/AB) [since June, 1984, the International Council for Scientific and Technical Information (ICSTI)] and the National Federation of Abstracting and Information Services (NFAIS). In addition, bilateral and trilateral agreements with BioSciences Information Service (BIOSIS), and BIOSIS and Engineering Information, Inc., (Ei), respectively, have provided for joint studies, conducted since 1970, on overlap of coverage, bibliographic standards, taxonomic vocabulary, and so on. Details on cooperative activities are given in the sections that follow:

A.   National Library of Medicine

In 1966, CAS developed a special registry of chemicals associated with foods, drugs, pesticides, and cosmetics under the joint sponsorship of the National Library of Medicine (NLM) and the National Science Foundation. This project established an experimental link between the CAS Chemical Registry and NLM's online *MEDLARS* service. Subsequently, CAS began providing information for NLM's online *TOXLINE* and *CHEMLINE* files and continues to do so.

B.   BioSciences Information Service and Engineering Information

In 1970, CAS, BIOSIS, and Ei undertook a study to determine overlap between the services in terms of journals monitored, amount and kind of documents selected for coverage, and differences and similarities in form and content of records published, and compatibility of machine-readable records. Results from studies of the overlap of the journals monitored and the articles selected for coverage, and a comparison of bibliographic descriptions used by the three services were published in 1972, 1973, and 1976, respectively. In 1975, BIOSIS, CAS, and Ei jointly produced the *Bibliographic Guide for Editors and Authors*.

In 1972, CAS and BIOSIS began working toward the development of common taxonomic indexing, common methods for indexing chemical substances, and common abbreviations, and CAS Registry Numbers began appearing in *BIOSIS*'s *Abstracts on Health Effects of Environmental Pollutants*. Since 1976, *Biological Abstracts* and several other BIOSIS publications have been photocomposed at CAS under contract.

In 1981, CAS and BIOSIS launched *BIOSIS/CAS SELECTS*, a series of biweekly current awareness publications that bring together abstracts from both the CAS and BIOSIS databases on specific research topics that span the biological and chemical disciplines.

C.   The Chemical Society—The Royal Society of Chemistry

An agreement was signed between ACS and The Chemical Society (now The Royal Society of Chemistry) in 1969 under which the British society provided CAS with abstracts and index entries for some British scientific papers and patents and marketed CAS publications and services in the United Kingdom and Ireland. While the agreement has seen several modifications, the general terms currently are approximately the same as the initial agreement.

Earlier, The Chemical Society's research unit had been one of the first organizations to work with CAS's experimental computer-readable files and developed some of the first search techniques and programs for these files.

D.   Gesellschaft Deutscher Chemiker—Internationale Dokumentationsgesellschaft fuer Chemie

When publication of *Chemisches Zentralblatt* was discontinued at the end of 1969, Gesellschaft Deutscher Chemiker (GDCh) began to provide abstracts for input to the CAS database and market CAS publications and services in West Germany. The German input ceased in 1975, when the agreement with GDCh was replaced by a broader agreement with Internationale Dokumentationsgesellschaft fuer Chemie (IDC), which continues in a somewhat modified form today. Marketing responsibility for CAS publications and services in West Germany is now handled by VCH Verlagsgesellschaft mbH.

E.   Japan Association for International Chemical Information

In 1977, an agreement was signed with the Japan Association for International Chemical Information (JAICI) under which the Japanese organization took over the marketing of CAS publications and services in Japan and was permitted to develop publications and services of its own from the CAS database for use in Japan. In 1984 JAICI, with the help of some outside collaborators, processed about 9000 Japanese patent documents for CAS.

F.   Centre Nationale de l'Information Chimique

A similar agreement was concluded with France's Centre Nationale de l'Information Chimique (CNIC) in 1978. CAS and CNIC had been cooperating informally on development of chemical information retrieval methods since 1972.

G. FIZ Karlsruhe

In 1984 the ACS initiated cooperative efforts with the West German organization Fachinformationszentrum Energie, Physik, Mathematik GmbH (FIZ Karlsruhe). The result is availability of *CAS ONLINE* to European users and eventual availability to American and Canadian users of all of FIZ Karlsruhe's files, including those relating to mathematics, chemical engineering, physics, nuclear research and engineering, and nuclear magnetic resonance spectra. The cooperative online system through which files are available in North America and Europe is STN[SM] (Scientific and Technical Information Network). See also Section 10.12.C.

Cooperative arrangements, such as those just described, permit CAS to more effectively keep up with literature from other countries and to add new databases or files not previously available. See also Chapter 8 for further descriptions of some of the abstracting services mentioned in this Chapter.

6.13. PRICING AND MARKETING

The changes in the pricing structure of *CA* (see Table 6.15) are fascinating. Impinging on it have been many complex factors, including increases in the cost of doing business, and in the number of documents covered and number of pages printed, as well as increasingly heavy reliance on computers as the principal tools for producing (typesetting and printing) abstracts and indexes.

When *CA* was first published, publication costs were subsidized from dues paid by ACS members, and members who wished to, received it free of charge. The basic subscription cost to nonmembers was $6.00. In 1956, it was decided that CAS was to be self-supporting, and in 1966 CAS dropped the individual member rate entirely. The subscription cost for 1986 was $9200 per year. (Colleges may purchase *CA* for $7800, and small undergraduate schools in the United States and Canada can purchase it for much less.) A decision was made by the ACS Board of Directors in 1978 that more than half of CAS future revenues must come from sources other than the printed *CA* by 1983. This was achieved. Thus, the full *CA* has evolved from what was once a very affordable tool for individual chemists to one that is now quite frankly and intentionally priced at the institutional level.

The dichotomy is that, at the same time, CAS does offer a number of newer, intensely personal tools, most notably *CA SELECTS* and some aspects of the online service, but these may carry significant prices—$120 for each topic of *CA SELECTS* (rate for non-*CA* subscribers) and from several to several hundred dollars for a search of *CAS ONLINE* depending on the type of question and searcher's experience. In addition, optimum use of *CAS ONLINE* requires training and equipment best afforded by well-financed organizations or individuals.

TABLE 6.15. Subscription Price[a] and Number of Subscriptions to *Chemical Abstracts*

Year	Annual Subscription Price[a]			Number of Subscriptions	
	Base ($)	College or University ($)	ACS Member ($)	Total	ACS Member
1934	6		6	11,883	9,768
1935	6		6	12,020	9,766
1936	6		6	12,463	10,007
1937	6		6	13,098	10,480
1938	6		6	13,408	10,595
1939	6		6	13,810	10,703
1940	12		6	13,934	10,780
1941	12		6	14,249	11,006
1942	12		6	13,605	11,228
1943	12		6	14,651	12,067
1944	12		6	15,428	12,463
1945	12		6	16,729	13,138
1946	12		6	19,428	14,622
1947	12		6	21,524	15,758
1948	15		7	21,705	15,586
1949	15		7	22,675	15,919
1950[b]	20		10	21,627	14,631
1951[b]	60		15	19,232	12,701
1952[b]	60		15	19,174	11,639
1953[b]	60		15	19,361	11,487
1954[b]	60		15	19,510	11,283
1955[b]	60		15	19,475	11,023
1956	350	80	20	16,907	11,688
1957	350	80	20	17,216	11,172
1958	350	80	20	17,301	11,036
1959	350	80	20	17,351	11,133
1960	570	150	32	16,107	10,254
1961	925	200	40	15,183	9,747
1962	925	200	40	14,271	8,901
1963	1,000	500	500	6,758	1,458
1964	1,000	500	500	6,802	1.426
1965	1,200	700	700	6,808	1,378
1966	1,200	700	[c]	6,672	[c]
1967	1,200	700		6,653	
1968	1,550	1,050		6,451	
1969	1,550	1,050		6,401	
1970	1,950	1,450		NA[e]	
1971	1,950	1,450		NA	
1972	2,400	1,900		NA	
1973	2,400	1,900		NA	

TABLE 6.15. Continued

Year	Annual Subscription Price[a] Base ($)	College or University ($)	ACS Member ($)	Number of Subscriptions Total	ACS Member
1974	2,900	2,400		NA	
1975	2,900	2,400		NA	
1976	3,500	3,000		NA	
1977	3,500	3,000		NA	
1978	4,200	3,700		NA	
1979	4,200	3,700		NA	
1980	5,000	4,500[d]		NA	
1981	5,500	5,000[d]		NA	
1982	6,200	5,200[d]		NA	
1983	6,800	5,800[d]		NA	
1984	7,500	6,400[d]		NA	
1985	8,500	7,200[d]		NA	
1986	9,200	7,800[d]			

[a] The nonmember subscription price of *CA* was $6 per year from 1907 through 1939. Prior to 1934, all ACS members who wanted *CA* received it free of charge. *CA* publication expense was subsidized from ACS member dues from 1907 through 1953. Subscription totals prior to 1934 are not available and since 1970 have not been made public.
[b] *CA* publication expense was subsidized in part from ACS member dues through 1955.
[c] Special subscription price for ACS members was eliminated in 1966.
[d] Since 1980 smaller U.S. colleges have been eligible for substantial grants toward *CA* subscriptions. The amount of the grant varies according to enrollment. Colleges in the smallest category (fewer than 500 students) were eligible for a total grant of $7800 in 1986.
[e] NA = not available from CAS (proprietary).

Without the changes in pricing structure just described, CAS would have been unable to continue its outstanding coverage of worldwide chemical literature, and indeed might well have been forced to sharply curtail or even to go out of business, a fate that was met by the U.K. counterpart of *CA*, *British Abstracts* in 1953, by its German counterpart, *Chemisches Zentralblatt* in 1969, and by the French *Bulletin Signalétique* at the end of 1983.

In any event, time has clearly demonstrated that high quality chemical information, delivered in large volume and on a prompt basis, is expensive. Most chemists seem to have fully accepted this fact.

No discussion of *CA* subscription costs would be complete without mention of the *CA* Section Groupings. Priced at $200 per year to ACS members (limit of two different Grouping subscriptions per member) and $1200 per year to nonmembers, for 1986, these continue to be one of the best technical information purchases for members of ACS.

Related to changes in pricing policies at CAS is the establishment of a strong Marketing Division in 1978. It has become clear that establishment of this division, although originally criticized by some as too commercial, was necessary as CAS products became more technologically sophisticated, as they continued to grow in size, and as economic pressures became more severe.

Achievements of the marketing staff include development of a complete battery of training programs and manuals. This division has played an important role in the development of a revised fee structure for CAS products. Most importantly, marketing has determined the need for new and improved products, especially those that are computer based, as mandated by the ACS policy decision of 1978 that the primary sources of future revenues are to be publications and services other than the printed *Chemical Abstracts*.

6.14. OTHER CHEMICAL INFORMATION SERVICES

Some aspects of CAS can be better appreciated when considered in the perspective of other major chemical information services. CAS is the only remaining major international literature and patent service of its type, with the exception of *Referativnyi Zhurnal, Khimiya,* published in the Soviet Union since 1953. The French *Bulletin Signalétique,* founded in 1940, was discontinued at the end of 1983. *British Abstracts (BA),* originally founded in 1871, ceased publication as such in 1953, and the venerable *Chemisches Zentralblatt (CZ),* founded in 1830, was discontinued as such in 1969. Both *BA* and *CZ* underwent a number of name changes over the years, and both, as well as the French publication, have been succeeded by publications with new names in greatly watered down, modified form. See also Chapter 8.

Discontinuance of the original German and British services has avoided much unnecessary duplication of *CA*. Both are, however, still valuable today for years not covered by *CA,* or for those subject areas or types of documents where they may complement *CA*. For example, *CZ* is noted for its coverage of some earlier patents, especially German patents, and for the detail of its abstracts compared to those of *CA*.

The only other major chemical abstracting and indexing service published in the United States is *Current Abstracts of Chemistry and Index Chemicus®,* published in the United States since 1960 by the *Institute for Scientific Information® (ISI®)* in Philadelphia. This publication does provide only partial coverage of journals and does not cover patents, but it does have a number of attractive and unique features. *ISI®* also offers a broad range of other important chemical information services. See also Chapter 8.

The leading British chemical information service is believed by many to be the Derwent *Chemical Patents Index,* issued in one form or another since 1951 by Derwent Publications, Ltd. This worldwide patent information service has an excellent reputation, especially among industrial chemists and engineers. See also Chapter 13.

The Royal Society of Chemistry publishes a series of rather specialized abstracting and indexing services primarily in such fields as analytical chemistry and safety. Unfortunately, these are probably not as well known in the United States as they should be. See also Chapters 8 and 18.

6.15. OTHER REMARKS

CAS has emerged as by far the most dominant factor in the chemical information industry worldwide. As with any other situation of comparable reputation, power, and influence, there are those who are somewhat uneasy about the outlook for the future. For example, will there continue to be room for small innovators in this field—can they hope to get into the arena now? Can (or should) additional healthy competition be nurtured to provide novel and diverse approaches? Would additional chemical information sources, if not really needed, simply dilute talent and financial resources and confuse unnecessarily?

What kind of influence can users expect to have in the future on shaping of CAS programs and policies? Are the user committees (or councils) that have been established by CAS a strong enough link to "grass roots" opinions of chemists? Now that chemical information has become so complex and sophisticated, how important are grass roots opinions—perhaps what count most are opinions of a relatively few who make the effort to become highly informed. How can average chemists most effectively express their opinions and recommendations on handling of literature and patents that they and their colleagues generate and use? How can CAS best be persuaded to do something promptly if a significant segment of the chemical community thinks it should be done? Have all of the major advances at CAS made chemical information access significantly easier, or "merely" helped chemists keep their heads above the waters of the flood of chemical publications?

These are complex and controversial issues. ACS leadership and CAS staff will surely welcome chemical community input on all of these questions. A brief comment on one of the issues is provided at the end of the next chapter.

REFERENCES

1. R. R. Freeman, J. T. Godfrey, R. E. Maizell, C. N. Rice, and W. H. Shepherd, "Automatic Preparation of Selected Title Lists for Current Awareness Services and or Annual Summaries," *J. Chem. Doc.,* **4,** 107–112 (1964).

2. D. J. Gluck, "A Chemical Structure Storage and Search System Developed at du Pont," *J. Chem. Doc.,* **5,** 43–51 (1965).

3. H. L. Morgan, "The Generation of a Unique Machine Description for Chemical Structures—A Technique Developed at Chemical Abstracts Service," *J. Chem. Doc.,* **5,** 107–113 (1965).

SELECTED PAPERS OF HISTORICAL INTEREST RELATING TO *CA* (in chronological order)

1. Bernier, C. L. and E. J. Crane, "Indexing Abstracts," *Ind. Eng. Chem.,* **40,** 725–730 (1948).

2. Emery, A. H., E. J. Crane, and G. G. Taylor, "The Future of *Chemical Abstracts,*" *Chem. Eng. News,* **28**(30), 2517–2520 (1950).

3. Crane, E. J., "Scientists Share and Serve (The Priestley Medal Address)," *Chem. Eng. News,* **29**(42), 4250–4253 (1951).

4. Crane, E. J. "Chemical Abstracts," in *A History of the American Chemical Society;* C. A. Browne and M. E. Weeks, Eds., American Chemical Society, Washington, D.C., 1952, pp. 336–367.

5. Volwiler, E. H. and A. C. Cope, "Chemical Abstracts—Millstone or Milestone," *Chem. Eng. News,* **33**(25), 2636–2639 (1955).

6. Crane, E. J., "The Chemical Abstracts Service—Good Buy or Good-by," *Chem. Eng. News,* **33**(26), 2752–2754 (1955).

7. "CA Today: The Production of Chemical Abstracts," American Chemical Society, Washington, D.C., 1958.

8. Dyson, G. M., "Research Expansion at the Chemical Abstracts Service," *Chem. Eng. News,* **37**(36), 128–131 (1959).

9. Dyson, G. M., "Closing the Gap in Chemical Documentation," *Chem. Eng. News,* **38**(19), 70–72 (1960).

10. Baker, D. B., "Growth of Chemical Literature," *Chem. Eng. News,* **39**(29), 78–81 (1961).

11. Crane, E. J. and C. C. Langham, "On-the-Job Training at Chemical Abstracts Service," *J. Chem. Doc.,* **2,** 199–204 (1962).

12. Freeman, R. R. and G. M. Dyson, "Development and Production of *Chemical Titles,* a Current Awareness Index Publication Prepared with the Aid of a Computer," *J. Chem. Doc.,* **3,** 16–20 (1963).

13. Dyson, G. M. and M. F. Lynch, "Chemical-Biological Activities—A Computer-Produced Express Digest," *J. Chem. Doc.,* **3,** 81–85 (1963).

14. Freeman, R. R., J. T. Godfrey, R. E. Maizell, C. N. Rice, and W. H. Shepherd, "Automatic Preparation of Selected Title Lists for Current Awareness Services and as Annual Summaries," *J. Chem. Doc.,* **4,** 107–112 (1964).

15. Cossum, W. E., M. L. Krakiwsky, and M. F. Lynch, "Advances in Automatic Chemical Substructure Searching Techniques," *J. Chem. Doc.,* **5,** 33–35 (1965).

16. Morgan, H. L., "The Generation of a Unique Machine Description for Chemical Structures—A Technique Developed at Chemical Abstracts Service," *J. Chem. Doc.,* **5,** 107–113 (1965).

17. Baker, D. B., "Chemical Literature Expands," *Chem. Eng. News,* **44**(23), 84–86 (1966).

18. Siegel, H., D. C. Veal, and D. A. McMullen, "Polymer Science & Technology (POST): A Computer-Based Information Service," *J. Polym. Sci., Part C,* (25), 191–196 (1968).

19. Baker, D. B., "World's Chemical Literature Continues to Expand," *Chem. Eng. News,* **49**(28), 37–40 (1971).

20. Donaldson, N., W. H. Powell, R. J. Rowlett, Jr., R. W. White, and K. V. Yorka, *"Chemical Abstracts* Index Names for Chemical Substances in the Ninth Collective Period (1972–1976)," *J. Chem. Doc., 14,* 3–14 (1974).

21. Proceedings of the Symposium on Chemical Abstracts in Transition, Chicago, IL, August 28, 1973, Chemical Abstracts Service, Columbus, OH, 1974.

22. Baker, D. B., "Recent Trends in Growth of Chemical Literature," *Chem. Eng. News,* **54**(20), 23–27 (1976).

23. Chemical Abstracts Service, in *A Century of Chemistry: The Role of Chemists and the American Chemical Society,* H. Skolnik and K. M. Reese, Eds.; American Chemical Society, Washington, D.C., 1976, pp. 126–143.

24. Baker, D. B., J. W. Horiszny, and Metanomski, W. V. "History of Abstracting at Chemical Abstracts Service," *J. Chem. Inf. Comput. Sci., 20,* 193–201 (1980).

25. Baker, D. B., "Recent Trends in Chemical Literature Growth," *Chem. Eng. News,* **59**(22), 29–34 (1981).

26. Zaye, D. F., W. V. Metanomski, and A. J. Beach, "A History of General Subject Indexing at Chemical Abstracts Service," *J. Chem. Inf. Comput. Sci.* **25,** 392–399 (1985).

7 ESSENTIALS OF *CHEMICAL ABSTRACTS* USE

7.1 INTRODUCTION

As indicated in Chapter 6, by far the most heavily used and valuable information tool in chemistry, and the one best known to chemists and chemical engineers, is *Chemical Abstracts (CA)*, the chief product of CAS (Chemical Abstracts Service). CAS is a division of the American Chemical Society (ACS).

CA is so important—and so complex—that there are audio courses on its use (1). CAS publishes a workbook and offers other user aids and seminars on the use of *CA* (2). Even though in the introductions to *CA*, its indexes, and the *Index Guide* (see Section 7.5.A) there are many pages explaining in detail policies and proper use, especially for the indexes, the key to successful use of *CA*—as for any tool—is *hands-on* use on a regular basis. It requires both study and experience to use *CA* to full advantage.

A lengthy book could be written about the use of *CA* alone, much less the balance of the important tools on chemical information. Accordingly, this chapter focuses on essentials, with emphasis on points less well understood by many chemists, on newer developments, and on important recent changes. In addition, understanding of the historical aspects, as described in the previous chapter, is essential to use of *CA*.

7.2. COVERAGE

The official statement of what *CA* covers is important and worth quoting in its entirety (3).

> It is the careful endeavor of *Chemical Abstracts* to publish adequate and accurate abstracts of all scientific and technical papers containing new information of chemical or chemical engineering interest and to report new chemical information revealed in the patent literature, but the American Chemical Society is not responsible for omissions or for such mistakes as may be made in abstracts and index entries.

100

Although the implication is that *CA* covers only new information, its editors cannot possibly check all material included to see if it is new. They can only make the *careful endeavor* noted in their policy statement. *CA* editors necessarily rely to a large extent on original statements as to novelty of information in the literature.

7.3.   ABSTRACT AND INDEX CONTENT

CA contains informative abstracts of original documents. As mentioned in Chapter 6, these abstracts are not intended to replace the original, nor are they critical or evaluative. Rather, the abstracts are filters, with the intent of providing the user with enough information on content to permit one to determine whether one wants to consult the originals. Any scientist who repeats a procedure based purely on the *CA* abstract is making a mistake. Moreover, not all new substances and subjects reported in the original appear in the abstract; all are, however, covered by the Volume and Collective Indexes, according to *CA* policy.

Recent (Vol. 101, July–December 1984) figures from CAS show that the overall subject and substance index "density" (number of index entries per abstracted document) is 7.4 for all 80 sections. Density for sections that pertain to synthetic chemistry—such as Section 28 (heterocyclic compounds)—is over 28 index entries. These figures are for general subject and chemical substance index entries only; they do not include additional access points such as author names and molecular formulas.

CA has been significantly increasing index density since 1972; earlier years were not as deeply indexed. Beginning with the ninth collective period (1972–1976), indexing of reactants and intermediates was extended to all subject areas covered by *CA,* thereby further strengthening the indexes. This was, of course, in addition to the products of reactions that are the usual focal points of interest.

7.4.   SPEED OF COVERAGE AND INDEXING

The median time lag for the appearance of the abstract following issuance of the original publication has shown good improvement. It is less than 3 months as compared to about 3.5 months in the late 1970s. This includes completion of the indexing process that is important to online use (see Chapter 6). For the 700 journals covered by *Chemical Titles,* this median is less than 2 months. Some chemists may believe that *CA* is still too slow, but the abstracts may be available before the original journals are on library shelves, especially in the case of journals published abroad.

CA editors are making determined (and in many cases successful) efforts to further improve timeliness for all parts of *CA.* One notable example of success is speed of issuance of the semiannual (6 month) Volume Indexes. Speed of index appearance has improved dramatically to within about 3 months after the completion of the semiannual volumes, as compared to about 6 months in the late 1970s. This is attributed in large measure to the massive computerization program initiated by CAS staff in the 1960s. The speed of appearance of both abstracts and indexes is probably approaching the optimum that can be achieved with contemporary technology, human capabilities, and management skills.

7.5.　*CHEMICAL ABSTRACTS* INDEXES

The following is a broad summary of the contents of the Volume, weekly and Collective Indexes to *CA* (see also Chapter 6):

A.　Index Guide (new edition every 18 months since 1982)

Details the major points of *CA* indexing policy for the appropriate collective period and provides cross-references from chemical substance names and general subject terms used in the literature to the terminology used. The *CA Index Guide* contains the following parts:

Introduction describes cross-references, homograph definitions, and indexing policy notes listed in the main portion of the *Index Guide*.

Alphabetical sequence of cross-references and indexing policy notes is the main portion of the *Index Guide*. New for the 1985 edition is the addition of all the valid general subject index headings (excluding latinized genus and species names) and diagrams for stereoparents.

Volume Indexes to Chemical Abstracts: Organization and Use discusses the relationship between and use of the chemical substance, general subject, and formula indexes and index of ring systems.

Selection of General Subject Headings discusses the content of the general subject index.

Chemical Substance Index Names summarizes the rules used in deriving the *CA* index names for chemical substances, and includes a section on converting *CA* index names to structural diagrams.

Hierarchies of General Subject Headings aids in generic searching.

B.　Chemical Substance Index　(semiannual)

This relates the *CA* index names of chemical substances and their CAS Registry Numbers to *CA* abstract numbers for documents in which the substances are mentioned. Included with the *CA* index name is a brief description of the document's context. This index was initiated with the ninth collective period (1972–1976), prior to which chemical substances and general subjects were in a single index.

C. General Subject Index (semiannual)

This relates index entries that do not refer to specific chemical substances to the corresponding *CA* abstracts. These entries include concepts, general classes of chemical substances, applications, uses, properties, reactions, apparatus, processes, and biochemical and biological subjects.

D. Formula Index (semiannual)

This index relates the molecular formulas for chemical substances with the *CA* chemical substance index names, CAS Registry Numbers, and corresponding *CA* abstract numbers.

E. Index of Ring Systems (semiannual)

Ring composition, ring size, and number of rings are listed providing a means for determining the systematic *CA* index names for specific ring systems as well as the nonsystematic *CA* index names for cyclic natural products containing these ring systems.

F. Author Index (semiannual)

Authors, patentees, and patent assignees are listed in alphabetical order with titles of their articles or patent specifications and *CA* abstract numbers.

G. Patent Index (semiannual)

This index includes (1) entries for all newly abstracted patent documents; (2) cross-references to the first-abstracted patent on an invention when more than one patent describes that invention and (3) a listing at the first-abstracted patent of all the equivalent patents. The first *CA* numerical patent index was introduced in 1912, but was discontinued in 1915. It was reintroduced in 1935. A *Patent Concordance* was started in 1963, and the present combined numerical and concordance index begins coverage with the tenth collective period (1977–1981). In addition, the Special Libraries Association has published a collective numerical patent index to *CA* covering 1907–1936.

H. Weekly Issue Indexes

Each weekly issue of *CA* contains several indexes. The issue indexes include (a) keyword index, (b) patent index, and (c) author index.

Chemists can use the keyword index to search specific issues of *CA* for subject content. As mentioned in Chapter 6, keyword index vocabulary is uncontrolled, that is, nonsystematic and not standardized. It reflects author terminology and the content of the abstract. This permits the index to be produced quickly and published as a part of each issue, but it makes for substantially less detail and much more superficial index coverage than is found in the printed semiannual volume indexes, which appear some three months after completion of the volume. For online searching, both the keyword index terms and the systematic, standardized volume index entries are available through *CA SEARCH* or *CAS ONLINE* at the time of abstract publication.

The patent index, which includes both abstracted and equivalent patents, is cumulated semiannually in the Volume Indexes and every five years in the Collective Indexes. The issue author index, consisting of author name and initials and the *CA* abstract number, is less detailed than the semiannual volume author index and is designed to serve the issue in which it appears.

I.   *CA* Collective Indexes

Tables 6.4, 6.5, and 7.1 depict the development of *CA* Collective Indexes.

7.6.   *CHEMICAL ABSTRACTS* NOMENCLATURE AND SEARCHING *CHEMICAL ABSTRACTS* FOR CHEMICAL SUBSTANCES

As previously mentioned, beginning with Volume 76 (1972), CAS made a significant change to systematic IUPAC nomenclature. Note that this change applies to the boldface index headings, not to index modifications in which original author nomenclature is used.

Examples of the old and new policies follow:

Old Policy	New Policy
Acetic acid	No change
Aniline	Benzenamine
Carbonic acid	No change
Ethylene	Ethene
Ethylenediamine	1,2-Ethanediamine
Ethylene glycol	1,2-Ethanediol
Ethylene oxide	Oxirane
Ethylenimine	Aziridine
Ethyl ether	Ethane, 1,1'-oxybis-
Ethyl methyl ketone	2-Butanone
Formic acid	No change
Phenol	No change

TABLE 7.1. Development of *Chemical Abstracts* Collective Indexes

- 1st–4th Decennial Indexes each contain author and subject indexes.
- 1st Collective Formula Index covers 1920–1946.
- Ten-year Numerical Patent Index covers 1937–1946.
- 5th Decennial Index contains author, subject, numerical patent, and formula indexes.
- 6th Collective Index contains author, subject, numerical patent, and formula indexes and index of ring systems.
- 7th Collective Index contains author, subject, numerical patent, and formula indexes; index of ring systems; and patent concordance.
- 8th Collective Index contains author, subject,[a] numerical patent, and formula indexes; index of ring systems; patent concordance; and index guide.
- 9th Collective Index contains author, chemical substance, general subject,[b] numerical patent, and formula indexes; index of ring systems; patent concordance; and index guide.
- 10th Collective Index contains author, chemical substance, general subject,[b] patent, and formula indexes; index of ring systems; and index guide.

[a] The *Subject Index* includes *Registry Handbook–Number Section,* which were issued as part of the Eighth Collective Index.
[b] The *Subject Index* was subdivided into *General Subject* and *Chemical Substance Indexes* beginning with the Ninth Collective Index period.

Some chemists believe that this policy change makes it more difficult for them to use *CA*. They contend that such heavily studied chemicals as aniline, for example, should be indexed under that name and not under *benzenamine*, which few chemists use in conversation, normal reporting, or commerce. Although the fully systematic name for aniline is cross-referenced from aniline in the *CA Chemical Substance Index,* most trivial names are not similarly cross-referenced, except in the *Index Guide*.

CA editors believe that their index name policy change is fully justified.

The primary intent of this effort is to group structural families together in the alphabetically ordered *CA Volume Chemical Substance Index*. Other reasons behind the use of more fully systematic index names include:

1. They are generated more consistently.
2. They are translated more readily to complete structures.
3. They are edited and verified more rapidly and consistently.
4. They reduce manual and machine search efforts.
5. They employ fewer and more systematic nomenclature principles.

A simple example of past confusion is 4-methoxytoluene, also known as 4-methylanisole. Anisole and toluene are both descriptive *trivial* names.

Under the new fully systematic index policies, there is no confusion. The compound is now designated by *CA* as *1-methoxy-4-methylbenzene.*

CA editors have provided several powerful tools to help chemists adjust to the new nomenclature and make most effective use of the general subject and chemical substance indexes published under the new policies. These include (a) the *Index Guide (IG)*, (b) the *Formula Index,* and (c) the Chemical Registry System and *Registry Handbooks*.

As previously indicated, the *IG,* which is reissued in its entirety every 18 months, is a collection of cross-references and explanations of headings, which should be consulted if the current *CA* index policy for a substance is not known. New for the 1985 edition is the addition to the alphabetical sequence of all the general subject index headings (except latinized genus and species names) and diagrams for acyclic and cyclic stereoparents. (Prior to the initiation of the *IG,* the chemist found such cross-references and notes scattered throughout the *Subject Index.*) The *IG* is the indispensable key to efficient use of the *General Subject* and *Chemical Substance Indexes,* with special value in helping the chemist *translate* trivial names to the more fully systematic index names used by *CA*. Chemists using *CA* should use the more recent *IG* first and then the *IG* for the tenth (1977–1981) or ninth (1972–1976) collective period when they are searching retrospectively in these periods. There is also an *IG* for the eighth collective period (1967–1971), but much of the latter *IG* does not apply to current practice.

The *Formula Index,* published semiannually and cumulated every five years, provides the chemist with the quickest and most reliable route of access to names of individual substances as listed in the *Chemical Substance Index*. The *Formula Index* is easier and more straightforward to use than the *IG,* especially if an accurate molecular formula is already in hand. Under each molecular formula, the chemist will find the *CA* systematic name for a compound with that empirical formula, the CAS Registry Number (for positive identification), and an abstract number in which the compound is mentioned in *CA* issues. A physical property, such as boiling point, is given when an original document does not provide enough information for an unambiguous name. Chemists should, however, not let these comments discourage them from use of the *IG;* calculation of molecular formulas can also be error prone.

The next step after use of the *Formula Index* is to consult the *Chemical Substance Index* if textual information is desired to narrow a search for a given substance. The *Chemical Substance Index* also provides references for derivatives of the substance in question.

It is helpful to understand the relationship between the *CA Formula Index* and the *Chemical Substance* and *Subject* Indexes.

Currently, as stated in the Introduction to the *CA Formula Index:*

Cross-references to the *Chemical Substance Index* appear in the *Formula Index* for very common substances where long lists of *CA* references would otherwise be necessary. The user is best served in these cases by reference to the *Chemical Substance Index,* where study-related descriptions assist in narrowing the search to items of particular interest.

For practical purposes, CAS considers *a common substance* a substance that in a given six-month volume has more than 20 abstract references. Thus, in Vol. 99 *Formula Index* (1983), **Silver chloride (AgCl)** and **Silver iodide (AgI)** have cross-references, but under **Silver fluoride (AgF),** six *CA* abstract numbers are listed.

A similar policy applies to a five-year collective period. The difference is that *a common substance* is one that would have more than 50 abstract references in the *Collective Formula Index.* Thus, in the *Tenth Collective Formula Index* (1977–1981), **Silver, isotope of mass 105,** has a cross-reference, but under **Silver, isotope of mass 106,** 50 *CA* abstract numbers are listed.

Historically, the same yardstick has not always been used. The introduction to the first *CA Formula Index* in Volume 14 (1920), stated:

Cross-references to the *Subject Index* have been used for all simple inorganic compounds, for all minerals of definite composition, and for the organic compounds more commonly met with, in general whenever it seemed likely that users of *Chemical Abstracts* would predominatingly refer to the *Subject Index.* The *Subject Index,* because of the modifying phrases entered there, is better to use in such cases.

Thus, for certain compounds such as *simple inorganic compounds* [e.g., AgF (silver fluoride), BaN_2O_6 (barium nitrate), and O_4SZn (zinc sulfate)] and *minerals of definite composition* [e.g., Ag_3S_3Sb (pyrargyrite), $CCa_5O_{11}Si_2$ (spurrite), and $Pb_2S_5Sb_2$ (jamesonite)], cross-references had been made a priori even when the number of entries for each such compound was small. It was believed that the average searcher would turn to the *Subject Index* for information on such compounds rather than to the *Formula Index.*

As far as *common organic compounds* were concerned, for the first *Collective Formula Index* (1920–1946), it was assumed that approximately 50 entries for a compound would justify a cross-reference to the *Subject Index,* but that arbitrary decision was applied with discretion. Entries were made in that *Formula Index* when it seemed likely that the average individual would turn to the *Formula Index* rather than to the *Subject Index* for information.

As far as the early annual *Formula Indexes* were concerned, a cross-reference to the *Subject Index* was usually substituted for entries in the *Formula Index* when the same formula and name appeared more than 10 times on cards that were edited for a given *Volume Formula Index.* Derivatives such as esters and oximes were exceptions to that rule. Their *Formula Index* entries often contained more than 10 *CA* references, and no cross-references were inserted.

Prior to Volume 71 (1969), there existed a list of permanent *Formula Index* cross-references that corresponded to the categories listed previously. The use of that list was discontinued in 1969. For the remainder of the eighth collective period (1967–1971) and for the collective periods that followed, the distinction between *simple inorganic compounds, minerals of definite composition,* and *common organic compounds* was no longer made.

Since Volume 71 (1969), each specific chemical substance with its assigned Registry Number has been treated in the same way by subjecting it to a count of entries in a given Volume or Collective Index. Whenever more than 20 entries in a Volume or 50 in a Collective Index are encountered, a cross-reference to the *Chemical Substance Index* is inserted.

Thus, as a general rule of thumb, to ensure complete coverage of *CA,* the *Chemical Substance* or *Subject Index* needs to be consulted. To try to remember the limits of *Formula Index* coverage for all of the different time periods, as just described, is too confusing.

As mentioned in Chapter 6, the CAS Chemical Registry System, developed by CAS beginning in 1965, reached 7 million records, corresponding to specific chemical substances, in early 1985. In this system, each substance is assigned a unique, unambiguous computer record that is independent of the vagaries of nomenclature. For example, the Registry Number for benzene is 71-43-2.

CAS Registry Numbers have been widely accepted by the chemical community and are used in many journals, in such tools as the *Dictionary of Organic Compounds* (see Section 12.4.B), *Merck Index* (see Section 15.10), *HEEP* (see Section 14.7), *TOXLINE* (see Section 9.6), and in many other places.

With a Registry Number in hand, the chemist can consult CAS *Registry Handbook–Number Section,* and its annual supplements, to locate the *CA* index name and the molecular formula.

A few Registry Numbers may change, particularly when the original document (article, patent, etc.) supplied incomplete or partial information. In some cases, multiple Registry Numbers may be assigned to what later proves to be the same structure. As CAS learns more about such structures, old numbers are replaced by the preferred numbers. These are reported in the cumulative *Registry Number Updates,* issued annually.

Note that it is not possible to find associated *CA* abstracts for all CAS Registry Numbers, because some of these numbers resulted from the registration of special data collections such as *The Colour Index.*

7.7. *RING SYSTEMS HANDBOOK*

Another new tool available from CAS is the *Ring Systems Handbook*. This tool, which first appeared in 1984, updates and expands the coverage of the

Ring Index (and its supplements) and the *Parent Compound Handbook,* which it replaces. Uses include the following:

- The *Ring Systems Handbook* contains information about all ring and cage systems presently in the CAS Chemical Registry System. The entries are in ring analysis order.
- Cumulative supplements to the *Ring Systems Handbook,* issued every six months, provide a current awareness service for alerting to new ring and cage systems entered into the CAS Chemical Registry System during processing of the primary literature for *CA*. A new, updated complete edition of the *Handbook* will be issued every three to four years.
- It helps determine the *CA* index name of a ring system or a cage system; the *CA* index name can then be used to enter the *Chemical Substance Index* to *CA* to find references for documents reporting substances containing the ring or cage system.
- It helps determine whether a ring system exists and what similar rings there may be.
- It provides a bridge between CAS systematic nomenclature and Wiswesser Line Notations (see Section 10.11B).
- When searching computer-readable files, the chemist can use information from the *Ring Systems Handbook* to formulate substructure search strategies—that is, strategies that allow one to search for a group of substances that have in common certain ring systems or products of interest.

The *Ring Systems Handbook* includes, for a ring system:

1. Ring analysis data: number of rings, size of rings, and elemental analysis of rings.
2. Chemical structural diagram illustrating the nomenclature locant numbering system.
3. CAS Registry Number.
4. Current *CA* index name used in *Chemical Substance Index* to *CA*.
5. Molecular formula.
6. Wiswesser Line Notation.
7. *CA* reference [if the ring system was referenced in a *CA* abstract after Volume 78 (1973)].

A sample entry is shown in Figure 7.1.

7.8. *REGISTRY HANDBOOK—COMMON NAMES*

Beginning in 1977, CAS initiated the *Registry Handbook—Common Names* on microform. This tool provides access to CAS Registry Numbers through

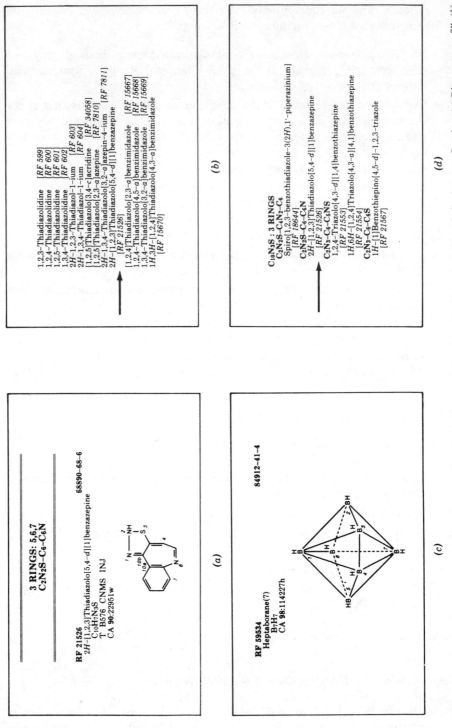

Figure 7.1. Sample Entries from the *Ring Systems Handbook*. Source: Advertising Brochure from Chemical Abstracts Service. (*a*) Ring systems fill; (*b*) ring name index; (*c*) cage systems; and (*d*) ring formula index.

the variety of substance names that are commonly used in the chemical literature and in chemical and allied industries. Molecular formulas are also given when known.

There are two parts: the name section and the number section. Emphasis is on simple, less systematic names, such as so-called author names, literature names, common names, and trivial names. The chemist can use the number section to identify sets of synonymous common names for the same substance.

7.9. USE OF SUBDIVISIONS

For hundreds of commonly reported chemicals and some classes of chemicals—such as alkanes, benzene, boric acid, carbohydrates and sugars, propane, 1-propanol, pyridine, steel, sulfuric acid, and tellurium—for which there are many index references, the chemist will find the searching task simplified by subdivision of the index heading into these general categories (called *qualifiers*) in *CA* indexes:

Analysis
Biological studies
Occurrence
Preparation
Properties
Reactions
Uses and miscellaneous

This practice was initiated with the eighth collective period (1967–1971).

For years prior to 1967, and for compounds for which the index is not subdivided as shown, searching the indexes can become both onerous and time consuming. The chemist or engineer who desires to identify all pertinent references in *CA,* should read all index entries under headings that are not subdivided. This advice holds true for important searches, but does not necessarily apply to *everyday* searches.

Even subsequent to use of subdivisions of the heading, the chemist can never be completely sure that a search under any one of the subdivisions will yield every item of desired information. A document that emphasizes one aspect of a subject may also have valuable data on other aspects. For example, in the indexing of a document on *preparation* and *properties* of a substance, either of these subdivisions may be used. The *CA* document analyst uses judgment as to which subdivision best fits emphasis of the document. Accordingly, the user may need to consult both subdivisions to retrieve all of what is needed. (The subdivision *preparation* spans the range from laboratory synthesis through commercial-style manufacturing processes.) The subdivision system just described can be a valuable time saver. Hopefully, the system can eventually be extended to include virtually all chemicals indexed by *CA*.

In addition to the subdivisions described, index headings with large numbers of entries are also divided into chemical functional group subdivisions. These include:

Acetals	Hydrazones
Anhydrides	Lactones
Anhydrosulfides	Mercaptals
Compounds	Mercaptoles
Derivatives (general)	Oxides
Esters	Oximes
Ethers	Polymers
Hydrazides	

7.10. SIZE

The sheer physical bulk of *CA* is imposing, almost forbidding to some. The indexes, in particular, are so large that considerable time can be required to locate what is needed in the printed volumes. For example, 75 books made up the *Tenth Collective Index* (1977–1981), and 87 books are estimated for the *Eleventh Collective Index* (1982–1986). A single issue of *CA* can total some 900 pages. But if *CA* is to maintain its objective of covering virtually all new chemical information, the current massive issues and indexes need to be maintained. Use of *CA* online (see Chapter 10) or in microform can help cope with the size/bulk situation. Familiarity with *CA,* and experience in its use greatly facilitates speed and ease of use.

Tables 7.2 and 7.3 summarize some trends in the growth of *CA*. Over the last few years, the rate of growth has either slowed or decreased significantly as compared to earlier years.

Part of the decline or slowdown may be due to a relatively new Soviet practice of depositing manuscripts in lieu of publishing. A counterbalance to this may be found in the significant increase in Japanese publications, especially patent applications.

7.11. COMPLEXITY

The development of the science of chemistry, coupled with the use of large numbers of automatic laboratory devices, have combined to produce substantially more data in recent years. This requires additional routes of access by CAS if the total literature is to be covered. *CA* today contains more data,

TABLE 7.2. Rate of Growth of *CA*

Years	Approximate Average Annual Rate of Growth (%)
1951–1960	8.2
1961–1970	8.4
1971–1975	6.9
1976–1980	4.6
1981	−5.3
1982	1.6
1983	−1.3
1984	2.0
1985	−0.6

and more indexing results, than ever before. The tools now published by *CA* are, of themselves, not necessarily more complex. It is simply a matter of more tools the user needs to master that, in the long run, may make the task easier.

7.12. PATENT COVERAGE

CA does not cover all chemical patents in its abstracts (see also Section 13.12). For example, if a patent is published first in a country other than the United States, and subsequently issued in the United States (or any other country), these equivalents (see Section 13.17) will not be reabstracted. Rather, they appear only in the *Patent Index* to *CA*. This policy resulted in

TABLE 7.3. *CA* Index Growth

Collective Index	Years Covered	*CA* Volumes	Number of Documents Referenced Including Equivalent Patents Cited
1st	1907–1916	1–10	192,000
2nd	1917–1926	11–20	218,000
3rd	1927–1936	21–30	544,000
4th	1937–1946	31–40	513,000
5th	1947–1956	41–50	663,000
6th	1957–1961	51–55	637,000
7th	1962–1966	56–65	1,024,000
8th	1967–1971	66–75	1,466,000
9th	1972–1976	76–85	2,024,000
10th	1977–1981	86–95	2,602,000
11th	1982–1986	96–105	(2,900,000)[a]

[a] Number of documents estimated for 1982–1986.

the provision of cross-references in the numerical patent index for 1961–1962, and in issuing of a *Patent Concordance* beginning in 1963—an index that was eventually integrated into the *Patent Index* in the *Tenth Collective Index* (1977–1981). Prior to 1961, there was full bibliographic reference in *CA* issue text to the abstract of an equivalent patent. This policy (of not reabstracting equivalent patents) is intended to reduce costs. But it makes for considerably more effort on the part of the chemist trying to locate equivalents in printed issues of *CA,* and, unlike Derwent, (see Section 13.11.A) it does not provide *automatic* notification when equivalents appear.

CA patent coverage has improved significantly, especially since about 1960. Prior to that, patent coverage is spotty, and, at least for composition of matter patents, *Chemisches Zentralblatt* is a better source. Despite improvements in patent coverage by *CA,* Derwent is the preferred source for this type of chemical document. See also Section 6.7.

7.13. REVIEWS

CA provides excellent coverage of review papers. The *CA* definition of a review is based primarily on whether the author of the original document so designates it. If new experimental information is also provided, however, the document is not treated as a review, but as a regular research report and is indexed thoroughly.

7.14. CHEMICAL MARKETING AND BUSINESS INFORMATION

This is not usually included directly in *CA*. It is *relegated* to the subsidiary publication issued by CAS known as *Chemical Industry Notes* (*CIN*) (see also Sections 14.7 and 16.6). *CIN* is excellent, but its abstracts (better termed extracts) and indexes are not as good as the full *CA*. Material appears in *CIN* quickly—within just a few weeks after the original. A typical issue contains the following sections:

Production	Corporate activities
Pricing	Government activities
Sales	People
Facilities	Keyword index
Products and processes	Corporate name index

CIN is not as complete as some other marketing and business information sources, such as those published by Predicasts, Inc. (see Chapter 16).

7.15. DISSERTATIONS

Some chemists believe that an example of an area for improvement is that of dissertations. *CA* has provided coverage of American and Canadian dissertations since 1960. Rather than provide an abstract of the dissertation, the user is usually referred by *CA* to *Dissertation Abstracts* (see Section 4.5). This requires an often inconvenient second step, since *Dissertation Abstracts* is not readily available in many chemistry libraries and information centers, except potentially in the online version. But there is no other efficient way. CAS staff could not easily assemble copies of all dissertations for their own abstracting—and particularly, indexing. The content of *Dissertation Abstracts* is too long for *CA* use, and *CA* editors are reluctant to modify abstracts without having the originals in front of them.

7.16. COVERAGE OF DOCUMENTS FROM THE SOVIET UNION

Many Soviet documents are fully abstracted and indexed by *CA*. Some, however, are accessible only through the abstracting and indexing service *Referativnyi Zhurnal* (*Ref. Zh.*). Because the abstracts in *Ref. Zh.* are copyrighted, only the titles and bibliographic information (including the *Ref. Zh.* data) are published in *CA*. The statement *Title only translated* is used instead of an abstract. Keywording and indexing are done from the title only.

7.17. MISTAKES

Mistakes, though infrequent, will be found in the abstracts, indexes, and other publications of CAS. Its staff members are human, and perfection is too much to ask for in such a large and complex work. Also, huge volumes of literature must be handled in relatively short periods of time. Correspondence with the editors of *CA* can almost always clear up any suspected mistake or other discrepancy.

Whenever possible, CAS staff handles mistakes in abstracts by republishing the corrected abstracts in later issues with reference to the earlier versions. Index references are then made to the corrected abstracts only.

Errors in semiannual indexes are corrected in the printed Collective Indexes and online. This provides another excellent reason for discarding semiannual indexes when corresponding Collective Indexes appear. The only reason for retaining the semiannuals could be the convenience of using smaller volumes with fewer pages when date of coverage can be well identified, but this leaves open the increased possibility of being misled by any uncorrected errors.

REFERENCES

1. J. T. Dickman, M. O'Hara, and O. B. Ramsay, Eds., *Chemical Abstracts: An Introduction to Its Effective Use* (ACS Audio Course), American Chemical Society, Washington, D.C.; The Royal Society of Chemistry, *Using CA,* Letchworth, England, 1984.

2. Chemical Abstracts Service, *Printed Access Tools Workbook,* Columbus, OH, 1984. In addition, the CAS catalog lists and describes available CAS Workshops and lists search aids for both the printed and online tools, for example, *CA Index Guide, CA Subject Coverage Manual, CA General Subject Index Heading List, Natural Language Term List,* and *Rotated Title Phrase List.* CAS also offers a free educational package for use with the printed *CA.* This includes: a film on searching *CA;* a booklet summarizing the film; a booklet *How to Search Printed CA;* summary sheets with brief tips on searching; and a 1981 edition of *CAS Printed Access Tools.*

3. This statement appears on the table of contents page of each issue of *Chemical Abstracts.*

8 OTHER ABSTRACTING AND INDEXING SERVICES OF INTEREST TO CHEMISTS

8.1. TRADITIONAL BRITISH, GERMAN, FRENCH, AND SOVIET SOURCES

In addition to *Chemical Abstracts* (*CA*), a number of very important chemical abstracting and indexing services have been published for many years in countries other than the United States. These have been described in detail by Skolnik (1) and others and include, for example, *British Abstracts (BA), Chemisches Zentralblatt (CZ), Bulletin Signalétique,* and *Referativnyi Zhurnal, Khimiya (RZh* or *Ref. Zh., Khim)* (see also Section 6.14).

Of these, only *CA* and *RZh* are still published as such. However, for complete coverage of early literature and patents, both *BA* and *CZ* need to be consulted, in addition to use of the *Beilstein* and *Gmelin Handbooks* (see Chapter 12), especially *CZ* for the literature published prior to the start of *CA* in 1907.

There is a question as to ready availability of these discontinued and older, but still important abstracting services. It is to be hoped that most better and larger libraries have space and funds needed to retain the volumes on their shelves. Even if a set of *CZ* is located, there can be an important, but unfortunate, language barrier for many English reading chemists; this barrier does not, of course, apply to use of *BA*.

Besides the advantage of earlier years of coverage, there are factors relating to content and scope of coverage. For example, as indicated in Section 6.14, *CZ* emphasizes earlier patents, especially from Germany, more than *CA* does, even for some years when the two services overlapped. In any event, because the abstractors and indexers of *BA* and *CZ* were usually different than those of *CA*, additional important information may be found in these abstracts and indexes. For example, *CZ* abstracts were noted for their length and detail as compared to other services. As could be expected, both *CZ* and *BA* were especially outstanding in coverage of publications from Germany and the United Kingdom, respectively, although their scope was international.

The following paragraphs from the book by E. J. Crane (editor of *CA* for many years), Austin M. Patterson, and Eleanor B. Marr (2), pp. 133–134 give a good comparison of *CA* and *CZ* and outline some of the history of the latter, but do not mention discontinuance of *CZ* in 1969 because it occurred after these words were written (1957):

This important German abstract journal [*CZ*] has value because of its early appearance, almost continuous publication, and good abstracts. There have been several changes of name, the original one being *Pharmaceutisches Centralblatt* (1830–1850). This was followed by *Chemisches–Pharmaceutisches Centralblatt* (1850–1856) and *Chemisches Zentralblatt* (1856–). *Zentralblatt* was spelled *Centralblatt* from 1856 to 1897. The Deutsche Chemische Gesellschaft published it from 1897 to 1945. Publication was suspended for a short time in 1945. Then two editions were issued in parallel from 1945 through 1950, one from the Eastern and the other from the Western Zone of Germany; the two editions were combined in 1951 to form a journal sponsored jointly by several scientific societies in each zone. It appears weekly. The usual form of reference to it is *Chem. Zentr.* or *Chem. Centr.*

Coverage. Up to 1919, *Chemisches Zentralblatt* was not a comprehensive abstract journal because it limited its abstracts to papers dealing directly or indirectly with pure chemistry, and it covered German chemical work more thoroughly than that of other countries. In 1919, the abstract section of *Angewandte Chemie* (then called *Zeitschrift für angewandte Chemie*) was made a part of *Chemisches Zentralblatt* and ever since it has endeavored to cover the world's periodical literature thoroughly for both pure and applied chemistry, though with some delays and omissions as an effect of World War II.

Before 1919 only German patents were abstracted by *Chemisches Zentralblatt,* and then only part of them, the leaning being toward patents on organic chemical substances. From 1919 to date *Chemisches Zentralblatt* has covered the world's chemical patents quite thoroughly. Throughout the years its patent coverage has not always been the same as that of *Chemical Abstracts* because the two journals have not made abstracts from the patents of all of the same countries. For example, *Chemisches Zentralblatt* does not make abstracts of Japanese patents whereas *Chemical Abstracts* has done so since 1917, and *Chemical Abstracts* has not always covered Russian patents, even in part. Because neither journal attempts to abstract all of the chemical patents of the world, the two sets of abstracts supplement each other in this respect.

Since 1926 new books have been announced by title in each of the divisions of the abstracts, usually at the beginning of the division or subdivision, but *Chemisches Zentralblatt* does not abstract or review books. The total number announced each year is much smaller than reported in *Chemical Abstracts*.

Chemisches Zentralblatt's coverage of the world's chemical literature was adversely affected by World War II and its aftermath, from 1939 through 1951. By 1952 it was making abstracts from papers in 4,925 periodicals. It is now [1957] approaching its prewar standard and the abstracts are reasonably prompt. Abstracts of some papers originating in the USSR and its satellite countries appear sooner in *Chemisches Zentralblatt* than in *Chemical Abstracts*.

It is very unusual to find an abstract of a paper in *Chemisches Zentralblatt* that is not also in *Chemical Abstracts* from 1907 on because of the care that the editors of the latter use to prevent omissions. If it does occur, it is not likely to represent a paper of major chemical importance but rather one of little chemical interest in a borderline field between chemistry and another science.

The same book gives a good description of *British Abstracts* (pp. 135–136):

Both the Chemical Society (London) and the Society of Chemical Industry (London) published abstracts in their representative journals, the *Journal of the Chemical Society* (London) and the *Journal of the Society of Chemical Industry,* until the end of 1925 (see below). In 1926 the two societies combined their abstracting activities and founded *British Chemical Abstracts,* which continued under this name through 1937. In 1938 *Physiological Abstracts* was merged with *British Chemical Abstracts* and the name was changed to *British Chemical and Physiological Abstracts*. In 1939 a section on anatomy was added. The name was changed again in 1946, *British Abstracts* becoming the new name.

Coverage. Up to 1938, as its history would lead one to expect, *British Abstracts* covered only the chemical field with considerable thoroughness, but after that date it became something more than a chemical abstract journal, with anatomy and nonchemical phases of physiology as the principal additional fields of coverage. The coverage of chemistry was extensive but not as comprehensive as that of *Chemical Abstracts* for the same period, 1926–1953. Applied as well as pure chemistry was abstracted, and many patents were covered. The excellent, promptly appearing abstracts, once informative, tended to become descriptive about the time of World War II.

Classification of Abstracts. From 1926 through 1944 there are two parts, A (Pure Chemistry) and B (Applied Chemistry). From 1944 through 1953 there are three parts: A, B, and C (Analysis and Apparatus). The parts have been subdivided in somewhat different ways at different times; they were issued and bound separately.

Indexes. Annual indexes cover subjects, authors, and patents. Collective indexes for parts A and B (1923–1932 and 1933–1937) cover authors and subjects. The 1923–1932 index includes abstracts that were published in the *Journal of the Chemical Society* (London) and in *The Journal of the Society of Chemical Industry* (1923–1925). The subject indexes are somewhat less extensive than those of *Chemisches Zentralblatt* and of *Chemical Abstracts*.

As previously mentioned, *CZ* ceased publication as such in 1969 and *British Abstracts* in 1953.

The Russian abstract journal *RZh, Khimiya* (Chemistry) started publication in 1953 and is one of a series of abstract journals published by VINITI (All-Union Institute for Scientific and Technical Information). Its stated goal has been the coverage of all the world's chemical literature.

Abstracts of chemical interest also appear in other journals of the *RZh* series, such as *Biologicheskaya Khimiya* (Biological Chemistry) [changed in 1984 to *Fiziko-Khimicheskaya Biologiya i Biotekhnologiya* (Physicochemical Biology and Biotechnology)], *Metallurgiya* (Metallurgy) and *Korroziya i Zashchita ot Korrozii* (Corrosion and Corrosion Protection).

There are major impediments to the use of *RZh* in many Western countries, because of the lack of ready availability (except in major libraries) and the widespread lack of understanding of the Russian language. Comparison with *CA* shows that *RZh* coverage is not as extensive. *CA* does cover most of the available Soviet chemical literature. Accordingly, it is believed that most American, British, and Canadian chemists can now safely ignore *RZh*.

The balance of this chapter deals with relatively smaller abstracting and indexing services. Potential benefits of use of some of these services, as compared to *CA,* may include one or more of the following:

1. Broader depth of coverage in the area of concentration, for example, more specialized publications, very brief reports, abstracts of presentations, letters to the editor, or industry news and developments that may not be covered by *CA* despite its broad scope. However, when taken as a whole, coverage of *all* publications of chemical interest by *CA* is superior to that of any other service.

2. More detailed abstracts that may be slanted to and written from the perspective of the particular need of a special interest group within chemistry or of another discipline outside chemistry. In some cases, however, abstracts may be identical with *CA* abstracts because of special arrangements that may have been made.

3. Indexes that are easier to use because they have more limited scope, are less complex, and are smaller than those of *CA*. On the other hand, indexing may be more specialized and in depth than that of *CA* in fields covered and may use jargon or technical terms more familiar to chemists and others working in the fields.

4. Individual issues that are easier to visually scan than those of *CA* because of brevity.

5. Unique presentation of abstracts and/or indexes that may be particularly well suited to more specialized purposes. A good example is in publications of the *Institute for Scientific Information*® as described later in this chapter.

6. Staffs that may be able to provide specialized services on demand including, for example: assisting in and performing of evaluations of state of the art; obtaining translations of documents that may not be readily available elsewhere; and referrals to outside practitioners and other experts who may be able to help solve laboratory and other problems.

For reasons such as these, chemists should determine whether or not abstracting and indexing services exist in their fields of specialization. This can usually be done with the assistance of the local chemistry librarian. Many such services are readily available online and, if so, an appraisal of any benefits is made easier. A full appraisal, however, can best be made by study of both printed and online versions. Parallel comparative evaluation with *CA* should help make a decision.

Criteria recommended for evaluating and selecting abstracting and indexing services include:

1. Reputation in scientific community
2. Quality of staff and sponsoring organization or publisher
3. Quality and depth of abstracts and indexes
4. Depth of coverage, including number and kinds of publications covered
5. Ease of use
6. Speed of coverage (abstracts and indexes)
7. Reasonableness of price
8. Availability of online version
9. Assistance to users including knowledgeable *help* (toll free) telephone lines, well-written user guides, copy service, translation service, and other user aids.

8.2. INSTITUTE OF PAPER CHEMISTRY, FOREST PRODUCT SERVICES, AND COMMONWEALTH AGRICULTURAL BUREAUX

Consider several examples of "smaller" services, published in the United States. One is found in the work of the Institute of Paper Chemistry.* The primary functions of the Institute are to conduct research and offer education in pulp and paper technology and science. The Institute's Division of Information Services, now under the leadership of H. S. Dugal, issues the following publications:

- *Abstract Bulletin* (available in printed, microform and online editions)—The printed index started in the 1930s, and the online version goes back to 1967. About 12,000–13,000 abstracts are published each year. There are monthly and annual subject indexes, a patent number index, and a specific name index, which lists trade names and names of companies, organizations, and products. The online version is known as *PAPERCHEM* and can be searched via *SDC®* Information Services or *DIALOG®* Information Services (see section 10.12A).
- *Keyword Index to the Abstract Bulletin*—this is based on a controlled vocabulary of key terms selected from the *Thesaurus of Pulp and Paper Terms* (published and sold by the Pulp and Paper Research Institute of Canada).*

- *Bibliographic Series*—approximately 300 annotated bibliographies many with supplements, have been published on key topics. Customized bibliographies are available for a fee.
- *Conference Proceedings*
- *Retrospective computer searches*
- *Translations*
- *Courses* (on slides or videocassettes)

From these listings, it should be clear that the Institute offers a complete array of information services especially geared to chemists and engineers doing research and other work in pulp and paper chemistry.

In addition to services of the Institute of Paper Chemistry, an example of still further specialization in the broad field of forest products is found in the abstracting and indexing service of the Forest Products Research Society.* This provides good coverage of papers relating to wood products, such as plywood, oriented strand board, wafer board, and flake board (among other topics), including chemical aspects. In its most up-to-date form, it is online in a database known as *FOREST,* which is available through *SDC*® Information Services. The printed version had appeared much later as part of *Wood Industry Abstracts* published by Washington State University, but the latter publication was discontinued in 1985.

In the field of agricultural science, an outstanding abstracting and indexing service, in addition to *CA,* is the Commonwealth Agricultural Bureaux (CAB).* Abstracts cover virtually every branch of agricultural science and are available in both printed and online forms. An expert staff also provides literature searches, bibliographies, translations, and document copies. Printed abstract journals are published covering highly specialized areas, for example, rice or seeds or soybeans. Broader areas are also covered, for example, soils and fertilizers, or weeds, or applied entomology. Many of these publications are succinct enough to be very conveniently scanned. In all, there are 23 abstract journals, 2 primary journals, 17 specialist abstract journals, and 12 serial publications. Online access to the comprehensive CAB database is available from 1973 to the present via *DIALOG*® Information Services, Information Retrieval Service of the European Space Agency (Rome), and DIMDI (Bonn); see also Chapter 10 on online searching. CAB is an international nonprofit organization owned by 28 Commonwealth governments. John R. Metcalf is CAB director of information services. CAB North American Office is at the Arid Lands Information Center in Tucson, AZ.*

There are a number of other "smaller" services, especially in the more applied or technical aspects of chemistry. These are too numerous to mention here, but this Section helps to illustrate potential advantages. Several important relatively large services that deserve the attention of the chemist are noted in the pages that follow.

8.3.   *INSTITUTE FOR SCIENTIFIC INFORMATION*®

The *Institute for Scientific Information (ISI*®) is an outstanding for-profit organization that offers a wide variety of services to the chemist. *ISI*® is located in Philadelphia, PA.* The success of this organization is due in large measure to the innovativeness of its founder and chief executive, Eugene Garfield.

ISI® was founded in 1960. It has approximately 650 employees worldwide with annual revenues forecast to be about $35 million in 1986. Over 6000 libraries and 30,000 individuals subscribe to *ISI*® publications. About 50% of users are academic, and the balance are equally divided between industry and government and other nonprofit organizations (about 25% each).

Some *ISI*® publications and other services are listed throughout this book. Several related to abstracting and indexing are described in the following paragraphs.

The cornerstone of *ISI*®'s coverage of chemistry is *Current Abstracts of Chemistry and Index Chemicus*® (*CAC&IC*®). This is a highly regarded weekly abstracting and indexing service, published since 1960, which tells what *new* organic compounds and syntheses are reported in over 110 of the *most important* journals to organic chemists. About 180,000 new compounds (including intermediates) are reported annually; over 3.9 million compounds have been recorded since 1960. *ISI*® says that this is over 90% of all new organic compounds reported in the journal literature. Coverage is prompt; *ISI*® says that articles appear in *CAC&IC*® within 45 days (on the average) after the article is published.

CAC&IC® provides the following information:

1. Structural diagrams for new organic compounds and reaction flows. This graphical information is the most significant feature of the service.
2. Complete bibliographic data, including language if other than English.
3. Indicator for presence of labeled compounds.
4. Indicator for presence of new synthetic methods. (Experimental details are available in the companion service known as *Current Chemical Reactions*®; they are not given in *CAC&IC*®.)
5. Analytical techniques used (expressed in codes).
6. Degree of experimental detail (full, part, none).
7. Applications or use profile, primarily biological activity, for example, *herbicidal activity* or *carcinogenic activity*.
8. *ISI*® accession number that permits ordering of document via *ISI*'s *The Genuine Article*™ service, formerly *OATS*® (*Original Article Text Service*); see Section 5.2 for details.
9. Author abstract (if available).

A typical abstract is shown in Figure 8.1.

Each issue contains indexes by subject, molecular formula, author, corporate (organization) source, journal title, labeled compounds, and biological activities. These indexes are cumulated quarterly and annually, the cumulations including a permuted (rotated) molecular formula index (Rotaform Index®).

The strong points of *CAC&IC®* are exceptional speed of coverage and easy-to-read, highly graphic abstracts that include structural diagrams and reaction flows. The indexes are another unique feature. Chemists working on synthesis of organic compounds will find this a tool of special utility although subject indexing is not as deep and systematic as that of *CA*. *CAC&IC®* is, however, limited to journals. This policy excludes compounds and syntheses reported in patents only.

As noted, although a large number of important journals are covered, some journals are not covered. However, studies at *ISI®* say that most new compounds are reported in only a relative handful of journals. One further point of interest is that much new information, such as new uses for *old* compounds is excluded from *CAC&IC®*. The reason for this is that the focus of *CAC&IC®* is primarily on *new* compounds, and *new* reactions and syntheses.

Current subscription costs (1986) for *CAC&IC®* are $3190 per year for industry and are considerably less for most educational institutions (rate varies with enrollment). Additional copies are available at reduced rates. Like *CA,* this is not intended for purchase by individual chemists.

In 1984, there was a major step forward in more rapid and efficient searching of *ISI®* files when *Index Chemicus® Online* became available through the online facilities of *Questel–DARC*. This is the online version of *CAC&IC®* and is intended to be primarily an organic structure database that offers both graphic and text input and retrieval (see also Section 10.11).

By early 1986 the file contained data on over 3.5 million structures as reported in the literature covered by *ISI®* since 1962.

Output from the *DARC* file includes: compound structure in graphic form, molecular formula, number of atoms, and abstract number. Output from the *Questel* file includes: abstract number, full bibliographic data, scientific data alerts (instrumental method, biological activity, explosive reaction, new synthetic method, level of experimental detail), molecular formula, specific isotope, Wiswesser Line Notation (WLN), and microform coordinates.

Structure access can be for specific compounds or substructures, or, it is said, on a generic (Markush) basis (see Section 10.11.A). WLN can be "string searched" (not the most satisfactory way of searching WLN) in the *Questel* bibliographic files; in the *DARC* file, WLN is not searchable at all.

The files are distinguished by rapid coverage and by exceptional attention to the inclusion of intermediates. Extension of structure coverage back to 1962, and of bibliographic coverage back to 1960, are also noteworthy, as is the scientific data alert feature. Unfortunately, abstracts are not available at

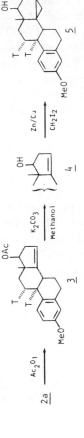
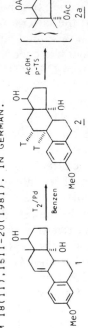

321448

TRITIUM LABELING OF A NEW STEROID WITH FERTILITY INHIBITING PROPERTIES:
3-METHOXY-14BETA,15BETA-METHYLENE-(9ALPHA,11ALPHA-T)-
ESTRA-1,3,5(10)-TRIEN-17BETA-OL.

WAGNER H, ROMER*J, KASCH H.

AKAD WISSENSCH DDR, FORSCH ZENT MOLEKBIOL, DDR-69 JENA, E GERMANY.

J LABEL COMPOUND RADIOPHARM 18(11),1511-20(1981). IN GERMAN.

The synthesis of a [9α,11α-]-tritium-labelled new
steroid with interceptive - post-coital antifertility -
activity, 3-methoxy-14β,15β-methylene-estra-1,3,5(10)-
triene-17β-ol, starting with the catalytic tritiation of
3-methoxy-estra-1,3,5(10),9(11)-tetraene-14α,17β-diol, is
described.
3-Methoxy-14β,15β-methylene-[9α,11α-³H]-estra-1,3,5(10)-
triene-17β-ol was obtained in a four-step synthesis with
high specific activity (3,8 Ci/mmol₂) and radiochemical
purity better than 99%. Radiolysis and storage conditions
are examined.

USE PROFILE

ANTIFERTILITY ACTIVITY

Figure 8.1. Sample *Current Abstracts of Chemistry and Index Chemicus*® Abstract. Reproduced with permission and copyright owned by the *Institute for Scientific Information*®, Philadelphia, PA, U.S.A.

this time, although consideration is being given to their potential inclusion in the future. Because of differences in indexing policies, fewer journals are included than in *CAS ONLINE,* and the latter covers about twice as many structures. Patents are not included at this time. Unfortunately, it is believed that response to this important product has been lukewarm so far. This may be because of intense competition from *CAS ONLINE.*

The *ISI®* service known as *ANSA®* (*Automatic New Structure Alert®*) permits chemists to be notified automatically each month when new structures or other data of interest are located by *ISI®* and reported in *CAC&IC®.* Users may describe interests in the form of structures and substructures, molecular formulas or elements therefrom, applications and biological activities, new synthetic methods, analytical techniques, keyword phrases, bibliographic data, or any combination of these interests. Output consists of the following information (when present): subject listings, applications and biological activities; labeled compounds; new synthetic methods; instrumental data; molecular formula; explosive reaction alert; complete bibliographic citation and Wiswesser Line Notation if the search query contains a structure. *ISI®*-assigned abstract numbers, from *CAC&IC®* and author-assigned compound numbers are also included. For typical output, see Figure 8.2. The cost of this service (1986) was $500 per substructure profile for United States subscribers. Retrospective capability is also possible for the full *ISI®* files back to 1962. It is not necessary to be a subscriber to *CAC&IC®* to use *ANSA®.*

Some organizations subscribe to the *ISI® Index Chemicus Registry System®* magnetic tape service, which is available on a monthly or annual basis. This is the machine readable version of *CAC&IC®.* The complete file covers over 3.7 million organic compounds from more than 300,000 articles. Organizations that subscribe to these tapes usually search it on their own computers.

Another way of searching of *ISI®* files is through use of the *Chemical Substructure Index®.* This monthly microfilm product is an index to *CAC&IC®.* It is cumulated monthly and annually and is intended to facilitate quick manual searching of substructures. Substructures or fragments can be identified through rotated (permuted) alphanumeric WLN entries (see Section 10.11.B) supplied for compounds covered (see Figure 8.3).

Current Chemical Reactions® (*CCR®*) is another *ISI®* tool, first published in 1979. This contains details of newly modified synthetic methods reported in over 120 journals and some books. Data presented include flow diagrams of new and newly modified reactions and syntheses, abstracts when available, product yields when available in original papers, citations, and other available information such as experimental details, advantages (or limitations) claimed, and an explosive reaction indicator (see Figure 8.4). Monthly and annual indexes are provided for new methods including products and starting materials, and there are author and corporate indexes. More than 300 new methods are usually reported in each monthly issue. Cost to United States subscribers is $540 (1986).

```
ACCT # 00016   PROFILE# 0012   REPORT FOR: SEP 1985
                                      RUN:11/03/85
                                      PAGE  0001

              SHELLY RAHMAN
              ISI
              3501 MARKET ST
              PHILADELPHIA


CAC ABSTR.# 378491

     ORGANOGERMANIUM COMPOUND.
     NEW SIMPLIFIED SYNTHESIS OF GERMATRANES SUBSTITUTED
     WITH
     NOVEL FUNCTIONAL GROUPS.
             KAKIMOTO N  SATO K  TAKADA T  AKIBA M
     ASAI GERMANIUM RES INST, TOKYO 201, JAPAN.
     HETEROCYCLES 23(6),1493-6(1985).
SUBJECT LISTINGS:
     ORGANOGERMANIUM CPD
     *GERMATRANE, 5-SUBST-, SYN FROM A,B-UNSATD ACID
        DERIVS & GECL3
     GERMATRANE
     *5-SUBST, SYN FROM A,B-UNSATD ACID DERIVS GECL3
USE PROFILES:
     ANTITUMOR ACTIVITY
INSTRUMENTAL DATA:
     NEW SYNTHETIC METHOD
     MASS SPECTRA
     INFRARED SPECTRA
     NUCLEAR MAGNETIC RESONANCE
     FULL EXPTL DETAIL

          TO ORDER THE CAC/IC ABSTRACT CHECK HERE ( )
          TO ORDER THE ORIGINAL ARTICLE CHECK HERE ( )


CAC ABSTR.# 378489

     THIENOPYRROLIZINES. 2.
     SYNTHESIS OF 4-IMINO (AND 4-AMINO)-4H-THIENO(2,3-B)
     PYRROLIZINES.
             LAPORTEWOJCIK C  GODARD A M  RAULT S  ROBBA M
     UNIV CAEN, LAB CHIM THERAPEUT, 14000 CAEN, FRANCE.
     HETEROCYCLES 23(6),1471-7(1985).
SUBJECT LISTINGS:
     THIENOPYRROLIZINE(2,3-B)
     *4-IMINO DERIVS, SYN & REDUCTN TO AMINO DERIVS
     *4-AMINO DERIVS, SYN VIA IMINIUM SALT
USE PROFILES:
     ANTITUMOR ACTIVITY
INSTRUMENTAL DATA:
     NEW SYNTHETIC METHOD
     FULL EXPTL DETAIL
     INFRARED SPECTRA
     NUCLEAR MAGNETIC RESONANCE

          TO ORDER THE CAC/IC ABSTRACT CHECK HERE ( )
          TO ORDER THE ORIGINAL ARTICLE CHECK HERE ( )
```

Figure 8.2. Sample *ANSA*® Output. Reproduced with permission from *ANSA.*® Copyright owned by the *Institute for Scientific Information*®, Philadelphia, PA, U.S.A.

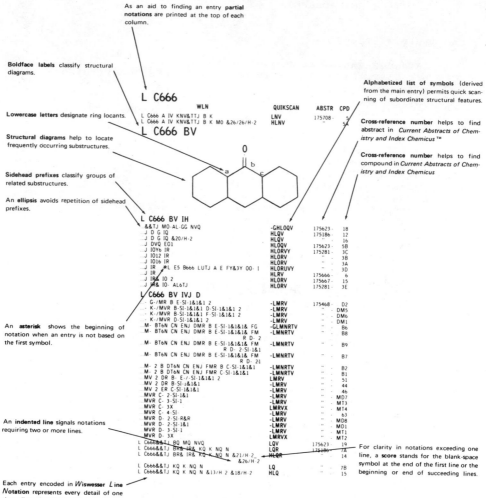

As an aid to finding an entry **partial notations** are printed at the top of each column.

Boldface labels classify structural diagrams.

Lowercase letters designate ring locants.

Structural diagrams help to locate frequently occurring substructures.

Sidehead prefixes classify groups of related substructures.

An **ellipsis** avoids repetition of sidehead prefixes.

An **asterisk** shows the beginning of notation when an entry is not based on the first symbol.

An **indented line** signals notations requiring two or more lines.

Each entry encoded in *Wisswesser Line Notation* represents every detail of one three-dimensional structure.

Alphabetized list of symbols (derived from the main entry) permits quick scanning of subordinate structural features.

Cross-reference number helps to find abstract in *Current Abstracts of Chemistry and Index Chemicus ™*

Cross-reference number helps to find compound in *Current Abstracts of Chemistry and Index Chemicus*

For clarity in notations exceeding one line, a **score** stands for the blank-space symbol at the end of the first line or the beginning or end of succeeding lines.

Figure 8.3. *Chemical Substructure Index®* Sample. Reproduced with permission and copyright owned by the *Institute for Scientific Information®*, Philadelphia, PA, U.S.A.

① J. Org. Chem. 49(24), 1984 ② TU 532

③ 023982 DISPLACEMENT ④ ⑩ EXPLOSIVE RXN

NITROCARBONS. 4.

⑤ REACTION OF POLYNITROBENZENES WITH HYDROGEN HALIDES. FORMATION OF POLYNITROHALOBENZENES.

⑥ NIELSEN*A T, CHAFIN A P, CHRISTIAN S L.

⑦— NAVAL WEAPONS CTR, CHEM DIV, RES DEPT, CHINA LAKE, CA 93555.

⑧ J ORG CHEM 49(24),4575-80(1984).

⑨ Hexanitrobenzene and pentanitrobenzene in benzene solution react with hydrogen halides (HCl, HBr, HI, but not HF) at 25 °C to produce high yields of pentanitrohalobenzenes and 2,3,4,6-tetranitrohalobenzenes, respectively. Pentanitrofluorobenzene also reacts readily with HCl to yield 3-chloro-2,4,5,6-tetranitrofluorobenzene, but the other pentanitrohalobenzenes are much less reactive. 1,2,3,5- and 1,2,4,5-tetranitrobenzenes react rapidly with concentrated hyrochloric or hydrobromic acids at reflux to form picryl halides and 1-halo-2,4,5-trinitrobenzenes, respectively; pentanitrotoluene reacts very slowly under these conditions to form 3-halo-2,4,5,6-tetranitrotoluenes in lower yields. The scope and limitations of this regioselective reaction are defined, and comparison is made to related reactions. A mechanism is presented and discussed.

⑬ **Exptl**:

a: HX(aq), C_6H_6, 25°, 0.1-4 hrs

b: HX(anh), C_6H_6, 25°, 16-18 hrs

c: HX, reflux, 15 min

d: HNO_3, H_2SO_4, 95° (72%); N_3Na, 20% oleum, 25°; reflux

e: HNO_3, H_2SO_4, 25-30° (64%); $MeOC_6H_5$, H_2SO_4, 25°

f: 98% H_2O_2, 20% oleum, 25°

⑪

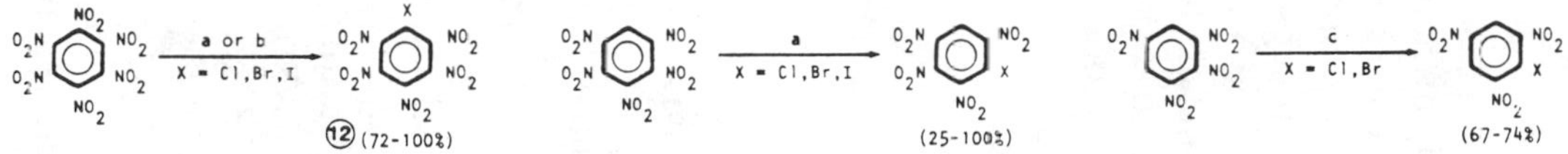

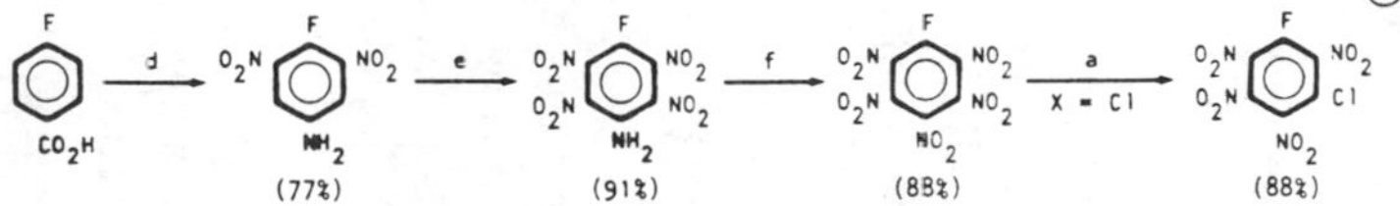

⑭ Advantages:
- regioselectivity
- products difficult to obtain by other routes
- high yields

Limitations:
- the rxn occurs only in acid media
- no rxn occurs with HF
- only one halogen introduced independent of the halogen and nitro groups present
- reactivity of nitroaromatic is affected by substitution

CAUTION: Polynitro compounds are explosives and should be handled with care.

① JOURNAL TITLE - ABSTRACTS ARE GROUPED BY JOURNAL	⑤ ARTICLE TITLE	⑨ AUTHORS' SUMMARY
② ISI ACCESSION NUMBER - USED FOR ORDERING COPIES OF THE ORIGINAL ARTICLES	⑥ AUTHORS' NAMES (WHEN USED BY SOURCE JOURNAL, ASTERISK WILL INDICATE CORRESPONDENCE AUTHOR)	⑩ EXPLOSIVE RXN INDICATOR
③ CCR REGISTRY NUMBER	⑦ CORPORATE ADDRESS	⑪ REACTION FLOW(S)
④ REACTION DESCRIPTOR	⑧ JOURNAL CITATION AND LANGUAGE OF ORIGINAL ARTICLE, IF OTHER THAN ENGLISH	⑫ PRODUCT YIELDS
		⑬ EXPERIMENTAL
		⑭ COMMENTS (e.g. ADVANTAGES, LIMITATIONS, HAZARDS, BIOLOGICAL ACTIVITIES)

Figure 8.4. Typical Abstract from *Current Chemical Reactions®*. Reproduced with permission from *Current Chemical Reactions®*. Copyright owned by the *Institute for Scientific Information®*, Philadelphia, PA, U.S.A.

Another *ISI®* publication is the multidisciplinary *Science Citation Index®* (*SCI®*), which is available in printed form and also in an online computer-based version known as *SCISEARCH®* through Lockheed *DIALOG®* Information Services (see Section 10.12.A). *SCI®* indexes material covering over 3300 journals representing more than 100 disciplines, including chemistry. Indexing is via:

1. *Citation Index*—this important tool lists documents referred to (cited) in the current literature alphabetically by name of author. It permits the user who already knows one or more authors who have written in a field of interest to identify other items citing the previously known pertinent work. This can lead to a network or chain of related references. For example, users may wish to know who has cited their own work or the work of colleagues.

2. *Source Index*—this index is alphabetically arranged by author, and there is also a separate index by organization. It is a way of keeping track of work by specific authors and organizations.

3. *Permuterm® Subject Index*—this is an index based on the original words in the titles of items covered. All significant words in a title are coupled or paired together as shown in Figure 8.5.

SCI® is published six times a year and cumulated annually going back to 1961. There are also cumulations covering 1955–1964, 1965–1969, 1970–1974, and 1975–1979. This can be a unique and powerful tool when properly used. The most significant feature is the citation indexing employed.

Other important *ISI®* services include *ASCA®*, *ASCATOPICS®*, and *Current Contents®*, which are discussed in Chapter 4. In 1986, ISI® initiated an index to the chapter contents of multiauthored scientific books.

8.4. THE ROYAL SOCIETY OF CHEMISTRY

Many pages are taken up in this book with a discussion of ACS activities in abstracting and indexing (primarily CAS). Less well known in the United States are the important activities of The Royal Society of Chemistry. Abstracting and related publications of The Royal Society include:

Methods in Organic Synthesis. A monthly current awareness service on novel and interesting reactions and reaction schemes. Graphic reaction schemes are heavily used.

Index of Reviews in Organic Chemistry. First volume covers November 1980–December 1982 and now to be published annually.

Chemical Engineering Abstracts. Available both in printed form each month and online through Pergamon InfoLine®.

Analytical Abstracts (monthly). Available online via Pergamon *Info-*

```
FORMATION - - TOMLINSO.J      REGULATION  - LUMENG L        ACETAMIDE
GROWTH  - - - SOINI J         REINFORCING - AMIT Z                        ◆BOGOSZLO.VV
HELOBDELLA - CORNEC JP        RELATIONSH. - TRUITT EB                     ◆FACELLI JC
HIRUDINIDAE -  "              RESPONSE  - - GORDON BHJ                    ◆PUCCIARE.F
INDEX - - - - ◆SCHULTER.FP    RH  - - - - - ORITA H
LEFT - - - TOMLINSO.J         RIBOFLAVIN - PINTO J         ACETAMIDES
LUXATION - - -  "             ROLE - - - - ZOGANAS HC                    ◆SAMSKOG PO
NEW - - - - -  "              SELECTIVE - PINTO J
OSTEOCHOND. ◆NIETHARD FU      SERUM-PROT. - GILES HG       ACETAMIDINE
OSTEOCHOND. -  "              STABLE - - - - LUMENG L                    ◆VEDASIRO.JR
PREDICTING - - SCHULTER.FP    STAGES - - - - GUERRI C
PRODUCED - - SOINI J          STUDY - - - IVANOV CP         ACETAMIDOCINNAMIC
PUBIS - - - - - SCHULTER.FP   SUBCELLULAR SHIOHARA E                     ◆NAKAMICH.K
RACE - - - - -  "             SUBJECTS - - - TRUITT EB
REGENERATL - CORNEC JP        SULFATASE - BLOMQUIS.CH       ACETAMINOPHEN
ROOF - - - - - KASSELT MR     SUPPORTED - LUNSFORD JH       sa PARACETAMOL
       - - - - NIETHARD FU            - - ORITA H          AA - - - - - ◆LASKIN DL
RYNCHOBDEL. CORNEC JP         SYSTEM - - GARCIN F           ABSENCE - - -  "
SEX - - - - - - SCHULTER.FP   TERATOGENI. - BLAKLEY PM      ACCUMULATI. -  "
ULTRASTRUC. - CORNEC JP       TRANSMISSI. - MOURA D         ACIDS - - - - ◆ZERA RT
YOUNG-RABB. - SOINI J                 - SOARESDA.P          ACTIVITY - - ◆HAZELTON GA
                              TREATED - - - HELLSTRO.E      ADMINISTRA. ◆SKJELBRE.P
ACETAL                        TREATMENT - SHIOHARA E        AGING - - - - ◆RICHIE JP
ALDOSES - - ◆KISO M           TURNOVER - - GILL K           ALCOHOL - - ◆BANDA PW
AMINO-SUGA. ◆STEFANSK.B       WATER - - - - IMPROTA C       ALKYLATION ◆SMITH CV
AMYLOID-P - ◆HIND CRK         XANTHINE-O. - ZOGANAS HC      ANALGESIC - ◆COOPER SA@
ASYMMETRIC ◆JOHNSON WS                                               - ◆QUIDING H
BEHAVIOR - - - KISO M         ACETALIZATION                 ANTAGONISTS ◆MITCHELL MC
BINDING-SP. - HIND CRK                   ◆IMAI K            APAP - - - - ◆NAHATA MC
COMPONENT -  "                           ◆RAGHAVEN.P        ASPIRIN - - ◆BERGH JJ
DIMETHYL - - STEFANSK.B                                     ASPIRIN-IN. - ◆STERN AI
EXCHANGE - KISO M             ACETALS                       RABOON - - - ◆ALTOMARE E
GALACTOSE - - HIND CRK        ACETYLENIC ◆ALEXAKIS A        BILIARY-EX. - ◆HJELLE JJ
N,N-DIMETH. - STEFANSK.B      ACID - - - - ◆MIYASHIT.M               - ◆SIEGERS CP
PYRUVATE - - HIND CRK         ACID-CATAL. ◆LEE DS@          BROMOBENZE. SMITH CV
REACTION - - - STEFANSK.B     ACID-ESTERS ◆SHONO T          CAFFEINE - - - COOPER SA
REAGENTS - - KISO M           ACIDS - - - - ◆BELLASSO.M     CELLULAR - - LASKIN DL
SERUM - - - - HIND CRK        ACTIVATED - MIYASHIT.M        CHILDREN - - NAHATA MC
SYNTHESIS - - JOHNSON WS      ADDITION - - ◆RAIFELD YE      CIMETIDINE - MITCHELL MC
TEMPLATES -  "                ADDITIONS - ◆YAMAMOTO Y                - ◆RIUO GD
                              ALCOHOL - - ◆GAZIZOV MB       CNS - - - - - ◆LEVY G
ACETAL-TYPE                   ALCOHOLS - ◆TSUOOI S          CODEINE - - - COOPER SA@
            ◆MUKAIYAM.T       ALDEHYDE - ◆SUGAI S                   - ◆QUIDING H@
                              ALDOL - - - - YAMAMOTO Y      COMBINATION ◆FRITZ AK
ACETALDEHYDE                  ALKYLARYLA. ◆LAMATY G         COMPARATIV. ◆JI S
ACTIVITIES - ◆BLOMQUIS.CH     ALKYLATION ◆WHEELER TN        COMPARED - - COOPER SA@
ACUTE - - - ◆GILL K                  - YAMAMOTO Y          COMPARISON ◆HJELLE JJ
ADDICTION - ◆EWING JA         ALLYL - - - - ◆DOYLE MP       CONCURRENT ◆MACDONAL.MG
ADDITION - - ◆MARTEN DF       ALPHA - - - - BELLASSO.M      CONJUGATION - LEVY G
ADDUCTS - - - ◆HOMAIDAN FR              ◆SHARMA KK          COSOLVENTS ◆PRAKONGP.S
         - ◆LUMENG L          ALPHA-METH. SHONO T           CSF - - - - - LEVY G
ADRENERGIC ◆MOURA D           ALUMINA - - - TSUBOI S        CURVE - - - - COOPER SA
         ◆SOARESDA.P          AMIDE - - - -  "              CYTOCHROME. ◆DAHLIN DC
ADSORPTION ◆MADIX RJ          ATOM-BRIDG. ◆PADIAS AB        DAMAGE - - - STERN AI
ALCOHOL - - ◆GUERRI C         BENZALDEHY. ◆LAMATY G@        DEFICIENCY - ◆REICKS M
ALDEHYDE - ◆SHIOHARA E        BENZYLIDENE ◆BINKLEY RW       DETERMINAT. - BERGH JJ
AMINES - - - - GILL K         BETA-AMINO - SHONO T                  - ◆HASSAN SM
AMINOANTHR. ◆DENISOV VY       BETA-BROMV. - RAIFELD YE
ATMOSPHERL ◆ORITA H           BETA-ETHYL. - BELLASSO.M
```

Figure 8.5. Example of *Permuterm® Subject Index*. Reproduced with permission from *Permuterm®*. Copyright owned by the *Institute for Scientific Information®*, Philadelphia, PA, U.S.A.

Line®. Covers the entire field of analytical chemistry. Printed indexes every 6 months (see also Section 18.4).

Mass Spectrometry Bulletin (monthly). Also searchable online on Pergamon *InfoLine®*, November, 1966 to date (see also Section 10.12B).

Current Biotechnology Abstracts (monthly).

Laboratory Hazards Bulletin

Chemical Hazards in Industry

Most of these publications are excellent examples of "packaging" of abstracts and related tools to meet the special needs of relatively concentrated groups of chemical specialists. The products can thereby each be tailored to meet these needs in the most efficient and economic manner.

The Royal Society has a number of other important publications and also offers such services as: supplying of CAS Registry Numbers, search of *TOXLINE* files, retrospective search of *CA* and *Biological Abstracts®*, and current awareness (individual customer service including abstracts) in the chemical and biological sciences. See also Section 18.4.

8.5.   RUBBER AND PLASTICS RESEARCH ASSOCIATION (RAPRA)

The abstracts, indexes, and online database of the Rubber and Plastics Research Association (RAPRA)* are important. The online file of *RAPRA Abstracts* starts in 1974 and is searchable through Pergamon *InfoLine®*. It provides excellent and detailed coverage of technology relating to polymers, plastics, and rubbers. An important feature is the special emphasis put on company and trade names that can be the best or sometimes the only way of identifying a key technology of more recent vintage. Sources include not only research and development reports, but also practical trade and commercial developments of the type not usually covered by *CA*. This means that, in searches relating to polymers, plastics, and rubbers, and especially in which trade/commercial information may be important, it is helpful to use *RAPRA* in conjunction with *CA*.

The printed version appears every two weeks and is also available through Pergamon. Sections include: general; commercial and economics; legislation; industrial health and safety, raw materials and monomers; compounding ingredients; intermediate and semifinished products; applications; processing and treatment; and properties and testing. Separate publications include an annual guide to trade names and an alerting service, *RAPRA EXTRACTS*. Other services include search service, magnetic tape leasing, translations, and photocopies.

8.6.   *CHEMISCHER INFORMATIONSDIENST*

Chemischer Informationsdienst is published weekly under the auspices of the Gesellschaft Deutscher Chemiker in cooperation with Bayer AG. This weekly service contains abstracts in German of about 20,000 papers per year from about 230 journals in the areas of general, physical, inorganic, and low-molecular weight organic chemistry.

8.7.   NATIONAL LIBRARY OF MEDICINE

In medicinal chemistry, major sources in addition to *CA* include the databanks and publications of the National Library of Medicine (NLM), Bethesda, MD. The two principal online files of NLM are *MEDLINE* (medical literature) and *TOXLINE* (published human and animal toxicity studies). The overall approach of NLM is, of course, from a medical perspective, but many chemists find its files, especially *TOXLINE,* helpful. One of the newer online files available from NLM is *DIRLINE,* which is a directory of information resources (organizations and experts) who can be contacted. This file was originally developed by the National Referral Center at the Library of Congress. Henry Kissman has been the leader of the efforts of chemical interest at NLM. These and other NLM online files are discussed in more detail in Section 9.6.

8.8. *BIOLOGICAL ABSTRACTS®*

This major abstracting and indexing service is a product of BioSciences Information Service (*BIOSIS*).* Companion products include *Biological Abstracts/RRM® (Reports, Reviews, and Meetings), BIOSIS/CAS SELECTS* and other awareness services, *Abstracts on Health Effects of Environmental Pollutants (HEEP), BIOSIS Previews®* (online database), *BIOSIS Information Transfer System (B-I-T-S™),* and others. For the biological aspects of a topic, *BIOSIS* products are an important source.

8.9. *CHEMICAL ENGINEERING*

Another source of engineering information is *Chemical Engineering,* which indexes and abstracts the contents of some 400 journals from 1976 to date. The producer is DECHEMA (Deutsche Gesellschaft für chemisches Apparatewesen) in cooperation with Fachinformationszentrum Chemie. Online accessibility is via *INKA* (see Section 10.12.C). As mentioned previously, The Royal Society of Chemistry has a similar publication.

8.10. *ENGINEERING INDEX®*

Engineering Index® (Ei®) was founded in 1884 and is now published by Engineering Information, Inc*. Types of literature included are journals, conference proceedings, conference preprints, technical reports, standards, and monographs. Patents are not included. The core areas of engineering covered include civil, mechanical, electrical, chemical, and mining.

The computer-readable form of *Engineering Index®* is known as *COMPENDEX®,* which covers literature since 1970. Online access is readily available through several major systems, such as some of those described in Chapter 10. A specialized publication, *Energy Abstracts,* provides information on conventional and alternative sources of energy. In conjunction with other *Ei®* activities, an exceptionally strong, full-service library, Engineering Societies Library, is operated at *Ei®* headquarters.

8.11. *INFORMATION SERVICE FOR THE PHYSICS AND ENGINEERING COMMUNITIES (INSPEC)*

Information Service for the Physics and Engineering Communities (INSPEC) is a product of the Institution of Electrical Engineers in the United Kingdom and is represented in the United States by the *INSPEC* IEEE Service Center.* *INSPEC* provides excellent coverage of the increasingly important and still emerging interface of chemistry with electrical and electronics engineering and computer technology, as well as physics. This source is especially recommended for any chemist or engineer interested in semiconductors or other electronics materials or chemicals. The major printed publications (also available online through a number of major systems as listed in Chapter 10) are *Electrical & Electronics Abstracts, Computer & Control Abstracts,* and *Physics Abstracts.*

Current awareness services, such as *Current Papers in Physics,* and selective dissemination of information are also available. In addition, a separate service, *Electronic Materials Information Service,* covers, in an online mode, hard data on properties of key electronics materials (see Section 15.16).

INSPEC abstracts are meaningful and complete, and indexing is excellent. One specific advantage of *INSPEC* is the indexing, which is from an electronics perspective; terms used are usually those familiar to workers in electronics. Unfortunately, however, patents are not included at this time, although they were covered in earlier years, until about 1976.

8.12. AMERICAN VERSUS BRITISH SPELLINGS

American chemists who use *INSPEC* (or any other British publication) need to constantly remind themselves of minute differences in spelling that could potentially result in missing significant information, especially when using printed or online indexes. For example, English (British) spelling of aluminum is aluminium, analog is analogue, color is colour, fiber is fibre, ionization is ionisation, molding is moulding, sulfur is sulphur, and vapor is vapour. Similarly, British chemists who use American publications need to remind themselves of these variations. Table 8.1 contains a more extensive list of spelling differences.

8.13.   CAMBRIDGE SCIENTIFIC ABSTRACTS

One of the largest private for-profit publishers of abstracts is Cambridge Scientific Abstracts.* Some of their publications of potential interest to chemists include:

> *Aquatic Sciences & Fisheries Abstracts®*
> *Cambridge Scientific Biochemistry Abstracts®*
> *Ecology Abstracts®*
> *Pollution Abstracts®*
> *Safety Science Abstracts Journal*
> *Science Research Abstracts Journal*
> *Solid State Abstracts Journal*
> *Toxicology Abstracts®*

8.14.   GOVERNMENT SERVICES

The U.S. Government is an important publisher of abstracting and indexing services. Some of these are summarized in Chapter 9.

8.15.   PATENTS; SAFETY AND HEALTH

A number of specialized services have been developed to cover information found in patents and on safety and health. These are discussed separately in Chapters 13 and 14.

TABLE 8.1.   American Variant Spellings for Terms in the INSPEC Thesaurus[a,b]

For a number of the terms contained in the INSPEC Thesaurus, the normal American spelling differs from the English form as used by INSPEC. The differences are generally small and almost all fall into the following categories:

> 'Z' in place of 'S'
>   e.g. 'ionization' and 'ionisation'
> 'or' in place of 'our'
>   e.g. 'color' and 'colour'
> 'er' in place of 're'
>   e.g. 'center' and 'centre'
> 'g' in place of 'gue'
>   e.g. 'analog' and 'analogue'

To assist users of the INSPEC database this appendix gives a list of all such terms contained in the Thesaurus, giving the American spelling and the equivalent English form.

It is planned to include the American variants as lead-in terms in future editions of the INSPEC Thesaurus.

NOTE: Both forms are likely to be used in the free-indexing.

TABLE 8.1. Continued

A-centers	use	A-centres
aluminum	use	aluminium
aluminum alloys	use	aluminium alloys
aluminum compounds	use	aluminium compounds
amorphization	use	amorphisation
analog computer applications	use	analogue simulation
analog computer circuits	use	analogue computer circuits
analog computer methods	use	analogue simulation
analog computer programming	use	analogue computer programming
analog computers	use	analogue computers
analog differential analyzers	use	differential analysers
analog digital computers	use	hybrid computers
analog-digital conversion	use	analogue-digital conversion
analog-digital convertors	use	analogue-digital conversion
analog memories	use	analogue storage
analog simulation	use	analogue simulation
analog storage	use	analogue storage
analog, direct	use	direct analogue
analyzer, differential	use	differential analysers
anodization	use	anodisation
anodized coatings	use	anodised layers
anodized layers	use	anodised layers
anodized thin films	use	anodised layers
apodization	use	acoustic imaging
		optic images
atomic polarizability	use	atomic polarisability
attenuation equalizers	use	attenuation equalisers
auroral ionization	use	auroral ionisation
autoionization	use	autoionisation
beta-ray polarization	use	beta-ray polarisation
biological effects of ionizing particles	use	biological effects of ionising particles
circular polarization	use	polarisation
carbon fiber reinforced composites	use	cabon fibre reinforced composites
carbon fibers	use	carbon fibres
channeling	use	channelling
chemical vapor deposition	use	chemical vapour deposition
code converters	use	code convertors
color	use	colour
color blindness	use	colour vision
color cameras, television	use	colour television cameras
color centers	use	colour centres
color display tubes, television	use	colour television picture tubes
color filters	use	optical filters
color model	use	colour model
color perception	use	colour vision
color photography	use	colour photography
color picture tubes, television	use	colour television picture tubes
color receivers, television	use	colour television receivers
color television	use	colour television
color television cameras	use	colour television cameras
color television picture tubes	use	colour television picture tubes
color television receivers	use	colour television receivers
color TV	use	colour television
color TV receivers	use	colour television receivers
color vision	use	colour vision

136

computerized air traffic control	use	air traffic computer control
computerized communications control	use	communications computer control
computerized control	use	computerised control
computerized industrial control	use	industrial computer control
computerized instrumentation	use	computerised instrumentation
computerized manufacturing control	use	manufacturing computer control
computerized materials handling	use	computerised materials handling
computerized monitoring	use	computerised monitoring
computerized navigation	use	computerised navigation
computerized numerical control	use	computerised numerical control
computerized pattern recognition	use	computerised pattern recognition
computerized picture processing	use	computerised picture processing
computerized power station control	use	power station computer control
computerized power system control	use	power system computer control
computerized process control	use	process computer control
computerized signal processing	use	computerised signal processing
computerized spectroscopy	use	computerised spectroscopy
computerized test equipment	use	automatic test equipment
computerized traffic control	use	traffic computer control
computerized transport control	use	transport computer control
converters	use	convertors
copolymerization	use	polymerisation
corporate modeling	use	corporate modelling
crystallization	use	crystallisation
demagnetization	use	demagnetisation
demagnetization, adiabatic	use	demagnetisation + magnetic cooling
deuteron polarization	use	deuteron polarisation
dialog programming	use	interactive programming
dielectric depolarization	use	dielectric depolarisation
dielectric polarization	use	dielectric polarisation
differential analyzers	use	differential analysers
digital-analog converters	use	digital-analogue conversion
digital-analog conversion	use	digital-analogue conversion
digital differential analyzers	use	digital differential analysers
digitizers	use	analogue-digital conversion
direct analogs	use	direct analogues
diverters	use	divertors
electron ionization	use	electron impact
electron microprobe analyzers	use	electron probe analysis
electron probe analyzers	use	electron probe analysis
electrosynchronization	use	synchronisation
elliptical polarization	use	polarisation
equalizers	use	equalisers
F-centers	use	F-centres
F_2-centers	use	M-centres
F_3-centers	use	R-centres
F_a-centers	Use	A-centres
fiber optics	use	fibre optics
fibers	use	fibres
fiber reinforced composites	use	fibre reinforced composites
field ionization	use	field ionisation
file organization	use	file organisation
flow visualization	use	flow visualisation
fluidized beds	use	fluidised beds
fluidized powders	use	powders
frequency converters	use	frequency convertors

137

TABLE 8.1. Continued

gamma-ray polarization	use	gamma-ray polarisation
glass fiber reinforced plastics	use	glass-fibre reinforced plastics
glass fibers	use	glass fibres
graphitization	use	graphitisation
graphitizing	use	graphitising
H-centers	use	H-centres
heat of crystallization	use	heat of crystallisation
heat of vaporization	use	heat of vaporisation
image converters	use	image convertors
impact ionization	use	impact ionisation
impedance converters	use	impedance convertors
information analysis centers	use	information centres
information centers	use	information centres
inverters	use	invertors
ion microprobe analyzers	use	ion microprobe analysis
ionization	use	ionisation
ionization chambers	use	ionisation chambers
ionization gauges	use	ionisation gauges
ionization of atoms	use	ionisation of atoms
ionization of gases	use	ionisation of gases
ionization of liquids	use	ionisation of liquids
ionization of molecules	use	ionisation of molecules
ionization of solids	use	ionisation of solids
ionization potential	use	ionisation potential
ionization time	use	ionisation
isobaric analog resonances	use	isobaric analogue resonances
isobaric analog states	use	isobaric analogue states
isomerization	use	isomerisation
latent heat of crystallization	use	heat of crystallisation
latent heat of vaporization	use	heat of vaporisation
lattice localized modes	use	lattice localised modes
light polarization	use	light polarisation
linearization techniques	use	linearisation techniques
liquid-vapor transformations	use	liquid-vapour transformations
localized modes in crystals	use	lattice localised modes
localized electron states	use	localised electron states
localized states, electron	use	localised electron states
M-centers	use	M-centres
magnetization	use	magnetisation
magnetization reversal	use	magnetisation reversal
magnetohydrodynamic converters	use	magnetohydrodynamic convertors
magnitude converters	use	magnitude convertors
mercury vapor lamps	use	mercury vapour lamps
mercury vapor rectifier	use	mercury vapour rectifier
metal vapor lamps	use	metal vapour lamps
metallization	use	metallisation
metallizing	use	metallisation
MHD converters	use	magnetohydrodynamic convertors
minimization	use	minimisation
minimization of switching nets	use	minimisation of switching nets
modeling	use	modelling
modeling, computer	use	computer-aided analysis
molecular electron impact ionization	use	molecular electron impact ionisation

molecular polarizability	use	molecular polarisability
monitoring, computerized	use	computerised monitoring
multichannel analyzers	use	pulse height analysers
negative impedance convertors	use	negative impedance convertors
network analyzers	use	network analysers
network equalizers	use	equalisers
neutron polarization	use	neutron polarisation
normalizing	use	normalising
nuclear isobaric analog resonances	use	isobaric analogue resonances
nuclear isobaric analog states	use	isobaric analogue states
nuclear particle track visualization	use	particle track visualisation
nuclear polarization	use	nuclear polarisation
nuclear polarization in solids	use	nuclear polarisation in solids
OH‾-centers	use	OH‾-centres
optical fibers	use	optical fibres
optimization	use	optimisation
orthogenelized plane wave calculation	use	OPW calculations
parametric up-converters	use	parametric devices
particle track visualization	use	particle track visualisation
Penning ionization	use	Penning ionisation
phase converters	use	phase convertors
photoionization	use	photoionisation
photoionization of gases	use	photoionisation
photon polarization	use	photon polarisation
polarizability	use	polarisability
polarization	use	polarisation
polarization in nuclear reactions and scattering	use	polarisation in nuclear reactions and scattering
polymerization	use	polymerisation
power converters	use	power convertors
power utilization	use	power utilisation
proton polarization	use	proton polarisation
pulse amplitude anlyzers	use	pulse height analysers
pulse analyzers	use	pulse analysers
pulse height analyzers	use	pulse height analysers
pulverized coal	use	pulverised fuels
pulverized fuel	use	pulverised fuels
quantization	use	quantisation
R-centers	use	R-centres
recrystallization	use	recrystallisation
recrystallization annealing	use	recrystallisation annealing
recrystallization texture	use	recrystallisation texture
renormalization	use	renormalisation
rotary converters	use	rotary convertors
Self-optimizing systems	use	self-adjusting systems
self-organizing storage	use	self-organising storage
signal processing, computerized	use	computerised signal processing
solid-vapor transformations	use	solid-vapour transformations
spectral analyzers	use	spectral analysers
spectrum analyzers	use	spectral analysers
spontaneous magnetization	use	spontaneous magnetisation
stabilization	use	stability

TABLE 8.1. Continued

stabilizers	use	controllers
standardization	use	standardisation
storage, analog	use	analogue storage
storage organization	use	file organisation
sulfur	use	sulphur
superconducting tunneling	use	superconducting tunnelling
surface ionization	use	surface ionisation
synchronization	use	synchronisation
synchronizing reactors	use	current limiting reactors
textile fibers	use	fibres
thermally stimulated depolarization	use	thermally stimulated currents
thermally stimulated polarization	use	thermally stimulated currents
track visualization, particle	use	particle track visualisation
transient analyzers	use	transient analysers
trapped electron centers	use	F-centres
trapped hole centers	use	V-centres
tunneling	use	tunnelling
tunneling spectra	use	tunnelling spectra
U-centers	use	U-centres
V-centers	use	V-centres
V_h-centers	use	H-centres
vaporization	use	vaporisation
vaporizing	use	vaporisation
vapor density	use	density of gases
vapor deposited coatings	use	vapour deposited coatings
vapor deposition	use	vapour deposition
vapor-deposited thin film	use	vapour deposited coatings
vapor-liquid transformations	use	liquid-vapour transformations
vapor phase epitaxial growth	use	vapour phase epitaxial growth
vapor pressure	use	vapour pressure
vapor pressure measurements	use	vapour pressure measurement
vapor-solid transformations	use	solid-vapour transformation
visualization, particle track	use	particle track visualisation
volatilization	use	vaporisation
wave analyzers	use	wave analysers
Z-centers	use	Z-centres

[a] While this list is intended for use with a database covering electronics and electrical engineering, it should also be of value in the use of some databases in chemistry.
[b] Reprinted by permission of *INSPEC*. The Institution of Electrical Engineers.

REFERENCES

1. H. Skolnik, *The Literature Matrix of Chemistry*, Wiley, New York (1982).
2. E. J. Crane, A. M. Patterson, and E. B. Marr, *A Guide to the Literature of Chemistry*, 2nd ed., Wiley, New York (1957).

LIST OF ADDRESSES

Arid Lands Information Center,
Office of Arid Lands Studies
845 North Park Avenue
Tucson, AZ 85719 (800-528-4841 or
602-621-7897)
Attention: Sandra Apsey

BioSciences Information Service
2100 Arch Street
Philadelphia, PA 19103-1399
(800-523-4806)

Cambridge Scientific Abstracts
5161 River Road
Bethesda, MD 20816

Commonwealth Agricultural
Bureaux
Farnham House, Farnham Royal
Slough SL2 3BN, United Kingdom
(011-44-2814-2281)

Engineering Information, Inc.
345 East 47th Street
New York, N.Y. 10017
(800-221-1044-Outside NY State)

Forest Products Research Society
2801 Marshall Court
Madison, WI 53705

ISI®
3501 Market Street
Philadelphia, PA 19104
(800-523-1850)

INSPEC IEEE Service Center
445 Hoes Lane
Piscataway, N. J. 08854
(201-981-0060, Ext. 380)

Institute of Paper Chemistry
P.O. Box 1039
Appleton, WI 54912

Pulp and Paper Research Institute
of Canada
570 St. John's Blvd.
Point Claire, Quebec,
Canada H9R 3J9

Rubber and Plastics Research
Association
Shawbury, Shrewsbury
Shropshire SY4 4NR, United
Kingdom

Washington State University
Pullman, WA 99164

9 SOME UNITED STATES GOVERNMENT TECHNICAL INFORMATION CENTERS AND SOURCES

9.1. INTRODUCTION

The research and development effort of the U.S. Government is huge. Its budget has been estimated at approximately $40 billion per year, and a large fraction of this is of direct interest to chemists and engineers. Thousands of reports and other publications are issued each year based on this massive government effort. However, many projects are not reported in forms readily available to the public. Fortunately, a number of important resources are available to help chemists and engineers access and use the results of federal R&D. Some of these are described in the pages that follow.

9.2. NATIONAL TECHNICAL INFORMATION SERVICE

The National Technical Information Service (NTIS) is the cornerstone of federal technical information efforts. All NTIS reports and services are readily available to the general public, usually at reasonable cost. Headquarters offices are located at Springfield, VA.*

Some of the functions of NTIS include:

1. Publishing biweekly abstracting and indexing bulletins (*Government Reports Announcement Abstracts & Index*) listing the most recent reports received by NTIS. Providing copies of all of these reports, usually designated by the prefix PB. Most, but by no means all, government reports are available through NTIS.

2. Making available of its abstracting and indexing files for online searching through such vendors as *DIALOG*® Information Services and others.

3. Compiling and publishing searches on topics (now over 3500) deemed important by customers and NTIS.

4. Publicizing of new government technology available for transfer to the private sector through *Government Inventions Licensing* (weekly) and *NTIS Tech Notes*. The latter are 1–2 page summaries of processes or products believed to have commercial potential. One of these categories is physical sciences. In addition, there are weekly *Abstract Newsletters* on 28 different topics. One of these is titled *Chemistry*. NTIS and other government documents are also included in *Chemical Abstracts* and other major information sources. In addition, NTIS handles the licensing of U.S.-owned patents developed at 8 Federal departments.

5. Acting as a clearinghouse for federally developed computer software and databases.

Search of the online NTIS databank is recommended because it includes some information of interest to chemists that is not readily found elsewhere. It is an especially simple file to use, and copies of almost all documents listed can be obtained from either NTIS itself or a variety of private document delivery services. Cost of use online is relatively low.

In 1986, NTIS published the *Directory of Federal Technology Resources* (PB86-100013). This is an invaluable guide, directory, and index to hundreds of federal laboratory and engineering facilities around the United States.

Despite all of its services and important advantages, users of NTIS should be aware that its document operation can be likened to that of a large, well organized, and intelligently managed mail order facility, except that it is not for profit. NTIS staff have neither the time nor the background required to evaluate the novelty or quality of all the many thousands of documents they handle nor to give expert referrals to the best sources of information. The quality of NTIS abstracts and indexes is much inferior to that of CAS or Derwent, and chemists using NTIS indexes must exercise ingenuity and creativity in finding what is needed.

A recent development is that the Administration is considering conversion of all or part of NTIS to private or quasi-public status.

9.3. NATIONAL REFERRAL CENTER

This often forgotten but important facility is located at the Library of Congress.* The purpose and function, as the name implies, is to provide referrals to sources of expertise nationwide. Referrals are made at no charge and can be to either government or private sources. The database of this organization has been made available online through the National Library of Medicine. This file is designated as *DIRLINE* (Directory of Information Resources) and is discussed further in Section 9.6.

9.4. DEFENSE TECHNICAL INFORMATION CENTER AND OTHER DEPARTMENT OF DEFENSE SOURCES

The Defense Technical Information Center (DTIC) is operated by the U. S. Department of Defense (DOD) to serve the information needs of DOD personnel and DOD contractors and potential contractors, and of other U.S. Government agency personnel and their contractors. DTIC users include universities and industries throughout the United States.

Because DOD is a huge generator of technical information, much of which is of interest to chemists, access to its data and documents through DTIC is important. DTIC headquarters is located in Cameron Station, Alexandria, VA, a short ride from Washington National Airport. Other locations are at Hanscom Air Force Base, near Boston, and Los Angeles, CA*. The services of DTIC include providing providing a *Technical Abstract Bulletin* (*TAB*) with indexes, full-text copies of reports, bibliographies, and current awareness and retrospective searches. Resources of DTIC are also searchable online. The total file at DTIC consists of over 1 million reports and documents. The categories of documents include those on: (1) completed research, (2) research in progress, (3) planned research, and (4) independent R&D.

Although DTIC could logically be expected to get copies of all technical reports sponsored by DOD, in fact it does not. For one reason or another, many thousands of reports are not available through DTIC, but rather only through the specific Defense Department agency or office that funded the work. In addition, DTIC is not well equipped to welcome visitors who may desire to do background reading on site within its specialized files. This is because DTIC headquarters houses almost all of its reports only in microfilm or microfiche form—a practice hardly conducive to intensive on-site study.

However, many unclassified reports originating from work sponsored by the Defense Department, are made widely and readily available by DTIC through the NTIS, which is described in Section 9.2. DTIC also manages and funds Information Analysis Centers (IAC's), which provide specialized and expert services. Some of these are of chemical interest, as described later in this section.

Like DTIC, the services of the IAC's are aimed primarily at government agencies and government contractors, but on occasion, the user community is more broadly defined. Services of the IAC's typically include analysis and evaluation of reports and other literature, abstracting and indexing, bibliographies, issuing of specialized compilations, and copies of documents. Some IAC's are not directly funded and managed by DTIC but are operated in cooperation with DTIC.

Except for the IAC's, which have some evaluative capabilities, most users can expect warehouse type of service from DTIC. This means that users are left with the task of pinpointing what is important and with interpreting results. Precise chemical nomenclature, or precise chemical abstracting and indexing, are not offered by DTIC.

DTIC and IAC services can be extremely valuable, but because much of the content of these centers is government classified, there is a complex procedure for use. This procedure requires that potential users fill out the appropriate registration forms that must then be approved by the government. Once approved, some DTIC services are free, although all IAC's operate on a cost-recovery (fee) basis.

From time to time, DTIC publishes a directory of its principal information sources and contacts, including many major federal laboratories and libraries. An example is report AD/A-095-600 *Defense Technical Information Center Referral Data Bank Directory* (August 1981). This directory is available through the NTIS. Compilations such as this are useful to anyone seeking contacts in the government.

A. Chemical Propulsion Information Agency

The Chemical Propulsion Information Agency (CPIA) has been in operation at the Johns Hopkins University Applied Physics Laboratory near Laurel, MD since 1946. In 1964, CPIA became one of the DOD Information Analysis Centers, and it is managed administratively by the DTIC.

CPIA produces *Chemical Propulsion Abstracts,* which contains abstracts of reports on government-sponsored R&D programs and is indexed by subject, author, corporate source, and contract number. In addition, CPIA publishes *Chemical Propulsion Technology Reviews* and a number of propellant manuals and other publications.

The staff is relatively small, but expert in appreciating major chemical propulsion trends. Expertise of its staff, however, is understandably not at the level of those chemists or engineers who do hands-on laboratory or pilot R&D in the fields covered. On the other hand, CPIA staff members can usually make good referrals to persons who can provide most needed detail.

For the most part, CPIA services are available only to organizations and individuals approved by the Defense Department, and there is a cost-recovery system of service charges.

B. Hazardous Materials Technical Center

Hazardous Materials Technical Center (HMTC) is a DOD Information Analysis Center operated since 1982 by Dynamac Corp., Rockville, MD. Principal emphasis is on hazardous waste technology (handling, transportation, storage, disposal) and regulations. Functions include: keeping track of pertinent literature, data, and regulations; performing literature searches; issuing a quarterly abstract bulletin and newsletter; and compiling special reports and handbooks. The principal function of HMTC is to serve DOD, and its very existence has been poorly publicized so far except within DOD. However, private firms or individuals can request data from HMTC and can subscribe to its abstracting service and newsletter—all for a fee. In addition, selected portions of the abstract bulletin are being added to the National Library of Medicine's online *TOXLINE* file (see Section 9.6) as the first step in an effort to augment *TOXLINE* so as to be responsive to the requirements of the Comprehensive Environmental Response, Compensation and Liability Act of 1980 (also known as the Superfund Act).

C. Plastics Technical Evaluation Center (Plastec)

This important Information Analysis Center is located at the U. S. Army Armament, Materials and Chemicals Command* and has been in operation since 1960. Some of the areas of specialization include plastics, adhesives, and organic matrix composites and laminates. A number of valuable compilations are made available to the public through NTIS, and literature searches of a specialized database can be performed for a fee. There is a strong collection of standards, specifications, and handbooks. See also Section 15.16.

9.5. ENERGY

A major source of energy information is the U. S. Department of Energy Data Base *(EDB)*. The principal printed product is *Energy Research Abstracts,* which is published semimonthly. The editorial staff of this publication is still located at Oak Ridge, TN, which reflects the original nuclear science orientation of this publication.

In addition to the printed form, *EDB* can be searched online through a government-sponsored system known as *RECON* and, since 1981, this file has also been available through commercial vendors such as *DIALOG*® Information Services.

Other major sources of energy information include *Chemical Abstracts; Engineering Index*® and *COMPENDEX*®; National Technical Information Service; and the databases and services of EIC/Intelligence, Inc., located in New York City (Section 14.5).

9.6. NATIONAL LIBRARY OF MEDICINE

The National Library of Medicine (NLM) is the strongest library of its type in the United States and possibly in the world. A complete range of information services is offered. Facilities are in Bethesda, MD.* Several very important online databases of interest to chemists are offered by NLM under the leadership of Associate Director Henry Kissman. These are described in the following sections.

A. *TOXLINE*

This is the NLM's extensive collection of computerized bibliographic information covering the pharmacological, biochemical, physiological, and toxicological effects of drugs and other chemicals. It contains approximately 1,600,000 citations, almost all with abstracts and/or indexing terms and CAS Registry Numbers. Records in the *TOXLINE* file have a primary date of publication of 1981 to the present. Older information is in the *TOXLINE* backfiles, *TOXBACK 76* (1976–1980), and *TOXBACK 65* (pre-1965–1975). The information in *TOXLINE* is taken from 11 secondary sources that formulate the following subfiles:

Toxicity Bibliography (TOXBIB)

Chemical-Biological Activities (CBAC)

Pesticides Abstracts (PESTAB) Terminated December 1981

International Pharmaceutical Abstracts (IPA)

Abstracts on Health Effects of Environmental Pollutants (HEEP)

Environmental Mutagen Information Center File (EMIC)

Research Projects (RPROJ)—(CRISP, a toxicology subset of the National Institutes of Health Grants Systems, was added in December 1984 [1983 FY])

Toxicology Document and Data Depository (TD3)

Toxic Materials Information Center File (TMIC)

Hayes File on Pesticides (HAYES)

Hazardous Materials Technical Center (HMTC)

International Labor Office (ILO)

B. *Registry of Toxic Effects of Chemical Substances*

The *Registry of Toxic Effects of Chemical Substances (RTECS)* is the online version of the National Institute for Occupational Safety and Health's (NIOSH) annual compilation of substances with "toxic activity". The original collection of data that makes up *RTECS* was known as the Toxic Substances List, compiled in 1971 by NIOSH in response to the Occupational Safety and Health Act of 1970. The printed version of *RTECS* is updated every year or two with quarterly microfiche editions that contain updated information. NIOSH is responsible for the file content in *RTECS,* and for providing the quarterly updates to NLM. *RTECS* currently contains toxicity data for some 60,000 substances. The information in *RTECS* is structured around chemical substances with toxic action, and thus provides a single source for basic toxicity information. *RTECS* includes some listings of basic toxicity data and specific toxicological effects that are searchable. The sources of toxicity data are identified, with the names of the journals, volumes, pages, and years given. Also included in *RTECS* are threshold limit values, air standards, NTP (National Toxicology Program) carcinogenesis bioassay information, toxicology/carcinogenic review information, status under various federal regulations, compound classification, and NIOSH Criteria Document availability. See also Section 14.3.

C. Chemical Dictionary Online

The *Chemical Dictionary Online (CHEMLINE)* is the NLM's online, interactive chemical dictionary file developed under contract with Chemical Abstracts Service (CAS). It provides a mechanism whereby over 600,000 chemical substances can be searched and retrieved online. This file contains CAS Registry Numbers; molecular formulas; *CA* chemical index nomenclature; generic and trivial names; and a locator designation that points to other files in the NLM system and the TSCA (Toxic Substances Control Act) Inventory. In addition, where applicable, each record contains ring information including: number of component rings within a ring system, ring sizes, ring elemental compositions, and component line formulas.

D. Directory of Information Resources Online

The *Directory of Information Resources Online (DIRLINE)* is an online, interactive directory that refers users to organizations providing information in specific subject areas. Records in *DIRLINE* come from two sources, the Library of Congress' National Referral Center (NRC) database, and the National Health Information Clearinghouse (NHIC) database. *DIRLINE* is updated quarterly. The NRC component contains information on over 13,000 public and private resource centers. Each record is updated by the NRC at least once every two years. The NHIC component contains 1500 records, which refer users to organizations providing information in areas of health and disease. Each record in NHIC is updated at least once a year. As an alternative to online use, clients can write or call NRC at the Library of Congress.*

   In addition, the NLM sponsors the Toxicology Information Response Center (TIRC). This was organized in 1971 at the Oak Ridge National Laboratory (ORNL) to serve as an international center for toxicology and related information. It provides information on individual chemicals, chemical classes, and toxicology related topics to the scientific, administrative, and public communities. As an information analysis center, TIRC synthesizes comprehensive literature packages according to a user's specific request. Formats may include, but are not limited to, custom searches of computerized databases, individualized literature searches, annotated and/or keyword bibliographies, SDI (selective dissemination of information) service, or written summaries of the material obtained. Published bibliographies are available from NTIS. Most services are provided on a fee (cost-recovery) basis. For more information, the contact is Toxicology Information Response Center, Oak Ridge National Laboratory.*

E. TOXNET

In 1985, NLM introduced its new TOXNET (Toxicology Data Network) data retrieval system. Currently available on TOXNET are the *HSDB (Hazardous Substances Data Bank)* and *CCRIS (Chemical Carcinogenesis Research Information System)* files. NLM's *TDB (Toxicology Data Bank)* was the forerunner of the *HSDB*. The passage of the Superfund Act (Comprehensive Environmental Response, Compensation, and Liability Act) provided the impetus to expand the rather pure toxicological scope of *TDB*. The resultant *HSDB* file took a somewhat broader look at hazardous chemicals. *TDB* was moved from NLM's ELHILL retrieval system to join *HSDB* in TOXNET. Both these files were TOXNET components until September 30, 1986, when the *TDB* file was permanently withdrawn and its entire contents subsumed within *HSDB*. *HSDB* is now a scientifically reviewed and edited data bank containing chemical and toxicological information strengthened with additional data related to environmental fate/exposure potential, emergency handling procedures, and regulatory requirements. Data are extracted from assorted monographs, government documents, special technical reports, and the primary journal literature. All data statements in each *HSDB* record are referenced and peer reviewed. *HSDB* contains records for over 4100 chemicals. The *CCRIS* file, sponsored by the National Cancer Institute, contains evaluated information, derived from both short and long-term bioassays, on 1200 chemicals. Studies relate to carcinogens, mutagens, tumor promoters, cocarcinogens, metabolites, and inhibitors of carcinogens, etc. Planning is under way to add additional files to TOXNET. Among these would be the National Institute for Occupational Safety and Health's *Registry of the Toxic Effects of Chemical Substances (RTECS)*. TOXNET offers a flexible and user-friendly approach to searching for toxicity data.

9.7. FEDERAL LABORATORY CONSORTIUM FOR TECHNOLOGY TRANSFER

Another significant way to access the know-how of the federal government in the United States is through appropriate contacts with the Federal Laboratory Consortium for Technology Transfer. The Consortium was established to help scientists in industry, and others in the United States, to identify and use the results of government research. Hundreds of federal laboratories and thousands of research workers are potentially accessible through contact with representatives of the Consortium. Under the Stevenson–Wydler Act of 1980 (Public Law 96-480), the government is mandated to strive to transfer its technology to the state and local governments and to the private sector. The Consortium has several regional representatives throughout the United States. The chairman at this time is Eugene Stark, who is located at Los Alamos National Laboratory. See also the discussion of Federal Research in Progress (Section 4.5).

9.8.   OTHER SOURCES OF GOVERNMENT INFORMATION

A number of private for-profit firms have been organized to help the public use government resources more effectively. These include, for example, Washington Researchers, Ltd.* They can perform investigations themselves or can refer clients to the proper sources. In addition, they have a number of publications intended to help identify key personnel and sources in Washington. One of their reports is a guide to some 3000 databases in the federal government, *The Federal Data Base Finder—A Directory of Free and Fee-Based Data Bases and Files Available from the Federal Government.*

It is often advisable to communicate directly and personally with federal officials. The best source for the names, addresses, and phone numbers of these officials is the *Federal Executive Directory,* published by Carroll Publishing Co.* Because this is updated several times a year, the frequent reorganizations and other personnel shifts in the federal government can be tracked using this tool. Even so, the federal establishment is so large and confusing, inquirers can often expect to be shunted around extensively until the proper contact is made. Local members of Congress can sometimes help make and expedite contacts with the most appropriate government employees and can help obtain copies of government reports and other documents.

Additional sources of technical information within the government are many and include, for example, the National Bureau of Standards and the Patent Office, both of which are covered in detail elsewhere in this book, as well as the Science and Technology Division of the Library of Congress and others too numerous to mention here.

LIST OF ADDRESSES

Carroll Publishing Co.
1058 Thomas Jefferson Street N.W.
Washington, D.C. 20007

Library of Congress
Washington, D.C. 20540
   (202-287-5680)

National Library of Medicine
8600 Rockville Pike
Bethesda, MD 20894

**Defense Technical Information
   Center-Field Locations**
Bedford, MA
and 11099 South La Cienega
   Boulevard
Los Angeles, CA 90045

**National Technical Information
   Service**
Springfield, VA 22161
   (703-487-4636)

National Referral Center, Library of Congress
Washington, D.C. 20540
  (202-287-5670)

Plastics Technical Evaluation Center
U.S. Army Armament, Materials &
  Chemicals Command
Dover, NJ 07801

Toxicology Information Response Center
Oak Ridge National Laboratory
P.O. Box X, Bldg. 2024
Oak Ridge, Tennessee 37831
  (615-576-1743)

Washington Researchers, Ltd.
2612 P Street N.W.
Washington, D.C. 20007
  (202-333-3499)

10 ONLINE DATABASES, CHEMICAL STRUCTURE SEARCHING, AND RELATED TOPICS

10.1. INTRODUCTION

The most significant development in chemical information since the first edition of this book is the continued growth and improvement of online databases as a source of information for chemists. Online databases have significantly transformed and enhanced the ways by which chemists can store, retrieve, and use chemical information. The impact has been virtually worldwide, and a slowdown of this transformation is nowhere in sight.

In recent years, existing databases have been improved, and important new bases and search methods have been added. Five developments are especially noteworthy:

1. The newer databases permit structure and substructure searching by both graphic and nongraphic methods. Structure searching was possible before, but not with nearly as much *power* as is now the case.
2. Much more data about patents, especially United States patents, are available online now than ever before (see Chapter 13).
3. Full-text online databases are more widely available.
4. Special attention is being given to facilitate access to databases through personal computers.
5. Numerical values (hard data) are increasingly available online.

The purposes of this chapter are to give chemists some appreciation of how online databases function, to review some of the advantages and limitations of online searching, to describe and explain some of the leading sources, to suggest additional strategic concepts that need to be considered in searching of both online and more conventional (printed) tools (see also Chapter 3), and to review chemical structure searching. Explanations of what to search and how are interspersed, as appropriate, with pertinent, selected lists of systems, centers, databases, vendors, and so on. Accordingly, the reader may find it helpful to quickly scan the table of contents for this chapter before reading it further.

10.2. THE BASICS

An online information retrieval system is one in which a user can, via computer, directly interrogate a machine-readable database such as *Chemical Abstracts*.

There is two-way (interactive) communication between user and computer through local input/output devices, such as a suitably equipped typewriterlike device and/or a cathode ray tube (CRT) display, connected to a main-frame computer center by a telephone line. The communication (input) from the user is the query. The communication from the computer (output) is the system's response to the query, or a prompt for the next query. For a flow chart of a typical online database and system see Figure 10.1.

Input devices may range from "dumb" or nonintelligent terminals (no memory or logic), to personal computers, to word processors. In all cases, a device called a modem (modulator–demodulator) is required for telephone communication unless the terminal is directly connected to the computer. In one option, an acoustic coupler may be needed to link the terminal to the telephone line. These devices may be either separate or already built in. Certain combinations of modems and computer terminals along with the appropriate software, permit users to automatically dial and log into preselected online systems. In such cases, a telephone instrument is not needed; a telephone line will suffice. Modems convert digital signals into analog signals in sending and analog into digital while receiving.

For databases that permit graphic input and display, special equipment is needed to take full advantage of the capabilities of these systems. This equipment can be specified kinds of graphics terminals, terminals with graphics adaptations, or microcomputers (personal computers) suitably modified as needed. In the case of text input, virtually any *dumb* (nonintelligent) terminal, word processor, or personal computer can be used, provided it is capable of ASCII transmission and reception. For such text terminals, structure searching is by means of alphanumeric input.

Online output usually consists of a list of references, including full bibliographic citations and indexing terms that may include CAS Registry Numbers. Some services, such as *CAS ONLINE* and *Questel–DARC,* offer structure diagrams for substances retrieved as answers. In some cases, full abstracts, hard data (numerical values, etc.), or full text (see Section 10.13) are also available as online output. Output format can sometimes be "tailored," and content can often be limited to what is of primary interest. In several systems, output can be sorted and ranked as needed. In many cases, copies of original documents can be ordered while the user is online. Future

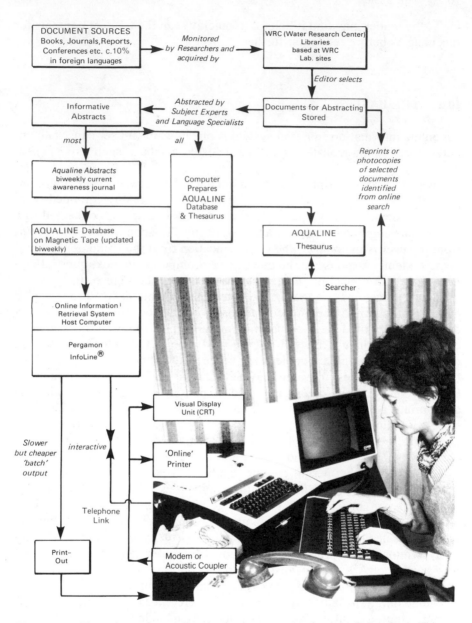

Figure 10.1. Flow chart of an online database system (production of *Aqualine*). This was derived from the *Aqualine* Online User Guide with permission from the Water Research Centre, U.K. DIALOG® is the registered trademark of DIALOG® Information Services, Inc.

expectations are for additional hard data and full text to be available as output.

The online concept, as applied to chemistry, is relatively new. It did not win truly widespread acceptance in the chemical community until the late 1970s and early 1980s. Not only are online databanks proliferating, but so are the vendors or service organizations that make the databanks available to chemists.

Several directories of online databases are available (1), and there are at least three English-language journals in this field (2). None of the directories, valuable though they are to chemical information specialists, make it significantly easier for chemists who are not chemical information specialists to learn details of how to use and evaluate competing systems and databases.

The principal public communication networks through which most online chemical information transactions in the United States are believed to be conducted, are *Tymnet®, Telenet®,* and *Uninet®.* In 1986, *Uninet®* became part of *Telenet.®* In addition, *DIALOG®* Information Services has established its own nationwide telephone network, which it designates as Dialnet®. Another network of increasing importance to chemists is *CompuServe®.**

These networks permit users to access computer-based databases with a local or toll free phone call in most metropolitan areas throughout the United States as well as in many foreign countries. Several of the networks publish valuable free directories of online systems (and databases) to which they provide access.

Most chemists applaud the availability of new databases that are both good and pertinent. What is more controversial is the increase in the number of specialized vendors. There are arguments both pro and con. Some believe that this trend is healthy competition that should yield the good results ordinarily brought about by competition. Others contend that they must now learn details of too many systems, and that this dilutes the opportunity to develop expertise in any of them. More vendors mean more systems, each of which may differ importantly in method of use. What will probably happen is that vendors who offer quality services that are needed, and that are easy to use, will survive, while the others will fall by the wayside.

10.3. WHY USE ONLINE SERVICES?

Online information retrieval offers chemists important advantages in searching of patent, journal, and report literature. Rapid, convenient, and fingertip access is made possible to millions of references and chemical substance structures as well as some hard data and full text of journal articles, patents, and so on. The systems available often permit users to identify recent pertinent references more completely, quickly, and effectively than is possible by using other techniques.

Access to printed indexes—whether these are alphabetical, classified, numerical, or specially organized—is inherently limited by index arrangement. In addition, users can find information in printed indexes only through a limited number of primary access points, and these are usually in a predetermined order. (More can be done if the user wants to spend the time.) In contrast, online services often afford many more access points and have much increased flexibility. The following are examples.

If the index heading is **Polymerization** and indexing policy is to index *oligomerization* under the more general heading, the user must turn to the heading **Polymerization** and look for *oligomerization* as a potential indenture or a term somewhere in a phrase (or modification) under that heading. Yet, in online access, *oligomerization* can directly be accessed no matter where it occurs in the file.

If the index heading is **Benzoic acid, 4-(1*H*-tetrazol-5-yl)-,** the user cannot usually find from the printed index, for instance, all (or certain) derivatives of *1H-tetrazole,* because the alphabetically arranged chemical substance indexes contain unique names constructed according to an order of precedence of chemical functions and compound classes.

If the index headings in the Formula Index are $C_{17}H_{18}IN_5O_4S$ and $C_{24}H_{20}I_6N_4O_8$, the user cannot usually find [unless the whole index is scanned (an impossible task), or unless it is a permuted index] other organic compounds each containing I, N, and O. Online access allows the user to go into *the middle* of a chemical substance name or formula, and then use combinations of terms such as described later.

The online approach has the capability to search for individual (or combinations of) access points that are not usually easily available in printed indexes. These may include, for example, not only index terms, keywords, authors, formulas, and patent numbers (these are readily available now in printed indexes), but also: parts of chemical substance names, formulas and notations; generic features; periodic groups; classes of substances; and formula weights. These are all relatively difficult to locate in most printed indexes, and combinations with one another, as well as concurrent limiting by language or type of document, are even more difficult or impossible in traditional indexes. However, some printed indexes with additional access points provided by permuted or rotated entries do exist, as, for example, in some publications and services of the *Institute for Scientific Information*® (see Section 8.3).

The online approach permits users to create and use, as necessary, search logic that is much more complex than can usually be handled manually. This is done by combining key terms (such as names of substances and other subjects) used in searching, as shown in the following examples:

A   or   B   or   C   or   D

A   &   B   &   C   &   D

(A & B)   or   (C & D)   not   (E & F)

See also Figure 10.2.

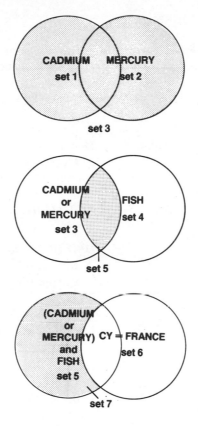

Figure 10.2. *Simplified example of Boolean Logic.* The first two combinations of circles indicate a Boolean Logic approach to obtaining references on the presence of cadmium or mercury in fish. Set 3 depicts the statement "cadmium *or* mercury" indicating that either or both may be present. In set 5, the results from set 3 are combined (in an *and* statement), which specifies that the word "fish" must also be present, thereby yielding set 5. In set 6, the country of France is excluded for purposes of this example (*not* statement) which results in set 7 (cadmium *or* mercury *and* fish, but *not* France). This was derived from the *Aqualine* Online User Guide with permission from the Water Research Centre, U.K.

Similarly, users can specify that only certain types of publications are wanted, such as: patents issued in a specific country; language within a given time; articles that contain one or more CAS Registry Numbers in combination with a specific end use; or review articles and books on a combination of topics. All other materials can be automatically excluded. Doing this kind of specification in a conventional manual search may sometimes be possible, but it is much more time consuming. In addition, some online systems facilitate ranking and sorting of results.

Online systems also make it possible for the chemist to search a wide variety of sources and tools that might otherwise not be conveniently available because of budgetary, geographical, or other limitations.

10.4. MAJOR SUPPLIERS

Fortunately for the chemist, all major information services of chemical interest are available online through several suppliers, most of whom provide
access throughout the United States and in other countries. Examples of
online suppliers (also called "host" systems or vendors) include:

BRS® **Information Technologies**
1200 Route 7
Latham, N.Y. 12110
(518-783-1161), (800-833-4707),
   (800-553-5566 in New York)

STN^{SM} **International**
(Cooperative venture of the ACS
   and Fachinformationszentrum
   Energie, Physik, Mathematik
   GmbH; the principal U.S.
   component is *CAS ONLINE*)
P.O. Box 3012
Columbus, OH 43210
(614-421-3600), (800-848-6533)
   (outside Ohio), (800-848-6538)
   (inside Ohio and in Canada)

Chemical Information System, Inc.
   (CIS)
Fein–Marquart Associates, Inc.
7215 York Road
Baltimore, MD 21212
(301-321-8440), (800-CIS-USER)
   (from outside Maryland)

   A historically related (but
      different) *Integrated Chemical
      Information System (ICIS)* is
      also available through

Information Consultants, Inc.
   (ICI)
1133 15th Street, N.W.
Washington, D.C. 20005
(202-822-5200)

**Defense Technical Information
   Center (DTIC)**
Cameron Station, Building 5
Alexandria, VA 22314
(202-274-6434)

DIALOG® **Information Services,
   Inc. (subsidiary of Lockheed
   Corporation)**
3460 Hillview Avenue
Palo Alto, CA 94304
(415-858-3785), (800-3-DIALOG)

**European Space Agency (ESA)
   Information Retrieval Service**
C.P. 64 Via Galileo Galilei
1-00044 Frascati, Italy
(06) 94011

   U.S. contact is

Information Services Associates
21 Maynard Court
Los Altos, CA 94022
(415-948-2326)

Mead Data Central
9393 Springboro Pike
Dayton, OH 45401
(513-865-6800), (800-621-0391)

National Library of Medicine (NLM)
8600 Rockville Pike
Bethesda, MD 20894
(301-496-6193)

Pergamon *InfoLine*®, Inc.
1340 Old Chain Bridge Road
McLean, VA 22101
(703-442-0900), (800-336-7575)

Home office is
12 Vandy Street
London EC2A 2DE, England
011-44-1-377-4650

SDC® Information Services
  (*ORBIT*®)*
2500 Colorado Avenue
Santa Monica, CA 90406
213-453-6194, (800-421-7229),
  (800-352-6689) (in California)

*In 1986, SDC® was purchased by Pergamon
which is also the owner of *InfoLine*.®

Telésystémés–Questel
  (*Questel–DARC*)
83–85, Boulevard Vincent-Auriol
75013 Paris
France
Telephone: 33 (1) 582.64.64

U.S. Office is
Questel, Inc.
Suite 719, 1625 Eye Street, N.W.
Washington, D.C. 20006
(202-296-1604), (800-424-9600)

10.5. DATABASES OF INTEREST TO CHEMISTS THAT ARE SEARCHABLE ONLINE

The brief list of online sources of interest to chemists in Table 10.1 is merely representative. A recapitulation of databases that emphasize patents is presented in Table 13.1, and other online sources are discussed throughout this book, especially in the chapters on physical properties and on safety. New databases (also called databanks or files) are added and others are dropped periodically, depending on availability, demand, and interest or need. The principal criterion as to the availability of a database to the public is usually financial (profit). Accordingly, the turnover in online databases is significant.

Chemists can consult their local chemical information specialists or chemistry librarians for full details on what is available or can refer to

TABLE 10.1. Some Online Databases of Chemical Interest

Chemical Abstracts (Registry; Bibliographic Files from
    1967 to date)
Derwent *World Patents Index* (*WPI*)
Index Chemicus® Online
IFI *Claims*™/*U.S. Patents*
Engineering Index® (*COMPENDEX*®)
Biological Abstracts (*BIOSIS PREVIEWS*)
Chemical Industry Notes (*CIN*)
Science Citation Index® (*SCISEARCH*®)
National Technical Information Service (NTIS)

published directories (1). A complete list of files of chemical interest might number more than 50–100 databases.

It is rare that the complete original printed source is available online from its beginning. Nevertheless, so much of importance is online that significance of this mode of search cannot be overemphasized.

10.6. USE OF ONLINE DATABASES—SOME IMPORTANT CAVEATS

The online database concept is not a panacea. For many online databases, coverage of the literature and patents does not begin until the late 1960s, although, for some, coverage begins much earlier, and for others, much later. For example, although the printed *CA* begins in 1907, the online version of *CA*'s bibliographic files does not begin coverage until the 1967 volumes. However, as of July 1, 1986, 67% of all *CA* substance indexing and bibliographic data since the beginning was online available for searching, and 49% of all *CA* abstracts since the beginning were contained in *CAS ONLINE*.

Most major online systems that offer online databases of interest to chemists did not start on a full commercial or public basis to the chemical community until the early 1970s or later. This means that for complete investigations—especially those that require going back to the *beginning*—the chemist also needs to conduct a conventional manual search. Accordingly, online services are often complementary to printed sources and manual search procedures. In addition, printed tools can have such important potential advantages over their online counterparts as: browsability, ease of use (no online instruction or equipment needed), and continuous availability (condition or availability of equipment, or online system schedules are not factors when using conventional printed sources). Finally, some important printed sources are not yet available online, and to ignore such sources, if they are pertinent, can be a significant error.

Availability of data through a computer, however sophisticated the hardware or software, does not in itself make that data accurate or reliable. This is a function of the original research that produced the data, the care with which these data are abstracted and indexed by secondary sources, and the care with which the data are entered into online files and systems. Data quality indicators have been suggested by the Chemical Manufacturers Association and others to help chemists evaluate online output from databases that provide hard (numerical) data (see Section 15.14).

The completeness that some users may attribute to online output per se can frequently be a myth. What count most are the intrinsic completeness and quality of the original sources, validity of search strategies, and the skill with which these are executed. Strategy is important because of the considerable chance for information loss due to such reasons as variations in: chemical names for the same substance, expression of concepts, and spelling of the names of organizations and people. Another example of the importance of strategy is in searching for patents assigned to a specific company. In such instances, it is important to look for patents to subsidiaries and affiliates in addition to simply looking under the name of the *parent* company and variations of that name.

As an example of skillful implementation of strategy, it is essential to understand and follow the protocol and logic of the database system. For example, a "minor" error such as a misplaced space or punctuation mark, or failure to use one when it is called for, can lead to a significant loss of information. An error in logic can be even more serious. The system may not signal when an error occurs and so the mistake can go undetected. A computer answer (output) showing zero "hits" always deserves especially careful scrutiny for this and related reasons.

Evaluation of online databanks is similar to evaluation of printed files (see Section 15.14). In assessing online services, criteria used include: scope; quality of the file and its originator or sponsor; promptness of updates and currency of content; cost (both connect time and cost of printing hits online and offline); response time (speed of response to online inquiries); schedule and availability; ease of use; quality (content and arrangement) of display; quality of instruction manuals and other documentation; and availability and quality of toll free *help lines*. In this regard, two of the services are distinguished by the excellence of their help desks. These are *CAS ONLINE* and Pergamon *InfoLine®*. Both provide expert and patient assistance. With regard to quality of user manuals, *DIALOG®* is particularly noteworthy.

Staff for some (but not all) online systems have advanced degrees in chemistry and can sometimes give valuable, informed advice in selection of the most appropriate databases and in formulation of online strategies. The quality of this advice, however, can vary sharply, depending on the vendor and on the person.

A thorough knowledge of the database being searched is as essential for online searching as it is for manual searching. The user needs to know as much as possible about the database, including scope, kind of indexing and abstracting used, and any special features or limitations. In addition, the user needs a thorough knowledge of the online system being used. This does not mean becoming a computer programmer, but it does mean study and knowledge of the manuals that suppliers of online systems make available. Further competence can be gained by attending training sessions held by suppliers at central locations, or as an option, at one's own facility. Expertisc in usc of systems can be further enhanced by online experience in real-life situations. A number of months may elapse before a user can be considered expert, although the key is regular and frequent use. Furthermore, online suppliers make frequent improvements and other changes in details of operation. Sources (databanks) that constitute input to the systems may also frequently change scope and other policies. The expert user needs to try to keep up with these changes by perusal of the newsletters that suppliers issue periodically, the notices that most systems supply online, and study of the sources in both printed and online forms.

In this connection, it is important to note that there may or may not be one-to-one correspondence between the printed and online forms of the same database. One may contain information or approaches that the other lacks. Accordingly, both may sometimes need to be consulted. It is often desirable to scan the printed form before going online to help develop the best strategy and for other similar reasons.

Expertise with any one online system does not assure competence in use of other systems, although all have fundamental similarities. For this and other reasons, user organizations should train more than one person, each of whom specializes in one or two systems. Basic typing skill is useful but not essential. This helps because communication with the computer (input) is by a typewriterlike keyboard. Ability to operate this equipment quickly and accurately saves times and money. From this, it should be clear that special training, and considerable study and effort are prerequisites to optimum use of online systems. Some laboratory chemists have the time and interest required, but most probably have their online searches conducted for them by chemical information specialists or chemistry librarians. The laboratory chemist should provide the person who does the searching with a complete description of what is wanted and should help evaluate strategy and output. Best results come from close partnership.

10.7.   COSTS

There are two basic components of online system use costs. The first is based on the time the user is connected to the central computer. This component, often referred to as "connect time," is typically about $100 per hour, but varies widely depending on the database and system used. Another $6–10 per hour needs to be added to cover use of one of the public data communication networks.

The second basic component is cost to display or print each hit or record, typically about 35–90 cents each (cost varies widely depending on the database, on how much of the record is displayed, and on whether or not it is printed offline). Additional costs may be accrued if, for example, strategies are saved online so that they may be executed in the future. Furthermore, a few systems require "up-front" fees independent of connect time and that must be paid by the user organization before any access is possible. A related cost that applies to some databases, is a basic subscription fee that the user organization must pay to have access to the database.

Up-front and subscription fees are not considered desirable by many users, and shopping around among various systems can sometimes identify suitable alternatives. A database or system for which users need only pay as it is used is usually preferable, assuming all else is equal.

Another cost factor is purchase, rental, or lease of the terminal required to access central computers. These costs vary widely, depending on manufacturer and special features selected. They can be as low as about $600 [1986 cost for purchase of a dumb (nonintelligent) printer terminal of good quality, but without all desirable features].

Examples of other factors that the user will want to take into account include:

1. Cost of equipment (modem or acoustic coupler) that may be needed to link the terminal to the computer (usually available for about $300–500), although some terminals and computers have this built in as part of their basic cost and configuration.
2. Cost of any required software (if a personal computer is used).
3. Cost of any authorized downloading (electronic capture) for subsequent use.
4. Any savings from requesting an *offline* print of abstracts or citations at a computer center and subsequent mailing of this output to the user— this approach can be advantageous when the user wishes to scan a large number of references.
5. Any savings from large volume or other types of discount plans.

10.8. ONLINE SEARCH STRATEGY

For most effective search results, strategy should be developed on paper or blackboard before going online. Search objectives, basic logic, and search terms should be specified clearly. Certain search aids, such as thesauri and word frequency count lists, are available specifically to help online searchers in their planning. Once online, users have at their disposal additional searching aids, some of which are not available manually.

The user will usually scan a small sample of initial results in an online search. If necessary, search strategy can then be quickly and easily changed while still online to achieve the most acceptable results. The capability of revising search strategy easily is one of the most attractive features of online searching. It is a major reason why this tool is used so widely and effectively.

In the use of chemical indexes, especially those searched online, it is important to look under as many logical term or chemical name variations as necessary. Thesauri or related types of guides are sometimes available to suggest or specify variants if any, to be examined. For *CA*, a basic tool for this purpose is the *Index Guide* (see Section 6.4).

One example of use of as many variations as possible in online searching occurs when the chemist desires references on ways of making a chemical. Some variations that need to be considered include:

1. a. manuf.   d. manufacture
    b. manufd.   e. manufactured
    c. manufg.   f. manufacturing

These usually denote references to methods of making chemicals on a commercial or plant scale.

One convenient shortcut in specifying name variations such as those mentioned is the use of truncation. For example, all of the variations just listed could be retrieved under the single truncation "manuf". Some online databases and systems can accept both right-hand and left-hand truncation.

Probably the most certain way that the chemist or engineer has of determining that a reference containing one of these variations *may* pertain to commercial manufacture of a chemical is identification of that reference as a patent (see Chapter 13 on patents). Many patents have at least the potential goal of commercial manufacture. In the printed *CA* indexes, association of an index entry with a patent is indicated by the symbol P preceding the abstract number; patents may also be specified in online searching of *CA*.

2. a. prep.   f. prepared
    b. prepd.   g. preparing
    c. prepg.   h. preparation
    d. prepn.   i. synthesis
    e. prepare

These almost invariably refer to methods of making chemicals on a laboratory or bench scale, but, again, when the reference is identified as a patent, commercial scale manufacture or use is a likely object.

3. formation

This usually signifies work in which the chemical occurs as a minor by-product, or only incidentally to the preparation of another compound that is the object of the investigation.

Other information pertinent to making a chemical could be found under such keywords or subject entries as:

4. fermentation (and related terms)
5. isolation (and related terms)
6. process (and related terms)
7. production (and related terms)
8. purification (and related terms)
9. recovery (and related terms)

But even inclusion of all terms 1–9, as noted, may not be enough. An article may relate to preparation of a compound without this information appearing in either the title or words used to index the article. For example, when a physical property of a substance is the principal object of the investigation, the article may include a report on preparation of a pure sample of the substance.

The compound may also be prepared or formed as a minor intermediate, and reported in an article dealing primarily with preparation of another compound, but not indexed. This depends on the policy of the database.

In the case of *CAS ONLINE*, Registry Numbers may occur with a suffix P to indicate preparation of that substance, a suffix D to indicate a generic or unspecified derivative, or a suffix DP to indicate the preparation of a generic or unspecified derivative. The key used by indexers for some of the most significant information tools is emphasis placed on the subject by the author of the original document.

As another example, consider the case of a chemical engineer concerned with selection of materials of construction to be used with a specific corrosive product. In addition to *materials of construction* as an entry in the online search, other entries to be searched could include:

anticorrosive (and variations)
compatibility (and variations)
corrosion (and variations)
inhibition (and variations)
noncorrosive (and variations)
prevention (and variations)
protection (and variations)
resistance (and variations)
specific materials, such as type of steel
stabilization (and variations)

Or consider the case of an analytical chemist searching for references on gases evolved on decomposition of a polymer. If this search were conducted online, an array such as listed in Table 10.2 could be constructed. Note that

TABLE 10.2. Examples of Term Variations Used in Online Searching

decomp.	decomposed	dissoc.	dissociation	oxidation
decompd.	decomposing	dissocd.	dissociating	oxidn.
decompg.	decomposition	dissocg.	dissocn.	pyrolysis
decompn.	degradation	dissociate	heat	temp.
decompose	degrdn.	dissociated	hot	temperature
				thermal

this array takes into account variant terms as well as abbreviations, all of which might be needed to ensure that the search is as complete as possible. However, considerable time and effort can be saved by truncation.

The chemist must also often include in searches variations in chemical names. A simple case is polyvinyl chloride, which might be found entered as any one or all of the following examples:

ethene, chloro-, polymers, homopolymer
PVC
polyvinylchloride (one word)
poly(vinyl chloride)
polyvinyl chloride (two words)
vinyl
vinyl compounds, polymers
vinyls

Note inclusion of both singular and plural variations of the same word. This list is not necessarily all inclusive; other variants may need to be used. Systems such as the *Chemical Information System* (see Section 10.11A.3) overcome this problem by conversion of all nouns to single form. Thus, hits are obtained in any combination of single/plural between query and stored text. The data displayed are as in the original record.

Fortunately, chemists performing online searches can frequently employ CAS Registry Numbers (see Section 6.10). These unambiguous identifiers can eliminate or minimize the need to memorize or use alternate chemical names in many searches. The Registry System is relatively simple to use, but it is a tool of great power, specifically in online searches of *CA*. In a search of *CA*, even if a Registry Number is not available or is inappropriate, use of variant chemical names is less important than in other sources because of uniform use of systematic nomenclature by *CA* and the availability of the *Index Guide* (see Sections 6.4, 6.6) (although indexing policies do change).

The Registry System is not perfect or all inclusive. The system is comprehensive, but did not begin until 1965. Consequently, many compounds are not yet in the system, although efforts are underway to take the file back through prior years (see Section 6.10). Older compounds can be especially troublesome because of nomenclature variations over the years. In addition, Registry Numbers are of little help in generic searching.

Another variation is due to changes in company or organization names over the years as shown in the example in Section 13.10. Failure to take such changes into account can result in loss of information. Fortunately, some database producers such as Derwent, IFI, and INPADOC (the International

Patent Documentation Center) (see Chapter 13), have attempted to standardize many company names. Such standardization is, however, incomplete and otherwise imperfect and does not usually take into account names of subsidiaries and affiliated companies.

In some cases, companies choose not to have inventions assigned immediately after issue. Such patents may be found only under names of individual inventors; none of the strategies described previously will help locate such patents quickly.

Another key point in online database searching relates to the importance of using more than one database, especially when looking for very critical information such as in a patent or safety investigation. The rationale is that even though some databases and systems may seem similar, or even identical, there are frequently important differences in coverage, in methods of indexing and abstracting, in system capability, or in mode of use. Although use of multiple files and systems usually results in overlap, it can frequently also result in location of important data that would not be found if just one system or database is used. For example, a search of *CA* on online system A may yield information not located on online system B. The online concept facilitates multiple database searching.

Finally, many experienced searchers believe that special attention needs to be paid to use of "not" logic. This capability is very powerful, but it must be used carefully to ensure that pertinent material is not accidentally excluded. Chapter 3 on search strategy contains some additional suggestions of specific interest to planning of online investigations.

10.9. INFORMATION CENTERS

There are several large *information centers* in the United States and other major industrial countries (for a partial list, see Table 10.3). These centers provide a wide variety of current awareness and retrospective searching services.

The services are usually computer based. Some provide primarily online service. Others are, however, *batch* or *offline*. Input (question) is transmitted by mail or ordinary voice telephone line, and output (answers) or search results are usually transmitted to users by mail. Response time is usually overnight at best, but more typically several days, depending on the mails and on workload.

Centers that provide offline or batch service continue to be a factor, despite the increasing popularity of online services. Information centers can offer such potential benefits as one or more of the following:

1. Providing expert assistance in correct structuring of inquiries, that is, concise and comprehensive statement of questions.

TABLE 10.3. Information Centers That Provide Service Based on Files Licensed from CAS and (in many cases) from Other Sources[a,b]

Information Centers Within the United States	

BRS® Information Technologies
1200 Route 7
Latham, N.Y. 12110
Phone: (518-783-1161, 800-833-4707),
  (800-553-5566) (in New York)
(Provides online service.)
	Mansfield Professional Park
	Storrs, CT 06268
	Phone: (203-429-3000)
	Telex: 710-420-0571
	Telecopier: 203-486-4586
	(Does not offer online services but will
	  conduct searches for a fee.)

***DIALOG*® Information Services**
3460 Hillview Avenue
Palo Alto, CA 94304
Phone: (415-858-3810), (415-858-3785),
  (800-3-DIALOG)
Telex: (DIALOG): 334499
(Provides online service.)
	System Development Corporation
	SDC® On-Line Bibliographic Search
	  Service
	2500 Colorado Avenue
	Santa Monica, CA 90406
	Phone: (213-829-7511), (800-421-7229)
	  (in United States except California),
	  (800-352-6689) (in California)
	(Provides online service.)

Institute for Scientific Information®
3501 Market Street
Philadelphia, PA 19104
Phone: (215-386-0100), (800-523-1850)
(This does not appear in the CAS list.
  It provides services based on its own
  files only.)
	Toxicology Information Program
	National Library of Medicine
	8600 Rockville Pike
	Bethesda, MD 20894
	Phone: (301-496-1131)
	(Provides online service. Does not offer
	  CA.)
	A related effort is

**Merit Information Center, Chemistry
  Division**
Director, Department of Chemistry
The University of Michigan
Ann Arbor, MI 48109
Phone: (313-764-7362)
(Does not offer online services but will
  conduct searches for a fee.)
	**Toxicology Information Response Cen-
	ter**
	P.O. Box X, Building 2024
	Oak Ridge, TN 37831
	Phone: (615-576-1743)
	(Not in the CAS list. Does not offer
	  online services but will conduct
	  searches for a fee.)

New England Research Application Cen- ter (NERAC) (Previously affiliated with the Univer- sity of Connecticut, but now inde- pendent.)	

Information Centers Outside Continental United States	
Belgium	*Brazil*
Royal Library of Belgium Director, CNDST Bd. de l'Empereur 4 B-1000 Brussels, Belgium Phone: (2) 513.61.80 Telex: 21157 BIBREG B	**Empresa Brasileira de Pesquisa Ag- ropecuaria** Head, Departamento de Informacao e Documentacao Venancio 2000 2° Sub solo Caixa Postal 11-1316 70.333 Brasilia, D.F Brazil

Canada

Canada Institute for Scientific and Technical Information (CISTI)
National Research Council of Canada
Head, Tape Services
Ottawa, K1A 0S2, Ontario, Canada
Phone: (613-993-1210)
Telex: 053-3115
(Provides online service.)

Czechoslovakia

Vyzkumny Ustav Technicko Chemickeho Prumysl^u
Ustredni Informacni Sluzba Chemie
VUTECHP-UISCH
P. O. Box 816 Stepanska 15
Prague 2000, Czechoslovakia
Telex: Praha 012114

England

Pergamon InfoLine® Ltd.
12 Vandy Street
London EC2A 2DE
ENGLAND
Phone: 01 377 4650
Telex: 8814614
[Provides online service; *Chemical Abstracts* is not online in this system.]
U.S. office is 1340 Old Chain Bridge Road, McLean, VA 22101
(800-336-7575)

France

Telésystémés–QUESTEL
40, rue du Cherche-Midi
75006 Paris
FRANCE
Phone: (01) 544.38.13
[Provides online service.] U.S. office is Suite 719, 1625 Eye Street, Washington, DC 20006
(800-424-9600)

Germany

FIZ Chemie
Chefredakteur
Postfach 12 60 50
Steinplatz 2
1000 Berlin 12, Federal Republic of Germany
Phone: (030) 31 05 81
Telex: 181 255
[Fachinformationszentrum Chemie offers a complete range of information services including online databases (*INKA*).]

Hungary

Veszpremi Vegyipari Egyetem Kozponti Konyvtara
Central Library of the Veszprem University of Chemical Engineering
Director of the Library
Schonherz Z.u.10.
8201 Veszprem, Hungary
Phone: 12-550
Telex: 32-397/Hungary

Israel

National Center of Scientific and Technological Information
Head, Department of Users Liaison
Hachashmonaim Street 84
P. O. Box 20125
Tel Aviv 61 201, Israel
Phone: 03-297781
Telex: 03-2332
Cables: COSTI

Italy

European Space Agency
Information Retrieval Service
Via Galileo Galilei
00044 Frascati, Italy
Phone: Italy (06) 94011
Telex: 610637 ESRIN I
(Provides online service.)

TABLE 10.3. Continued

Japan

The Japan Information Center of Science and Technology
Head, Literature Search Section
52, Nagatacho, 2 Chome, Chiyoda-ku
Tokyo 100, Japan
Phone Tokyo 03-581-6411
(Provides online and offline services.)

Japan Association for International Chemical Information (JAICI)
Gakkai Center Building
2-4-16 Yayoi, Bunkyo-ku
Tokyo 113, Japan
Phone: 03 (816) 3462
Telex: 272-3805 (JAICI J)

Korea

Korea Institute for Industrial Economics and Technology (KIET)
Director
206-9, Cheongryangri-dong, Dong-daemun-ku
CPO Box 1229
Seoul
Federal Republic of Korea
Phone: 96-6501-6

South Africa

South African Council for Scientific and Industrial Research
Centre for Scientific and Technical Information

Head, SASDI
P. O. Box 395
Pretoria
0001 Republic of South Africa
Phone: (012) 86-9211
Telex: 3-630

Spain

Instituto de Informacion Y Documentacion en Ciencia y Tecnologia
Director
Joaquin Costa, 22
Madrid 6, Spain
Phone: 2614808
Telex: 22628 CIDMD E

Sweden

Royal Institute of Technology Library
Head, Information and Documentation Center, IDC
S-100 44 Stockholm 70, Sweden
Phone: 08 7878950
Telex: 103 89 KTHB

Switzerland

Data-Star
12 quai de la Poste
CH-1204 Geneva
Switzerland
Phone: (33) 1 766 0226
Telex: 290393
(Provides online service.)

---

[a] This list is believed to be up to date as of 1986. It was compiled by CAS and is reprinted here through the courtesy of CAS. Those U.S. centers that are primarily vendors of online services were previously listed in Section 10.4.

[b] CAS itself provides to the general public most of the services associated with an information center, including searching of the literature as based on its own files, both online and the printed *CA* back to 1907, and providing copies of many recent documents covered by CAS.

2. Assisting in developing search strategy that will be most likely to yield desired results.

3. Selecting those files or sources that are most likely to yield fruitful output.

4. Organizing, screening, and evaluating for pertinent output or results.

5. Supplementing results with critique and data, when the centers have on their staffs persons who possess the required expertise.

6. Outputting in a variety of forms, such as 8½ × 11 in. paper or index card form.

7. Eliminating of need for user to rent or buy special equipment (such as interactive terminals) to use this kind of service.

8. Eliminating of need for user to learn special search languages to communicate with the computer.

9. Assisting in obtaining copies of pertinent documents identified and translations, if needed.

10. Referrals to, and assisting with, making contact with experts in outside organizations who may have knowledge on the question at hand.

Some persons question whether batch or offline information centers will survive in the wake of the tidal wave of online services. But it is believed that there will probably be a long-term need for offline information centers, such as those just described, especially for smaller organizations and for independent chemists such as consultants. Such factors as economics, availability of user and computer time, and capability will determine relative popularity of off- and online services and information centers in the future.

As shown in Table 10.3, Chemical Abstracts Service (CAS) has licensed a number of organizations to use one or more CAS files from which these organizations derive marketable services for their respective customers. These information centers provide a variety of services, among which are current awareness services and retrospective searches. Centers often provide an economical means for individuals and organizations to obtain needed information immediately while minimizing cost since charges are generally based on the specific service rendered.

Individual centers operate differently, providing different services and using different CAS files. Many offer services based on files from other suppliers. Inquiries regarding operation of a specific center listed in Table 10.3 should be directed to the center.

10.10. INFORMATION "BROKERS" AND CHEMICAL INFORMATION CONSULTANTS

There are also a number of independent "brokers" who can conduct, or arrange to have conducted, both on- and offline computer-based searches of the literature. These brokers often offer a wide variety of information services (including manual searches) and can be of particular value to smaller organizations or to overloaded larger organizations.

One good source for obtaining the names and locations of many of the most capable of the brokers is the Information Industry Association.* This organization includes as its members most of the profit-making organizations that produce databases and systems. Two of the larger, better known information brokers are listed here by way of example and illustration:

Information on Demand, Inc.
P.O. Box 9550
Berkeley, CA 94709
(415-644-4500), (800-227-0750)

This organization is associated with the Pergamon group of companies. A complete range of information services is offered.

FIND/SVP
500 Fifth Avenue
New York, N.Y. 10110
(212-354-2424)

In addition to a broad range of information services, this firm publishes special market studies and reports.

There are also independent chemical information consultants, for example,

Technology Information Consultants
163 Hearn Lane   5 Science Park
Hamden, CT 06514   New Haven, CT 06511
(203-281-3693)   (203-786-5030)

This firm was founded by R. E. Maizell, previously manager of Business and Scientific Intelligence Centers and consulting scientist for New Technologies at Olin Corporation.

Info/Consult®
P.O. Box 204
Bal Cynwyd, PA 19004
(215-667-0266)

This firm was founded by Gabrielle S. Revesz, a former executive at the *Institute for Scientific Information.*®

In addition, the Gale Research Co., Detroit, Michigan has published a comprehensive directory of online database retrieval services offered by private information companies and by libraries of all types (3). This is expected to help chemists and other users select outside sources to do online searching for them, if this is decided on as the option of choice.

10.11. CHEMICAL STRUCTURE SEARCHING USING COMPUTERIZED SYSTEMS

In recent years, the capability to search for and locate chemical substances by specifying structural characteristics has been significantly improved and expanded especially in online systems. Chemical structure and substructure searching approaches have been categorized by Witiak (4), Heller and Milne (5), and others, into several principal approaches.

The approach exemplified in Table 10.4 employs listing of chemical substance names. In its most refined form, this approach is based on systematic nomenclature such as that employed by CAS.

Fragmentation codes (see Table 10.5) are alphanumeric systems that provide the indexing needed to assist in searching for the structural aspects that are especially crucial for patents. They are intended to help provide the most complete recall possible, a feature important to searching of chemical patents.

A connection table is compiled by humans or computers. It shows numerically, and usually unambiguously, the sequence and ways in which specific atoms are connected to one another, and their relative stereochemistry. (See Table 10.6 for examples of connection tables.) Input of connection tables into a computer facilitates location by specified combinations of atoms. This is known as topological searching. The best known and the biggest single group of connection tables is probably that of CAS. The *MACCS* connection table system is closely related to that of CAS in that it employs what is claimed to be an improved (added stereochemistry) version of the so-called Morgan algorithm employed at CAS (see Section 6.10). The CAS system is used for both public online access (CAS Registry), as well as for some private structure files of individual companies (6–8). The *MACCS* approach

TABLE 10.4. Examples of Nomenclature or Dictionary-Type Approaches

File	Vendors or Developers
Chemdex® (online)	SDC® Information Services
CAS ONLINE	Chemical Abstracts Service
CHEMLINE (online)	National Library of Medicine
Chemname® (online)	*DIALOG®* Information Services
Toxicology Data Bank (online)	National Library of Medicine
NIOSH *Registry of Toxic Effects of Chemical Substances* (online)	National Library of Medicine and Chemical Information System

TABLE 10.5. Examples of Fragmentation Coding

File	Vendors or Developers
Derwent *World Patent Index* (online)	Derwent Publications, Ltd.
IFI *Claims*™ (online)	IFI/Plenum

is widely used by private companies on their own computers for proprietary chemical structures.

Notation systems typically use combinations of letters, numbers, and punctuation marks to represent chemical formulas in simplified form (see Table 10.7).

Classification approaches are hierarchical in nature and are typically based on grouping together of related chemical structures (see Table 10.8).

As indicated previously, private files are those maintained by individual companies or other organizations on their own computers for their own proprietary (in-house) use. Some of these approaches may also be used for public or external files. In addition to those in Table 10.9, CAS provides *private registry* capability through its facilities in Columbus, OH.

Tables 10.10 and 10.11 summarize the features of representative computerized systems for searching chemical substances. The data in these tables are as supplied by personnel connected with each of the systems covered and are up to date as of the end of 1984 and, in some cases, up to early 1986. These are basically partial revisions of tables originally developed by Heller and Milne (5).

A.   Connection Tables or Topological Systems Using Graphics and Related Approaches

CAS ONLINE and *DARC* are by far the largest and most important online structure search systems. Both permit access to the contents of the CAS Registry System since its beginning in 1965 and now (1986) cover over 7.4 million compounds registered by CAS. In addition, *DARC* permits graphics

TABLE 10.6. Examples of Connection Tables or Topological Approaches

File	Vendors or Developers
CAS ONLINE (Registry file)	Chemical Abstracts Service
DARC (includes certain files of *Chemical Abstracts* and *Index Chemicus*® *Online*)	Questel, Chemical Abstracts Service, and Institute for Scientific Information®
MACCS/REACCS	Molecular Design, Limited
SANSS (online)	Chemical Information System

TABLE 10.7. Examples of Notation Systems

File	Vendors or Developers	Chapter References
Current Abstracts of Chemistry and Index Chemicus® (Wiswesser Line Notation) (online)	Institute for Scientific Information®	8,10
GREMAS	Internationale Dokumentationsgesellschaft für Chemie	10

access to other important files such as *Index Chemicus*® *Online* (*Institute for Scientific Information*®).

CAS ONLINE is offered directly by CAS through STN[SM] (Scientific and Technical Information Network) and employs computers located in that organization's Columbus, OH, headquarters. *DARC* is a system developed and supported by the French government and employs computers located in Valbonne in the south of France. These systems are readily available to users throughout North America, Western Europe, and Japan.

In both cases, the need to know *correct* chemical names or special chemical codes is minimized. Chemists need know only how to use the systems and how to *draw* structures or substructures of interest correctly. Ambiguity associated with some chemical names or coding systems is essentially eliminated for most structures. Specifications of structures of interest can be done by simply drawing on a graphics terminal or input through a text terminal. The total approach is "user friendly."

It is not, however, a completely simple matter to use these systems. At

TABLE 10.8. Examples of Classification Approaches

File	Vendors or Developers	Chapter References
Beilstein Handbook of Organic Chemistry	Beilstein Institute	12
Gmelin Handbook of Inorganic Chemistry	Gmelin Institute	12
International Patent Classification System (online)	Derwent, Pergamon, *Questel, INPADOC,* other	13
U.S. Patent Office Classification System (online)	IFI/Plenum *Claims*™, Pergamon *PATSEARCH*®, Derwent files, other	13

TABLE 10.9. Examples of Systems That Can Be Used for Small or Large Private Files

File	Vendor or Developer	Chapter References
DARC–SMS	*Questel–DARC*	10
MACCS/REACCS	Molecular Design, Limited	10
Molecular Formula Indexes	Various	6,7,8
Wiswesser Line Notation; *CROSSBOW*	Fraser Williams (Scientific Systems) Ltd.	8,10

least one full day of specialized training, and additional days (or preferably weeks) of practice, are needed to utilize these systems to best advantage. Special equipment, beyond normal online requirements, is needed for graphic input and output, although text input of structures is adequate.

Although structure searching is a powerful tool, it still leaves something to be desired. Kaback (9) has pointed out that online files such as *CAS ONLINE,* and *DARC–Questel* can cope best with fully defined structures and less well with generics. Another disadvantage is that searches for generic data in these files may produce lists of compounds that can be too lengthy to be easily used.

Similarly, Kaback believes that fragmentation coding systems (e.g., Derwent and IFI/Plenum) fall short. Kaback states that these systems cannot presently differentiate as to whether two structural groups are both present in a compound or are alternatives. A typical generic search requirement could be, for example, that both groups are present.

Efforts are underway to further strengthen and improve both graphics and other structure systems to make them more suited for handling of generic searches, and chemists can expect to see important changes.

On March 26–28, 1984 an important conference, Computer Handling of Generic Chemical Structures, took place at the University of Sheffield, England (10). Proceedings of this meeting are an excellent review of the state of the art at the time of the meeting, which was held under the auspices of the Chemical Structure Association. An older book by C. H. Davis and J. E. Rush (11) covers information retrieval in chemistry with special emphasis on structure searching; this volume is highly recommended.

The advantage of graphic (structure) searching of chemical substances is that it is independent of any vagaries or eccentricities of nomenclature, which can, on frequent occasion, be quite arbitrary. Graphic searching provides freedom from nomenclature substitutes, such as classification or coding systems. All of these (nomenclature, and classification and coding systems) have a certain amount of artificiality and can be difficult to learn. In addition, graphic searching provides the potential for more effective searching for less well defined structures, including true generic searching.

TABLE 10.10 Comparison of Representative Computerized Systems for Searching of Chemical Substances—Principal Features[a,b]

Feature/Item	CAS ONLINE	Questel/DARC	CIS (SANSS)	MACCS	CROSSBOW
First available or operational	11/80	1/80	5/77	1979	1969
Approximate number of user organizations	3000	450	650	100	30
				in house	in house
Cost					
Online use, annual subscription fee ($)	0 (varies)	0	300	NA	NA
Online use, approx. hourly fee ($)	50	50	55–95	NA	NA
Online use, hit fee ($)	0.06–1.16	0.25	0	NA	NA
Online use, substructure search fee ($)	86–94[c]	40–80[c]	0	NA	NA
System size					
Number of compounds searched	7.5 MM (12/85)	7.5 MM[d] (12/85)	350 K (11/84)	600 K[l] in house (3/86)	10 K–2 MM in house (12/84)
Number projected (estimate)	7.9 MM (12/86)	7.9 MM[d] (12/86)	up to 400 K (1988)	NA	NA
Software available for purchase	Y	Y	NA	Y	Y
Hardware	IBM 3090 PDP-11	IBM, VAX	PDP-10, PDP-20	PRIME, VAX, IBM, PDP-10, PDP-20, Honeywell (Multics)	IBM 370 PDP-10, VAX, PRIME-400, ICL-1900 Burroughs-6700

177

TABLE 10.10 Continued

Feature/Item	CAS ONLINE	Questel/DARC	CIS (SANSS)	MACCS	CROSSBOW
Software	Assembler, PL-1	FORTRAN	FORTRAN	FORTRAN	COBOL, Assembler, FORTRAN
Batch (B) or interactive (I)	I	I	I	I	B and I
Structure Data Sources					
CAS connection table	Y	Y^e	Y^e	Y^e	$N^{e,f}$
Other connection table	N	Y	Y	Y	Y
WLN	N	Y^e	N^e	Y^e	Y
Fragment codes	Modified Swiss/CAS screens	DARC screens (FREL)	CIDS screens and CIS ring and fragment screens	MACCS and user defined screens	CROSSBOW fragments; user defined screens; WLN
Structure Input/Output					
Output display					
Teletype® (ASCII)	Y	N	Y	N	Y
Raster or vector graphics	Y	Y	Y	Y^i	Y

178

Structure input				
Teletype	Y	Y	Y	N
Vector graphics	Y	Y	Y	N
WLN	N	N	N	Y
Administrative				
Manuals English	Y	Y	Y	Y
Other	French	Japanese		Japanese
Training	Y	Y	Y	Y
Training cost ($) per person	40/220	50/125	100	Included[h]
Training required by supplier	N	N	N	N
U.S. Hotline/Assistance (business hours)	Y[g]	Y	Y	Y
Toll free phone	Y	Y	Y	N
Linking				
Bibliographic information	Y	N[j]	Y[k]	Y
Nonbibliographic information	Y	Y	Y	Y
Private files	Y	Y	Y	Y

[a] Sources of data in this table are usually the systems listed. Up to date as of 1984–1986. Source of data for *CIS* is Fein–Marquart Associates, Inc.

[b] NA = Not applicable or not provided; MM = Million; K = Thousand; N = No; Y = Yes.

[c] For *CAS ONLINE* and *DARC*, this is a fixed fee/search. Discounts apply in some cases.

[d] *DARC* says that it lags *CAS ONLINE* by about two weeks in entry of compounds; it says that its online processing of search queries is very rapid, however.

[e] Can be converted for use in the system.

[f] Connection table conversion for links to *CAS*, *DARC*, *MACCS*, and *CIS*.

[g] Plus Hotlines in West Germany, France, United Kingdom, and Japan.

[h] Included in yearly maintenance $2000.

[i] *MACCS* says it is the only system of those listed that has vector graphics.

[j] Indirect links provided to *DIALOG®*, *SDC®*, and NLM.

[k] Only if included in database.

[l] This is in just one of many systems. In addition, all or any subset of the structures and references in *Index Chemicus®* are available on *MACCS* via *ISI®*. Further, *ISI®* will have *Current Chemical Reactions®* available on *MACCS*.

TABLE 10.11 Comparison of Representative Computerized Systems for Searching of Chemical Substances—Searching Capabilities

Feature/Item	CAS ONLINE	Questel/ DARC	CIS (SANSS)	MACCS	CROSSBOW
CAS Registry Number	Y	Y	Y	Y[a]	N
Name	Y	N	Y	Y	Y
Synonyms	Y	N	Y	Y	Y
Name fragment	Y	N	Y	Y	N
Molecular formula	Y	Y[b]	Y	Y	Y
Molecular weight	Y	Y[c]	Y	Y	Y
Number of rings (d)	Y	Y	Y	Y[f]	Y
Ring structure	Y	Y[g]	Y[h]	Y	Y[d]
Augmented atoms	Y	Y[i]	Y	Y	N
Screens	Y	Y	Y	Y	Y[j]
Substructure	Y	Y	Y	Y	Y
WLN	N	N[k]	Y[l]	Y[a]	Y[e]
Fragment code	Y	Y	Y	Y	Y
Tautomers	Y	Y	N	Y	N
Identity search	Y	Y	Y	Y	Y
Stereochemistry (explicit)	N	Y[m]	N	Y	N
Atom by atom	Y	Y	Y	Y	Y

[a] If included in database.

[b] Molecular formula searching may be done in the *CAS* and *Merck* bibliographic files, for example.

[c] Some files (*Spectra, Merck*).

[d] Number of rings, size, heterocontent.

[e] Some question whether computer-based searching of WLN can be done effectively in any of these systems, with the exception of *CROSSBOW*.

[f] Available, but not automatic unless database set up with this feature.

[g] Ring skeleton or specific rings.

[h] Heteroatom and substituent pattern.

[i] Incorporated in FREL screen.

[j] Number of rings, size, heterocontent, WLN.

[k] Although some have said that *Questel/DARC* has had some capability in this regard, the operators of the system prefer the more conservative position.

[l] Only where WLN is present as synonym.

[m] If in database (not presently in *CA* or *ISI*®).

Whether this potential is fully realized in practice is a matter of debate, but if it is not, it is surely on the near horizon.

Thus, graphic searching is consistent with the work habits of most chemists. The graphics systems available permit input by drawing the structure much the same as it would be drawn on paper, or alternatively (or in addition), by means of text input. Either approach offers the potential of a high degree of user friendliness and of more power than would otherwise be available with a conventional nomenclature approach, such as search for a wide variety of substructures, and specification of substituents.

Despite the advances in graphic searching, many single-structure nomenclature queries can sometimes be most conveniently handled by nomenclature search techniques such as full or partial name match, providing the *correct* name is readily known. However, according to Molecular Design, Limited (see Section 10.11.4), users of their systems report that if chemists are also given the option of structure input, they usually prefer this over the nomenclature option.

1. **CAS ONLINE.** One leading example of structure searching of chemical substances is *CAS ONLINE*. Offered directly to customers by CAS, this offers the ability to search a file of over 7 million chemical substances cited in *CA* since 1965 on the basis of structural characteristics—to identify substances within the file that share particular arrangements of atoms and bonds and thus share properties of interest as well.

Questions can be phrased by constructing the structure diagram of the substructure to be searched. Two structure input methods (see below) are available, ensuring that the *Registry File* can be searched at virtually any type of terminal. Each answer retrieved includes CAS Registry Number, *CA* index name, molecular formula, up to 50 synonyms, structure diagram, and full bibliographic information for the 10 most recent references about the substance. The total number of references for the substance in the *CA File* and an indication of further references in the *CAOLD File* are also given. Synonyms, trade or common names, Registry Numbers, molecular formulas, formula weights, or element counts may also be used as search terms.

Full subject (controlled vocabulary, keywords, natural language, and Registry Number) and author and other bibliographic searching is readily available in the companion *CA File* that contains bibliographic data and *CA* index entries in a single file from 1967 to date. The *CA File* is the only online system on which it is possible to retrieve complete *CA* abstract texts for documents abstracted since mid-1975 (and for some earlier documents). At first, abstract texts could not be searched online, just displayed and printed, although *CAS ONLINE* has now begun to make all abstracts in the system searchable. In addition, all abstracts back to 1967 are in process of being made available. A third file, *CAOLD,* contains references to substances cited by *CA* prior to 1967. This file is being expanded incrementally as these "earlier" substances (many of which also have references after 1967) are added to the CAS Chemical Registry System. Registration by CAS of pre-1965 substances began in February 1984, with online access via *CAOLD* available soon thereafter. For pre-1967 references, information available is limited to *CA* Reference Number, Document Type (for patents), and Registry Numbers as index terms. The *CA* Reference Number is the reference to the *CA* issue indicating the volume number, column number, and column fraction where the abstract was printed; there were no abstract numbers prior to the *Eighth Collective Index. CAOLD* is available only on STN^{SM} (see Section 10.12).

In *CAS ONLINE,* the chemist can input a structure in various ways, depending on the type of terminal. If an intelligent graphics terminal is available, (specifically a Hewlett–Packard 2647), the structure may be input by using a menu. The menu is a list of basic rings, elements, bonds, and commands.

Structures may also be input on standard (nonintelligent) graphics terminals or on text terminals. With these types of terminals, chemists can build structures by typing keyboard commands that call up or modify structural components. A structure display on a standard graphics terminal is the same as on the intelligent graphics terminal.

In a text terminal that can be equipped with a graphics board, the chemist can also draw the structural skeleton using the terminal cursor buttons.

In 1985, *CAS ONLINE* began to offer the capability to search for chemical substances using complete and partial names and molecular formulas.

In the same year, CAS announced plans to introduce full Markush structure capability. When fully implemented, this should permit chemists to use shorthand notations to ask complex structural questions of a highly generic nature in the *Registry File* and get Markush-type structures as answers. At present, only specific compounds can be retrieved. Plans call for eventual introduction of a Markush structure file database, which is to be based on indexing and input of actual Markush-type structures that appear frequently in chemical patents. Markush language designates artificial groupings by wordings such as "material selected from a group consisting of A, B, C, or D" where A, B, C, and D are different, but can serve the same function. Possible variations in claims are almost infinite, and the examples just given are merely illustrative. The designation Markush comes from the name of the individual whose patent introduced this concept in 1925. Some use the term loosely to designate any type of generic claim or searching.

A further announced enhancement is a structure-based reaction search service for which database building has already begun. It is expected that about 300,000 reactions per year will be encoded, and that the service will be offered through STN^{SM} International. This will be the first entirely new database that CAS has offered in many years.

2.   QUESTEL–DARC.   This system was first conceived of by Jacques-Emile Dubois (University of Paris) in 1963 and was fully commercialized in 1980 with the "CBAC" portion of the CAS files. *DARC* thus claims to be the first *commercially* available online structure search system, if *CIS* (see Section 10.11.A.3) is regarded as originally a *public sector* operation. (For a brief description of the development of *DARC,* see reference 12.) *DARC* services are available in the United States through Questel, Inc., located in

Washington, D.C. The parent company is Telésystémés–Questel, which was founded in 1969 and has world headquarters in Paris, France.* Worldwide access is available through public data communications networks such as *Tymnet®, Telenet®, Transpac®,* and *Euronet. DARC* is the part of the system that permits structure searching. *Questel* is the system component that provides access to bibliographic databases.

The *DARC* System provides online access to *EURECAS,* a chemical structures database based on the CAS Registry File. *DARC* allows chemists to perform structure and substructure searches with straightforward alphanumeric (text) or graphic input of structures. With *DARC, screens* are automatically generated that will retrieve all structures sharing the same substructure. Atom-by-atom, bond-by-bond searching may then be used to complete the search. The result of a search is a list of CAS Registry Numbers which, when used with a simple command, leads to the closely related *Questel* System that contains bibliographic information from *CA* and other databases. This system competes directly with *CAS ONLINE.*

In addition to *CA,* approximately 35 other databases are available through *DARC* and/or *Questel.* (Language and orientation of many of these is French.) For example, *DARC–Questel* provides searching of the Mass Spectral Data Files of the U.S. National Bureau of Standards (see also Chapters 15 and 18). The latter file, consisting of 38,000 compounds and their mass spectra, can be searched by structure, substructure, molecular formula, mass spectral peaks and intensities, or by combinations of spectral and structural data. The very important Derwent *World Patent Index* (see Section 13.11) is also available on the *Questel* component of the system; *DARC* capability (graphic structure searching) is not available for Derwent as yet in 1986.

Questel–DARC has introduced initial capability to define a Markush (see p. 182) or generic query for the structure searching of specific compound files. However, they regard present capability as only a first step to truly generic searching. *Questel–DARC* is said to be actively working on solutions to the problem as defined by Kaback (see Section 10.11A). Cooperative development efforts are underway between Telésystémés, INPI (the French Patent Office) and Derwent on this point. Stated plans are to achieve fully generic searching (Markush searching of a Markush database) in stages.

DARC software is available for in-house use by private organizations. It has just begun to be marketed in the United States, although marketing in Europe has been in progress for several years.

DARC–SMS (Structure Management System) provides the same query input and structure capability as the online version of *DARC*. It also contains special features, such as report generation and structure registration, which are applicable to in-house systems. For input, users may employ a light pen or a graphics tablet to draw molecules onto the screen.

DARC–RMS (Reaction Management System) provides in-house capability to store and search reactions. Search can be done by reactant, by product, by combinations of reactants and products, by reaction conditions, by catalysts, or by related data that have been stored by the user in the system.

DARC–DMS (Data Management System) permits storage and searching of data related to chemical substances, such as, for example, biological activity.

3. **Chemical Information System (CIS).** The *Chemical Information System (CIS)* concept was originally developed in 1973 to aid federal agencies concerned with chemicals having an environmental or other regulatory impact. That has been a major thrust of its development, although NBS (National Bureau of Standards) has also played a key role. However, since 1979, *CIS* has been available to, and used by, the general public.

CIS provides single-point access to a number of largely independently generated files obtained from several different sources. Examples of system components are shown later in this Section. Physical property, safety, analytical, and regulatory data for some 350,000 compounds are included, as is strong capability for searching by structure for these compounds.

Much of the credit for original development of *CIS* should go to Stephen R. Heller (then with U.S. Environmental Protection Agency) and G. William Milne (U.S. National Institutes of Health).

In 1983 and 1984, the future of *CIS* became uncertain and government funding was reduced although there were about 600 subscribers, 40% of whom were federal agencies. On November 1, 1984, government involvement ended, and operation of *The CIS* was taken over by Chemical Information Systems, Inc., a wholly owned subsidiary of Fein–Marquart Associates,* and separately as *Integrated Chemical Information System (ICIS)* by Information Consultants Incorporated (ICI), jointly with Questel. ICI* is now owned by the Bureau of National Affairs (BNA) located in the same city. The two companies offer two competitive systems, each of which is expected to take on unique characteristics. For example, Fein–Marquart has added the online version of the *Merck Index* to its *CIS* system.

Fein–Marquart is the organization that originally developed and operated *The CIS* from 1975 through 1984, and ICI has also been closely involved with *CIS*. The original government-operated system software and the database it generates can now be licensed by any organization through the National Technical Information Service. *ICI–Questel* purchased such a license and began operating *CIS* on October 1, 1984. Fein–Marquart began operation of *The CIS* on a commercial basis on November 1, 1984.

Components of Fein–Marquart's *CIS* (as of 1986) were as follows:

1. Aquatic Information Retrieval (*AQUIRE*)—contains information on toxicity effects for freshwater and saltwater organisms. Over 4100 chemicals are included. Part of the *SPHERE* umbrella described in item (28).

2. Pattern Recognition (*ARTHUR*)—pattern recognition and data analysis program.

3. Chemical Activity Status Report (*CASR*)—lists over 8000 chemicals that the Environmental Protection Agency has studied.

4. Chemical Carcinogenesis Research Information System (*CCRIS*)—assay results of test conditions for about 900 substances.

5. Chemical Evaluation Search and Retrieval System (*CESARS*)—toxicity and properties for 190 chemicals of special interest to the Great Lakes Basin.

6. The Chemical Modeling Laboratory (*CHEMLAB*)—molecular structure calculation routines. A significantly improved version of this is claimed to be offered by Molecular Design, Limited (see Section 10.11.A.4).

7. Clearinghouse Technical Reports Database (*CLEAR*)—an index to research materials developed by or for the Environmental Protection Agency.

8. Carbon-13 Nuclear Magnetic Resonance Search System (*CNMR*)—over 11,600 compounds. The user may search this database for compounds having specific spectral characteristics, that is, values of shift, multiplicity, or intensity.

9. Cambridge X-Ray Crystallography Database (*CRYST*)—about 49,000 bibliographic citations.

10. Clinical Toxicology of Commercial Products (*CTCP*)—includes about 16,000 products (see also Section 14.3).

11. Dermal Absorption (*DERMAL*)—contains data related to dermal absorption of chemicals. Part of the *SPHERE* umbrella described in item (28).

12. Environmental Fate (*ENVIROFATE*)—has records on environmental fate or behavior for about 450 chemicals. Part of the *SPHERE* umbrella described in item (28).

13. Federal Register Search System (*FRSS*)—from January 1978 to November 29, 1983. Covers regulations, rules, standards, and guidelines involving chemicals.

14. Genetic Toxicity (*GENETOX*)—mutagenicity data on more than 2600 chemicals. Part of the *SPHERE* umbrella described in item (28).

15. Industry File Indexing System (*IFIS*)—information relating to regulation of chemicals under EPA's industry-specific legislation.

16. Infrared Search System (*IRSS*)—over 4500 spectra from two primary sources, the Boris Kidric Institute in Yugoslavia and the EPA.

17. Information System for Hazardous Organics in Water (*ISHOW*)—six types of basic physical properties for more than 5000 chemicals. Part of the *SPHERE* umbrella described in item (28).

18. Electronic Mail System (*MAIL*)—allows communication among *CIS* users.

19. Merck Index Online (*Merck*)—has about 12,000 compounds. (See also Section 15.10.)

20. Modeling Laboratory (*MLAB*)—(tool for experimenting with and evaluating mathematical models and functions).

21. Mass Spectral Search System (*MSSS*)—over 42,000 compounds. In addition to searches based on weight, molecular formula, and partial formula, the database may be searched on the basis of individual peak and intensity values and on the basis of the complete spectrum. Based on file from NIH/EPA/MSDC.

22. Nuclear Magnetic Resonance Literature Search System (*NMRLIT*)— about 43,000 references from 1964.

23. Nucleotide Sequence Search System (*NUCSEQ*)—over 200 major DNA sequences.

24. Oil and Hazardous Materials/Technical Assistance Data System (*OHM/TADS*)—over 1400 substances. Provides data to assist in case of discharge of materials listed.

25. Plant Toxicity Data (*PHYTOTOX*)—data on the biological effects of organic chemicals on terrestrial plants.

26. Acute Toxicity Data from the NIOSH *Registry of Toxic Effects of Chemical Substances* (*RTECS*)—over 77,000 compounds. See Section 14.3.

27. Structure and Nomenclature Search System (*SANSS*)— approximately 350,000 substances. See description at the end of this list.

28. Scientific Parameters for Health and the Environment, Retrieval, and Estimation (*SPHERE*)—detailed toxicity and related data for literature since 1970 plus other sources. This is the umbrella for several other files in the system.

29. Thermodynamic Property Values Database (*THERMO*)—values for more than 14,000 substances from the U.S. National Bureau of Standards *Tables of Chemical Thermodynamic Properties* and for several hundred compounds from the files of the Texas A & M University Thermodynamics Research Center.

30. Toxic Substances Control Act Plant and Production (*TSCAPP*)— production and plant site data (1975–1977) for some 53,000 commercial products.

31. Toxic Substance Control Act Test Submissions (*TSCATS*)— references unpublished health and safety studies submitted to Environmental Protection Agency.

32. Single Crystal Reduction and Search System (*XTAL*)—database consists of over 57,000 cells from the Determinative Tables (Vol. 1–4); new data collected since publication of the Determinative Tables will replace the current database.

33. Wiley Mass Spectral Search System (WMSS)—includes about 72,000 compounds (see also Chapter 18).

The hub of *CIS* is the *Structure and Nomenclature Search System* (*SANSS*), which is a file of some 350,000 chemicals containing formula, name, synonyms, and complete structural record for each chemical listed. CAS Registry Numbers are also available. The hub file can be searched using any of these elements. When the desired compound is located, the user is told which *CIS* files contain the compound, and transfer to these files can be accomplished to obtain the data such as a bar graph display of a mass spectrum. CAS connection tables are the source of substructure data.

SANSS is considerably smaller than other major online substructure search systems, although Fein–Marquart officials believe that the 350,000 compounds covered is a reasonable number and that the file includes most chemicals used in commerce. It concentrates on a relatively small number of chemicals that are of high interest to the U.S. Government, especially the Environmental Protection Agency (EPA), because of toxicological, environmental, and other properties of potential regulatory interest. One major feature is the capability to search not only by numbers and types of atoms and their connections, but also by complete or partial name or by chemical or trade name.

In addition to environmental and toxicity data, the *CIS* system contains a significant amount of valuable analytical and physical property data. A graphics terminal is not required for input of queries to *SANSS,* but specialized training and practice is essential to use the sometimes complex commands of the system. Also available is a stand-alone software package for personal computers that permits drawing of chemical structures for queries to *SANSS.*

4. *Molecular Design, Limited.* Molecular Design, Limited,* designs and offers a license of computer software for use by the chemical and pharmaceutical industry. These systems are intended for use on an organization's in-house computers so as to facilitate handling of data about proprietary compounds. Orientation of Molecular Design, Limited, is to computers made by Prime, Inc., Digital Equipment Corporation, and International Business Machines Corporation. The company was founded in 1977 by Stuart Marson and W. Todd Wipke. Wipke is a Professor of Chemistry at the University of California at Santa Cruz.

One of their software products is *MACCS* (*Molecular Access System*). *MACCS* is a database management system, first developed in 1979, for storage and retrieval of chemical compounds and related data. It is interactive, menu driven and user friendly. It provides graphic representation of chemical compounds, for both input and output, utilizing the natural language of the chemist—a structural diagram. There are no codes for the user to translate nor systems of nomenclature to learn. Any number of databases can be created, ranging from a few compounds of interest on a single project to hundreds of thousands of compounds for an entire organization.

To initiate a graphic search, a molecular structure or substructure is drawn on the screen with a light pen, tablet, joystick, or "mouse," just as it would appear on paper. Charges, isotopes, and asymmetric centers can be specified. *MACCS* can search for duplicates (exact match), stereoisomers, tautomers, or compounds that contain the graphically specified substructures. Search results are displayed immediately on the screen. These graphical search capabilities for compound substructures allow the chemist to hypothesize modes of biological activity based upon substructural fragments. A new capability reportedly provides improved generic searching.

Attached to this structural system is capability for data storage and retrieval. Related data such as physical properties, biological test results, compound history, literature references, company registry numbers, or other data from any number of user-defined categories may be searched and displayed. This allows the researcher to pose questions interrelating structural and biological properties.

MACCS "understands" the periodic table so that valences, atomic charges, and isotopic labels are handled correctly for all elements. It can "clean" drawings to produce textbook-quality structures. It also "understands" aromaticity, so that molecules may be located regardless of which Kekulé form is drawn.

Commonly used functional groups, ring systems, or substructural components may be stored as templates and recalled to be attached to a framework. This speeds construction of derivatives. Users can add or alter groups, enlarge, reduce or rotate the structure, or delete bonds and atoms. *MACCS* lets users store and search structures containing as many as 255 nonhydrogen atoms.

New compounds can be added by drawing the structure on the screen and transferring it into the database. *MACCS* updates the file instantly, making new information immediately available to all authorized users. These performance characteristics have made *MACCS* probably the most widely used system of its kind (over 90 separate installations worldwide).

The *Fine Chemicals Directory* (from Fraser Williams) is available for use with *MACCS*. This is a buyer's guide to approximately 70,000 research and other specialty chemicals typically used by laboratory investigators and made by over 30 manufacturers. Although this directory is searchable online elsewhere (e.g., Pergamon *InfoLine*®), the Molecular Design system permits graphical search and retrieval of structures.

Another Molecular Design Limited, system, *REACCS,* provides for graphical storage and retrieval of chemical reaction details. *REACCS* (Reaction Access System) is a database management system for both chemical compounds and reactions. *REACCS* allows a researcher to query synthetic methodology surrounding a compound or class of compounds. The system is used to catalog chemical reactions that have been utilized within a research organization, thereby providing the historical chemical experience of that organization to every present member. A graphic query is built by drawing reactants and products on the computer terminal screen with the same methods used in *MACCS*. They can be drawn as substructures or complete molecules.

REACCS searches and displays the reactions containing the product and/ or reactant represented by the query. Reactions appear on the screen in traditional form including the arrow. If desired, *REACCS* will highlight reacting centers automatically, making a distinction between groups that react and those that are incidental.

REACCS can also provide reaction conditions, physical properties, yield, name of synthesizing chemist, or any other data from any number of user defined data fields. Reactions can be searched through any of these data categories. Several search categories may be included in one query for more specific results. *REACCS* has the same search features as *MACCS,* including storage and retrieval of templates for attachment to a framework and the "clean" function to ensure uniform angles and bonds. It also lets users alter, add, or delete structures and data. New molecule and reaction information can be added. The user draws the structures or reactions on the screen, and it is transferred immediately into the database.

Furthermore, the *REACCS* system has available all reactions, and much related data, contained in *Organic Syntheses* from 1921 (about 5300 molecular structures contained in about 4500 unique reactions). Powerful reaction searching capability, including use of chemical fragments, is provided. In addition to reactions, the file includes: molecular structures; quantities of reactants, solvents, and catalysts; yields; reaction conditions; hazards; workup; and references to original literature and *Organic Syntheses*.

This system also includes a computer readable version of Volumes 1–35 of *Theilheimer's Synthetic Methods of Organic Chemistry* (originally published by S. Karger, A.G. Switzerland) with the capability of graphic searching. The database has approximately 70,000 molecules representing about 35,000 reactions as reported in the literature. For each reaction there is a page and volume reference to the original publication and to the Theilheimer volume. Other data available include reaction yields, catalysts, solvents, and reacting centers (bonds that change during course of a reaction). Annual updates of these databases, as well as introduction of additional databases, are also planned by Molecular Design, Limited.

The actual computer storage method for chemical structural information—the Stereochemically Extended Morgan Algorithm (SEMA)—is an algorithm that has been mathematically proven to uniquely name chemical structures according to Molecular Design, Limited.

The Data Access System (*DATACCS*) of Molecular Design, Limited, offers interactive design of report forms and tables and rapid collation of data from a variety of sources including other, non-Molecular Design, database management systems. In addition to display capabilities, this program can be used to move data between databases.

Also offered by Molecular Design, Limited are

GENOA—an artificial intelligence computer program to help determine structures of unknowns by drawing fragments believed to be present.

COMPAR—a display program for simultaneous study of up to six three-dimensional models graphically on the terminal screen.

ADAPT—a package of software tools for structure-activity and pattern recognition studies.

CHEMLAB-II—a molecular modeling software package for characterization of molecules and estimation of their physicochemical properties.

DRAWMOL—a drawing tablet for use with molecular modeling tools.

PRXBLD—for rapid molecular modeling.

MMZ/MMPZ—for more detailed molecular mechanics modeling of three-dimensional structures.

Various three-dimensional display packages to depict ball-and-stick and space-filling models, including some animation techniques.

MACCSLIB—library of FORTRAN-callable subroutines designed to help users develop their own specialized software.

Molecular Design, Limited refers to the preceding software offering as its *Laboratory On-Line*. These packages read and write the same common interface file, thus allowing information to be easily passed between them.

B.   Notation Systems

The best known chemical notation system is the Wiswesser Line Notation (WLN), which is an alphanumeric linear encoded representation of structures. This system is probably the most fully used by the *Institute for Scientific Information*® (*ISI*®), and has also been used by many other organizations and in numerous publications. It gave rise to a closely knit group of highly enthusiastic advocates of top caliber.

William J. Wiswesser developed this system in the 1950s; improvements have been introduced on an ongoing basis ever since. Some advantages claimed for WLN include: independence from the vagaries of some nomenclature systems; relatively easy search for structures and substructures; and use either manually or via computer. Decoding (reading) of WLN is said to be learned relatively easily. Encoding of structures is claimed to be possible after about a month of study and practice.

WLN is not always unambiguous. An excellent manual (13) is available for those interested in further details, and *ISI*® offers other useful assistance.

With the increasing use of graphic systems for searching of chemical structures and substructures, popularity and use of WLN is expected to decrease very significantly over the next three to five years, although WLN in the hands of those completely familiar with it can reportedly provide a more rapid *input* than other approaches. *ISI*® staff find that WLN is still a cost effective form of compound registry and storage. They report that the conversion rate of WLN's to connectivity tables (employing their special conversion program) is between 97–98% on current *ISI*® data (rate depends on file content and can be as high as 99%). Their "checker" program (which compares the molecular formula generated by WLN with that calculated by the encoder) is believed by *ISI*® to be one of the best in the world, if not the best.

Another system that employs WLN (as input) is *CROSSBOW,* originally developed by Ernest Hyde and colleagues at Imperial Chemical Industries. *CROSSBOW* is an acronym for Computer Retrieval of Organic Sub-Structures Based On Wiswesser. It is reportedly less cumbersome and more flexible than other approaches using WLN. *CROSSBOW* provides a three tier search system: fragmentation, WLN string, and atom-by-atom search. It uses connection tables based on WLN or atom connectivity tables produced by the DARING program. Line printer structure output with reporting facilities is available using *CROSSBOW.* Graphic structure output is available using the REWARD program. *CROSSBOW,* DARING, and REWARD are marketed by Fraser Williams (Scientific Systems) Ltd.*

1.   Internationale Dokumentationsgesellschaft für Chemie. Another example of use of notation systems is found in the work of IDC (Internationale Dokumentationsgesellschaft für Chemie). This is a consortium of large chemical companies organized to facilitate use of the chemical literature. Most of the member companies are West German firms, although there are also members from Austria and Japan. Headquarters of IDC are in Sulzbath, West Germany, near Frankfurt. Services are normally available only to members, but others may also participate under some circumstances on payment of an appropriate fee.

Functions of IDC include: (1) computer program development for data storage and retrieval; (2) procurement of foreign databases, including abstracts from sources such as CAS and Derwent; (3) preparation of IDC files and merging these with external input; (4) providing databases to members for online searching; (5) the conduct of searching by IDC staff; and (6) selective dissemination of information and published bulletins.

The IDC system comprises several approaches toward chemical structure searching. One is a method for topological representation of structures. There are also nontopological approaches, the principal of which is designated as Genealogisches Recherchieren durch Magnetbandspeicherung (*GREMAS*) or Generic Retrieval by Magnetic Tape Search. The system was originally described in 1961 by Fugmann and co-workers at Farbwerke Hoechst AG (14). It is a classification or notation system that can be used for describing both chemical structures and organic chemical reactions. IDC employs *GREMAS* in their indexing of patent and journal literature. This is believed to be one of the most powerful and sophisticated approaches to structure and substructure searching.

10.12.   DESCRIPTIONS OF SOME OTHER MAJOR ONLINE SYSTEMS

A.   *DIALOG®* Information Services, Inc.

One of the most popular online services of its type in the United States, and probably the world, is *DIALOG®* Information Services, Inc., also referred to as Lockheed *DIALOG®*.* The *DIALOG®* system became commercial in 1972 and now includes over 200 databases of all types, considerably more than SDC®, *BRS®*, Pergamon *InfoLine®* or *DARC–Questel,* who are the principal competitors.

The reasons why *DIALOG®* is so popular are the large number of databases offered, relative simplicity of system use, high-quality documentation, excellent user support, and competitive prices. *DIALOG®* includes a number of databases of chemical interest, most notably *CA* from 1967 to date.

Examples of other files of chemical interest that are on *DIALOG®* include: IFI/Claims™ files of U.S. patents; Derwent worldwide patent files; *Chemical Industry Notes,* which covers chemical business; *COMPENDEX®* engineering files; *PAPERCHEM* files of the Institute of Paper Chemistry; several files relating to chemical substance nomenclature; and a number of others.

DIALOG® staff point to a number of features of interest in searching of their online files of *CA:*

1. Online cross-referencing of all *CA* Index terms using the *CA Index Guide*.
2. Comprehensive chemical nomenclature searching, including especially designed segmentation of chemical names.
3. Capability for automatic transfer of CAS Registry Numbers in crossfile searches.
4. Full ring information to help distinguish substances or ring systems having identical molecular formulas, analysis of rings, and formulas of rings.
5. Searchable *CA* section codes, including automatic cross-referencing.

DIALOG® claims to facilitate substance identification in its files of *CA* by making it possible to employ combinations of nomenclature, ring data, molecular formulas, periodic group numbers, and other access points. *DIALOG®* lacks the graphics search capabilities of *DARC–Questel* and of *CAS ONLINE,* and *CA* abstracts are not displayable or searchable. Further, the *BRS®* and Mead *NEXIS/LEXIS®* systems surpass *DIALOG®* in number of major full-text files, and SDC® *ORBIT®* permits customer tailoring of print format. All of the online systems, notably Pergamon *InfoLine®*, have exclusive files, which the others lack, although there is also extensive overlap.

B.   Pergamon *InfoLine®*

InfoLine® was originally started in 1979 by Derwent and several other British organizations. In March 1980, it was purchased by Pergamon, a major scientific and technical publisher. In March, 1982 it began offering services from its main computers in London to the United States and elsewhere via satellite.

Pergamon is a British-managed organization, and this is reflected in *InfoLine*® databases, many of which are produced in the United Kingdom. Information covered, however, is worldwide and many of the databases are exclusive to *Infoline*®.

Some of the files of chemical interest believed to be unique to *InfoLine*® at this time include: the international patents equivalents file *INPADOC* (see Section 13.17), which is available elsewhere only directly from *INPADOC; RAPRA Abstracts* (indexing and abstracting service of the *Rubber and Plastics Research Association* of Great Britain, see Section 8.5); *Fine Chemicals Directory* [product of Fraser Williams (Scientific Systems) Ltd. of England]; *Chemical Engineering Abstracts* and *Mass Spectrometry Bulletin* (both products of The Royal Society of Chemistry); *World Textile Abstracts* (of the Shirley Institute of England); and *Zinc, Lead and Cadmium Abstracts* (product of the Zinc Development Association of London).

One of the most significant features of *InfoLine*® is the ability to rank and sort search results by frequency of occurrence. This is a powerful and useful tool. Another important feature of this system is the emphasis on safety-related databases.

C. STN^SM International

STN^SM (Scientific and Technical Information Network) was first announced in 1983, and it became operational in 1984. It is a cooperative venture of CAS with West Germany's FIZ Karlsruhe (Fachinformationszentrum Energie, Physik, Mathematik GmbH.)

The cooperative agreement links two major online systems. Through this agreement, users in North America have access not only to *CAS ONLINE*, but also to the databases of *INKA*, the online system of FIZ Karlsruhe. Users in Europe gain access to *CAS ONLINE*, offered exclusively from the Columbus, OH, headquarters of CAS, through the Karlsruhe node of the new network. The first *INKA* database to be made available to North American users through the arrangement with CAS is *Physics Briefs* (*Physikalische Berichte*). By mid-1986, *STN*^SM was scheduled to have about 10 databases available.

INKA has more than 40 files (see Table 10.12) available covering such topics as chemical engineering, nuclear magnetic resonance spectra, nuclear research and engineering, electronics and electrical engineering, and mathematics. All of these may now be readily accessed by users in the United States and other countries by direct telephone linkage with *INKA* computers. *INKA* is perhaps now better known as *FIZ*, but the former is used in this book to provide continuity.

Fachinformationzentrum Energie, Physik, Mathematik (FIZ) was originally part of the Gmelin Institute (see Section 12.5.A) in Frankfurt, West

TABLE 10.12. Selected List of Online Files of *INKA*

File Name	Description in *INKA* Catalog
C-13 NMR	C-13 NMR spectra: Numeric data and chemical formulas
COAL	Coal research and technology
COMPENDEX®	Engineering (general and special)
DECHEMA	Abstracts of publications on chemical technology and biotechnology (chemical engineering)
DETHERM	Data on substances (thermodynamic data)
DKI	Plastics, rubber, fibers (abstracts on properties, technology, and applications)
DOMA	Mechanical engineering
ENEC	Energy and economy: Numeric data on world energy balances, nuclear power plants, and world energy resources
ENERGY	Energy and all related fields (energy research and technology, energy economics and politics)
ENERGYLINE	Energy with the focal points energy, economics, and U.S. energy politics.
ENSDF	Nuclear structure and decay: Numeric data critically evaluated
ENSDF–MEDLIST	Nuclear structure and decay: Numeric data for medical and radiation biological applications
ENSDF–NSR	Nuclear structure and decay: Bibiographic part
GEOLINE	Geosciences, natural resources, and water supply
ICSD	Crystal structures of inorganic compounds: Numeric data
INIS	Nuclear research and technology
INSPEC	Physics and related fields, including astronomy and astrophysics Electrical engineering and electronics, control
METADEX	Metallurgy and metals
PATENTE	Patent literature from Austria, Switzerland, and the Federal Republic of Germany without any restriction to the *International Patent Classification*
PATSDI	*INPADOC Patent Gazette:* Patent literature without any restriction to the International Patent Classification
PHYS	Physics and related fields including astronomy and astrophysics
PHYSCOMP	Data compilations in physics: Bibliography
SDIM1	Metallurgy and metals
SDIM2	Metallurgy and metals
ZDE	Electrical engineering, control, data processing

Germany, and its scope focused initially on atomic energy. This arrangement was in place until 1968 when FIZ became a separate organization with its own facilities, and its interests subsequently expanded to include most of the physical sciences. FIZ provides both online and printed database services and can perform searches and supply copies of documents. The U.S. representative is Scientific Information Services,* under the direction of Dimitri R. Stein, who also represents *Gmelin Handbook of Inorganic Chemistry* in the United States.

10.13.   FULL-TEXT SEARCHING

This approach usually means that every significant word in the full texts of original documents covered is searchable and retrievable (displayable and/or printable). In contrast, most other online sources permit searching by such means as subject and author indexes, title words, keywords, Registry Numbers, patent numbers, classification or coding schemes, and abstracts, or some combination of the above. Users are "limited" to these alternatives.

Full text offers the potential (theoretical) advantage of not missing any information in the original, no matter how brief the detail or data. In addition, in the use of a full-text system, there is no requirement to learn complex, predetermined indexing terms, systematic nomenclature, or coding and classification systems, although this knowledge can be very helpful. Full-text users must instead guess accurately as to natural language words used by the original authors of documents. This means that, as in any other system, information can be lost. But a clear advantage over other systems is immediate availability of the full document, except for any drawings, curves, and similar graphics; this is document delivery in its most rapid form.

The question of the merits or disadvantages of full-text searching versus other systems has not been resolved. The issue is further briefly discussed in Chapter 13.

One good example of full-text systems is the *NEXIS/LEXIS®* service of Mead Data Central, part of the Mead Corp.* *NEXIS®* permits full-text searching of recent years of *Chemical Engineering, Chemical Week, Oil and Gas Journal,* and other publications of chemical interest, although the primary focus of *NEXIS* is not chemical. The Mead system includes also the full text of all U.S. patents since 1975 (see Section 13.14). In addition, the *BRS®* System (see Section 10.4) includes the full-text of American Chemical Society (ACS) journals since 1980 and of the *Kirk-Othmer Encyclopedia of Chemical Technology* (see Section 12.2.A.1). In 1986, ACS Journals became available on STN^{SM} and it was announced that *Kirk-Othmer* is to be available in systems beyond *BRS* such as *DIALOG®*. Also in 1986, *DIALOG®* expanded its full text coverage to include some McGraw-Hill publications such as *Chemical Week*.

ACS journals available online are *Accounts of Chemical Research, Analytical Chemistry* (research papers only), *Biochemistry, Chemical Reviews, Environmental Science and Technology* (research papers only), *Industrial and Engineering Chemistry—Fundamentals, Industrial and Engineering Chemistry—Process Design and Development, Industrial and Engineering Chemistry—Product Research and Development, Inorganic Chemistry, Journal of Agricultural and Food Chemistry, Journal of Chemical and Engineering Data, Journal of the American Chemical Society, Journal of Chemical Information and Computer Sciences, Journal of Medicinal Chemistry, The Journal of Organic Chemistry, The Journal of Physical Chemistry, Macromolecules,* and *Organometallics.* Online availability for most of these journals began in 1980, although in one case (*Journal of Medicinal Chemistry*), coverage began in 1976 and, in other cases, as late as 1982. A new ACS journal, *Langmuir,* was added in 1985.

In searching ACS journals online, users can choose from a variety of access points when planning search strategy: spectroscopic techniques; molecular formulas; thermodynamic data; biological data; chromatographic data; chemical names; CAS Registry Numbers; figure captions; footnotes; authors; employers; countries; and virtually any terms in the journals.

10.14.   CHEMICAL SUBSTANCES INFORMATION NETWORK

Chemical Substances Information Network (CSIN) is a federally funded attempt to link together many independent and autonomous data and bibliographic computer systems oriented primarily to chemical substances, thus in effect establishing a *library of systems.* It is claimed that users may converse with any or all systems interfaced by CSIN (see Table 10.13) without prior knowledge of, or training on, these independent systems, regardless of the hardware, software, data formats, or protocols of these information re-

TABLE 10.13. Systems Planned to Be Interfaced through CSIN[a]

Organization/System	Approximate Number of Files
BRS Information Technologies/*BRS*®	115
Chemical Abstracts Service/*CAS ONLINE*	15
Occupational Health Services/*Hazardline*[b]	3
Lockheed Information Systems/*DIALOG*®	275
National Library of Medicine/*MEDLARS*	25
System Development Corporation/*ORBIT*®	70

[a] Source: Data supplied by CSIN administration. CSIN can be regarded as both a front-end and a gateway.

TABLE 10.14. Objectives of CSIN

Provide a selection of databases to search.

Provide query lists to use in searching.

Provide a logical structure for Boolean operatives used to create query lists.

Automatically dial up and log on to selected databases.

Send or transmit to the central computer the user specified queries.

Search designated databases.

Retrieve, store and/or print results.

sources. However, the more the user knows about the various systems interfaced by CSIN, including details of operation, the more effective a job of searching can be done.

Benefits claimed for CSIN include: increased productivity of professional staff; improved responsiveness at low cost; high product quality; and effective data sharing in and among government, industry, and academic communities. There have been a few preliminary studies that attempt to document these claims, but so far there is nothing definitive. Table 10.14 lists the original objectives of CSIN.

In 1984, the minicomputer version of CSIN ceased because of lack of funding. A microcomputer version called "Micro–CSIN Workstation", has been developed by NLM, EPA, and the Council on Environmental Quality, with emphasis on responding to emergencies such as fires, explosions, and spills. This is to be made publicly available when tests are complete.

As can be predicted from its sponsorship, the orientation of CSIN is toward substances potentially subject to government regulation, either now or in the future. Principal interest is in the environmental aspects, although a number of other related topics are covered.

The CSIN system has evoked a moderate amount of interest and response, some of it favorable, but also some serious concern. The latter comments have come primarily from representatives of the chemical industry. Their concern centers about the perceived lack of adequate data quality indicators. The argument is that since CSIN makes large amounts of data very readily available, a serious possibility of error in regulatory and other decisions could arise unless the quality of data retrieved is carefully evaluated. As a result of this criticism, more attention is being paid to quality issues. See Section 15.14.

It is considered unlikely that the government will continue to fund CSIN as such for much longer. It is expected that the CSIN software will eventually be made available to industry, and the public generally, through the National Technical Information Service. The CSIN system itself may be continued in whole or part through an existing or new commercial information system, if it is believed that a profit can be made through private sector operation.

10.15.   PERSONAL COMPUTERS

Personal computers may be thought of in several ways with regard to chemical information as considered in this book. Two of the principal ways are (1) use as terminals to access databases in an online mode; and (2) use in freestanding modes to manipulate (search, organize, etc.) one's own databases obtained through previously authorized downloading of selected portions of an online file, or by direct receipt of preselected data sets (on disks, tapes, etc.), or by utilizing one's own data. The free-standing mode thus offers to the user the opportunity to create, maintain, manage, and use one's own database. This requires, of course, the knowledge and acquisition of appropriate software packages.

As expected, the upsurge in personal computers has prompted response from the vendors and producers of major online database systems, in addition to merely permitting downloading of some of the contents of some online files under specified conditions.

For example, to meet this upsurge, both *BRS*® and *DIALOG*® have initiated services operable mainly at night and on weekends with very low per hour costs compared to their normal rates. One *BRS*® system is called *BRS*®/*After Dark*™ and includes *CA* as well as other files of chemical interest such as ACS journals (full text) and *Kirk-Othmer Encyclopedia of Chemical Technology* (full text). A similar *BRS*® system is *BRS/BRKTHRU*™; standard rates apply during the day and reduced rates at night and on weekends. The comparable *DIALOG*® service is *Knowledge Index*®; this does not include *CA,* ACS journals, or *Kirk–Othmer*. Both systems are handy and inexpensive, but they do not have all capabilities or files of the full systems.

Another major development is an increasing number of software packages intended to facilitate logging on and using of the major online systems. These are aimed primarily at personal computer users.

Thus, a computer program offered by SDC® Information Services, *ORBIT*® *SearchMaster*™, permits such features as creating search strategies offline, automatic dialing up and logging on, automatic preformulated search execution, and reformulation of search results. In adition to *ORBIT*®, this program is said to be suited for the online systems of *DIALOG*®, the NLM, and *BRS*®.

A similar program, *Sci-Mate*™, is offered by *ISI*® (see Section 8.3). This permits automatic dial up and log on for the online services of *DIALOG*®, SDC®, *BRS*®, and the NLM. In addition, it is intended to minimize or obviate need to know the detailed search techniques of these host systems.

Programs such as *SearchMaster*™ and *Sci-Mate*™, and the previously mentioned CSIN, are examples of what have come to be called front-ends and gateways. They represent an increasing trend to capitalize on the need for both experienced and inexperienced users to navigate their way more smoothly through the increasingly confusing torrent of online databases and systems. An example of another gateway system is *EasyNet*, which is headquartered in Narbeth, PA. *Easy Net* provides access to a number of major online vendors, including *DIALOG,*® *Questel,* and *BRS.*® Other work on gateways and front-ends is underway, for example, by V. E. Hampel and coworkers at Lawrence Livermore National Laboratory.

Another major development is the increasing number of databases that can be accessed on local or personal computers without going online. For example, Fein–Marquart already offers (or soon plans to offer) several of the *CIS* databases for "stand-alone" use on personal computers (e.g., databases covering data on powder diffraction, thermodynamic properties, hazardous spills, and mass spectra). The databases are supplied on disks along with appropriate programs. As mentioned earlier (see Section 10.11.A.4), similar capability (with other databases) is available through Molecular Design, Limited. Much spectral data is available in disk form from sources such as Sadtler Research Laboratories (Section 18.1.B). The potential of optical disks is attracting considerable interest. *CA* now offers a stand-alone optical disk for many abstracts relating to health and safety (see Section 14.7).

Several computer programs are now available to help chemists input and manage their personal data files using micro (personal) computers. One of the best known of these programs is *Sci-Mate*™ Personal Data Manager, a product of *ISI*® (see Section 8.3), which is especially suited for management of reprint files.

As a major future trend, chemists can expect to access an increasing number of databases by direct use of personal computers without having to go online except for any needed updates covering the most recent material or perhaps for complex logical maneuvers.

REFERENCES

1. Examples of database directories: (a) Cuadra Associates, *Directory of Online Databases,* Santa Monica, CA. Updated several times each year; (b) M. E. Williams, Ed., *Computer-Readable Data Bases,* American Library Association, Chicago, IL, 1985; (c) L. Moore, Ed., *Data Base Directory,* Knowledge Industry Publications, White Plains, NY, 1984. There are several such directories, but these are probably the best known. The public data communications networks (Tymnet®, Telenet®, Uninet®) are sources of their own directories of the databases they carry; there is no change for copies at this time. The most important database directories are also available via online searching.

2. *Online* and *Database,* both published by Online, Inc., Weston, CT; *Online Review,* Learned Information, Oxford, United Kingdom and Medford, NJ.

3. J. Schmittroth, Ed., *Online Data Base Services Directory,* Gale Research Publishers, Detroit, MI, 1984.

4. J. Witiak, "Chemical Substructure Searching—Ready or Not, Here It Comes,"

Proceedings of the Online Conference, Dallas, TX, 1981, pp. 143–147; "Substructure Searching Using *CAS ONLINE,*" *Proceedings of the Second National Online Meeting,* New York, 1981, pp. 537–538.

5. S. R. Heller and G. W. Milne, *Online Chemical Information Workshop Notes,* 1982 (mimeographed).

6. P. G. Dittmar, R. E. Stobaugh, and C. E. Watson, "The Chemical Abstracts Service Chemical Registry System. I. General Design," *J. Chem. Inf. Comput. Sci.,* **16,** 111–121, (1976).

7. A. P. Lurie and A. S. Bernstein, "The Development, Implementation, and Operation of Eastman Kodak Company's Chemical Registry System," Abstract of presentation at the 188th ACS National Meeting, Philadelphia, PA, 1984.

8. W. J. Howe, M. M. Milne, and A. F. Pennell, Eds., "Retrieval of Medicinal Chemical Information," (ACS Symposium Series No. 84), American Chemical Society, Washington, D.C., 1978; pp. 132–143. Paper by R. G. Westland et al., "Warner-Lambert/Parke-Davis CAS Registry III Integrated Information System."

9. S. M. Kaback, "Online Patent Information," *World Patent Inf.,* **5,** (3)186–187, (1983).

10. J. M. Barnard, Ed., *Computer Handling of Generic Chemical Structures,* Gower Publishing Co., Aldershot, (U.K.), 1984.

11. C. H. Davis and J. E. Rush, *Information Retrieval and Documentation in Chemistry,* Greenwood Press, Westport, CT, 1974.

12. R. Attias, "*DARC* Substructure Search System: A new approach to chemical information," *J. Chem. Inf. Comput. Sci.,* **23,** 102–108 (1983).

13. E. G. Smith and P. A. Baker. *The Wiswesser Line-Formula Chemical Notation (WLN),* 3rd ed., Chemical Information Management, Cherry Hill, NJ, 1976.

14. R. Fugmann, W. Braun, and W. Vaupel, "Zur Dokumentation chemischer Forschungsergebnisse," *Angew. Chem.,* **73,** 745–751 (23) (1961).

LIST OF ADDRESSES

CompuServe
5000 Arlington Center Boulevard
Columbus, OH 43220

DIALOG® Information Services Inc.
3460 Hillside Avenue
Palo Alto, CA 94304 (800-227-1927)

EasyNet
134 N. Narbeth Ave.
Narbeth, PA (800-841-9553)

Fein–Marquart Associates
7215 York Road
Baltimore, MD 21212

Fraser Williams (Scientific Systems) Ltd.
London House, London Road South
Poynton, Cheshire, SK12 1YP, United Kingdom

ICI (Information Consultants Inc.)
1133 15th Street, N.W.
Washington, D.C. 20005

Information Industry Association
316 Pennsylvania Avenue, S.E.,
    Suite 502
Washington, D.C. 20003
    (202-544-1969)

Mead Data Central
9393 Springboro Pike
Dayton, OH 45401 (513-865-6800)

Molecular Design, Limited
2132 Farallon Dr.
San Leandro, CA 94577
    (415-895-1313)

Scientific Information Services
7 Woodland Avenue
Larchmont, N.Y. 10538

Telenet®
8229 Boone Boulevard
Vienna, VA 22180

Tymnet®
2710 Orchard Parkway
San Jose, CA 95134

Uninet®
7700 Leesburg Pike
Falls Church, VA 22043

11 REVIEWS

Comprehensive and evaluative reviews can minimize or obviate the need for the chemist to do further literature searching on the subject in question. Reviews save time and help provide perspective. Along with encyclopedia articles (see Chapter 12), they can be excellent starting points.

Some questions that can help the user evaluate the quality of a review include:

1. Is the author of the review a known expert in the field?
2. Is there a clear statement of scope, methodology, and purpose of the review?
3. What is the time frame covered? How up to date are the most recent sources included, and how far back in time does the author go? The more comprehensive the better.
4. Are all types of literature covered, including patents?
5. Is the author critical or evaluative? Are the most important and valid data identified?
6. Is the material well organized and presented?
7. Is there a complete bibliography with all sources clearly identified?
8. Is there any reason to expect significant lack of objectivity?

The best known review journal is *Chemical Reviews (CR)*. Initiated in 1925, and now published monthly by the American Chemical Society, *CR* contains critical, comprehensive reviews by experts in the field, with emphasis on the more theoretical aspects of chemistry.

A cumulative index to the volumes for 1925–1960 appears in Volume 60 (1960), and Volume 70 (1970) contains an index to the volumes for 1961–1970. These collective indexes are in addition to conventional annual indexes. Unfortunately, however, the practice of cumulative indexing has been discontinued, at least for the present, for this publication.

Other significant review tools include:

• *Annual Reports on the Progress of Chemistry* (published annually since 1904 by the Royal Society of Chemistry). Until 1967, *Annual Reports* was published in one volume. At that time, there was a split into separate volumes for Part A (General, Physical, and Inorganic) and Part B (Organic). Beginning in 1979, the Parts became: A (Inorganic), B (Organic), and C (Physical).

- *Chemical Society Reviews* (published quarterly since 1972 by the Royal Society of Chemistry).
- *Reports on the Progress of Applied Chemistry* (published annually for more than 60 years by the Society of Chemical Industry). The title is self-explanatory. Beginning in 1980, the Society discontinued these *Reports* and initiated a new series, *Critical Reports on Applied Chemistry*. Initial titles include: *Topics in Non-Ferrous Extractive Metallurgy; Chemical Engineering and the Environment; Progress in Pharmaceutical Research; Biophysical Methods in Food Research; Methods Used in Pharmaceutical Formulation; Developments in Chemistry and Technology of Organic Dyes;* and *Chemical Thermodynamics in Industry: Models and Computation.*
- *Russian Chemical Reviews* (English translation of *Uspekhi Khimii*).
- *Specialist Periodical Reports.* This series was initiated by the Royal Society of Chemistry in 1967 to provide "systematic and comprehensive review coverage of the progress in major areas of chemical research." The series has now reached over 40 titles, some of which are published annually and others as biennial volumes. Examples of some titles include:

Alicyclic Chemistry
Aliphatic, Alicyclic, and Saturated Heterocyclic Chemistry
Aliphatic Chemistry
Aliphatic and Related Natural Product Chemistry
The Alkaloids
Amino acids, Peptides, and Proteins
Aromatic and Heteroaromatic Chemistry
Biosynthesis
Carbohydrate Chemistry
Catalysis
Chemical Physics of Solids and Their Surfaces
Chemical Thermodynamics
Colloid Science
Dielectric and Related Molecular Processes
Electrochemistry
Electronic Structure and Magnetism of Inorganic Compounds
Electron Spin Resonance

Environmental Chemistry

Fluorocarbon and Related Chemistry

Foreign Compound Metabolism in Mammals

Gas Kinetics and Energy Transfer

General and Synthetic Methods

Heterocyclic Chemistry

Inorganic Biochemistry

Inorganic Chemistry of the Main-group Elements

Inorganic Chemistry of the Transition Elements

Inorganic Reaction Mechanisms

Macromolecular Chemistry

Mass Spectrometry

Molecular Spectroscopy

Molecular Structure by Diffraction Methods

Nuclear Magnetic Resonance

Organic Compounds of Sulphur, Selenium, and Tellurium

Organometallic Chemistry

Organophosphorus Chemistry

Photochemistry

Radiochemistry

Reaction Kinetics

Saturated Heterocyclic Chemistry

Spectroscopic Properties of Inorganic and Organometallic Compounds

Satistical Mechanics

Surface and Defect Properties of Solids

Terpenoids and Steroids

Theoretical Chemistry

- Annual Reviews* publishes a series of annual review volumes in various sciences. Those of most interest to chemists include volumes on physical chemistry, biochemistry, pharmacology and toxicology, biophysics and biophysical chemistry, and materials sciences.
- The CRC Press* has initiated a series of review journals that include the following: *CRC Critical Reviews in Analytical Chemistry; CRC Critical Reviews in Biochemistry; CRC Critical Reviews in Solid State and Materials Science;* and *CRC Critical Reviews in Toxicology.*

Chemical Abstracts (*CA*) editors have recognized the importance of reviews by designating these with the symbol R in the Volume Indexes to *CA* (beginning in 1967). Many chemists examine abstracts of reviews before they look at abstracts for other index entries, because a good review can provide all that is immediately needed on a topic. See also Section 7.13.

The Royal Society of Chemistry has now begun to publish *Index of Reviews in Organic Chemistry*. The first volume (*IROC 81–82*) covers review articles published in English, French, or German during the period November 1980–December 1982. This work is a successor to *Reviews in Organic Chemistry,* edited by D. A. Lewis, the last edition of which appeared in 1981. Each entry in the new series includes: index term; item number; title of original document; bibliographic details; a brief abstract if scope of review is not clear from title; and indication of the number of reviews cited in the original review. Future issues, comprising about 1200 entries each, are planned to be published annually. The Royal Society's index to reviews in analytical chemistry is mentioned briefly elsewhere in this book (Section 18.4).

The *Index to Scientific Reviews*™ (*ISR*)™ published by the *Institute for Scientific Information*® since 1974 (see Section 8.3), helps locate more than 26,000 review articles published each year in chemistry and other scientific disciplines. *ISR*™ is issued in two semiannual cumulations.

There are other review publications. For example, Academic Press (New York, Orlando, and London) publishes titles such as these "*Advances in*" publications: Chemical Engineering, Applied Biochemistry and Bioengineering, Catalysis, Coal Science, Energy Systems and Technology, and Fermentation Processes.

Not all reviews appear in journals and books specifically devoted to reviews. A good review can appear in almost any chemical journal or other information source. Reviews are frequently labeled as such or in similar terms, for example, *Recent Advances* or *Recent Trends.* However, any article or book with an extensive list of references, and a good discussion of these, is potentially a review, whether or not it is so labeled. Patent sources can also provide an excellent review of prior art (see Chapter 13).

Another way of reviewing technologies is through citations in articles and patents. Once the user has located a key reference, additional pertinent literature may be located, based on citations to or from this reference. There are those who are highly enthusiastic about the merits of this approach, while others believe that there are limitations. The *Institute for Scientific Information*® (see Section 8.3) has developed the use of citations in articles to its highest form. Sources of patent citations are discussed in Section 13.19.

LIST OF ADDRESSES

Annual Reviews, Inc.
4139 El Camino Real
Palo Alto, CA 94306-9981

CRC Press
2000 N.W. 24th Street
Boca Raton, FL 33431

12 ENCYCLOPEDIAS AND OTHER MAJOR REFERENCE BOOKS

12.1. INTRODUCTION

Journals and patents almost always contain the latest information, and that is frequently what chemists need. The chemist, however, cannot afford to ignore books. Books are usually the best starting point for searching the chemical literature. They can often provide quick, straightforward answers that could otherwise take many hours to locate in widely scattered journal articles and patents on the subject. Books provide the foundation from which to launch any further search of the literature required.

When the literature on a chemical or reaction is voluminous, use of an appropriate book is highly recommended to help sort out what is important and vital from what is not. Some *advanced* books are evaluative or critical and can help minimize further searching. Review journals and other review sources (see Chapter 11) perform a similar function.

The chemist will find some older books, seemingly obsolete, of value. As noted later in this chapter, older editions may contain some information not included in more recent editions, and this may be precisely what the chemist needs.

If a search is being made to see if an idea being considered for a patent is novel, consultation of pertinent books is mandatory.

Various kinds of books are referred to throughout this volume. This chapter focuses on some of the major reference works found in most good chemistry libraries.

First, there is a discussion of significant encyclopedias in chemistry and chemical engineering. This is followed by brief remarks on the German *Handbuch* concept and a review of some important reference works in organic and inorganic chemistry.

12.2. ENCYCLOPEDIAS—INTRODUCTORY REMARKS

The expert or specialist in a field will usually find treatment of this field by encyclopedias too general and elementary. Encycopedias are also quickly dated. By the time the last volume is completed, some of the articles in initial volumes may be obsolete. Nevertheless, most chemists and engineers beginning work on a new project about which they know little or nothing will find that a major chemical encyclopedia is a good starting point. For further details, users can then consult monographs, patents, and articles cited in the encyclopedia, and more recent sources as required. The relative conciseness, which is a feature of encyclopedias, is usually a plus because it gives the chemist a quick overview of a field.

Encyclopedias, however, are by no means all inclusive. For example, encyclopedia articles do not usually cover the recent process technology and economics for important commercial products with the depth of analysis and detail some users require. The special tools that provide this kind of information are described in Chapter 17.

The principal encyclopedias of interest to chemical engineers and chemist include:

> *Encyclopedia of Chemical Technology* (commonly referred to as *Kirk–Othmer* after the names of the original editors).
>
> *Ullmanns Encyclopädie der Technischen Chemie,* now called *Ullmanns Encyclopedia of Industrial Chemistry.*
>
> *Encyclopedia of Polymer Science and Technology—Plastics, Resins, Rubbers, Fibers* (frequently referred to by the name of its original editor, Herman F. Mark), now entitled *Eycyclopedia of Polymer Science and Engineering.*

Each is described in the sections that follow.

A. Specific Encyclopedias

1. **Kirk–Othmer (1).** Most chemists and chemical engineers agree that, with the exceptions of *Chemical Abstracts (CA)* and the desk handbooks, *Kirk–Othmer (KO)* is the most indispensable tool in any chemistry or chemical engineering library. There are now three complete editions of *KO.*

One of the most important uses of *KO* continues to be its value as a starting point for research workers and others beginning work in a less familiar field of chemistry or related technology. In this regard, *KO* is often unexcelled in its helpfulness.

KO editors place major, but by no means total, emphasis on the more applied aspects—the application of chemistry and chemical engineering to industrially important concepts, products, processes, and uses. Most articles in *KO* provide good coverage of significant aspects of the topics covered. Numerous references to patent, journal, and other literature permit chemists and engineers using *KO* to pursue fields of interest in more depth.

KO is not all inclusive and is characterized by the uneven treatment that is inevitable in an encyclopedia in any field. Also, if an article on a chemical is written by an employee of a manufacturer of that product, the article may provide strong coverage for properties and uses but weak coverage of manufacturing details. This approach helps users of the product but is otherwise incomplete.

KO is an American encyclopedia, written mostly by American authors for American chemists and engineers. Although there are discussions of, and references to, European and Japanese technology, emphasis is on U.S. practice. Omission of foreign practice is a shortcoming in some fields.

One of the most interesting and helpful features of the third edition of *KO* is that the entire set is searchable through several computer-based online systems such as those of *BRS®* (see Section 10.4) and IHS® (see Section 5.8). This online file permits chemists who have access to the online systems, but do not have a set of *KO* readily available, to determine whether and how the treatise covers topics of interest to them. The online file can permit more detailed, specific, and sophisticated searching than is possible with the conventional (printed) index to *KO*.

Another innovation with the third editon is that a one-volume version of the entire set was published in 1985 under the title *Kirk–Othmer Concise Encyclopedia of Chemical Technology*. This should prove to be an important addition to the daily working tools of chemists and chemical engineers who can afford the original purchase price of a little more than $100. In addition, selected volumes of reprinted *KO* articles, grouped by subject, have been published. An example is the collection of related articles on electronics chemicals (2). These articles are as they appear in the full original text of *KO*, except that they are conveniently grouped together instead of being scattered through multiple volumes. The collection of articles on electronics chemicals covers, in one volume, topics (articles) that appear throughout the original volumes in the appropriate alphabetical order as shown in Table 12.1. Other one-volume subsets of *KO* cover antibiotics, composite materials, glass, information retrieval, textiles, and recycling.

Kirk–Othmer contains over 1200 articles with about 9 million words, 6000 tables, and 5000 figures in the 24 volumes. A typical article is 15–20 pages in length. CAS Registry Numbers are useful additions to the third edition.

Table 12.2, which lists the topics covered by *Kirk–Othmer,* will help give the reader some idea of the comprehensiveness and scope of this set.

The initial volume of the first edition of *KO* was published in 1947 and the final volumes (in this case *Supplements*) were published in 1960. The first volume of the second edition appeared in 1962, and the *Supplement* and *Index* were published in 1971 and 1972. The first volume of the third edition appeared in 1978. This edition was completed in 1984, under the editorship of Martin Grayson, with the issuance of a complete index volume and a supplement volume.

The first and second editions of *KO* remain useful and are still consulted. They contain some information of value that does not appear in the third

TABLE 12.1. Contents of *Encyclopedia of Semiconductor Technology* (Collection of Articles on Electronics) from *Kirk–Othmer*[a]

Amorphous Magnetic Materials	Integrated Circuits
Arsenic and Arsenic Alloys	Ion Implantation
Ceramics as Electric Materials	Light Emitting Diodes and Semicon-
Chromogenic Materials—Electro-	ductor Lasers
chromic and Thermochromic	Magnetic Materials—Bulk
Cryogenics	Magnetic Materials—Thin Film
Digital Displays	Magnetic Tape
Electrical Connectors	Materials Reliability
Electroless Plating	Phosphorus and the Phosphides
Electromigration	Photodetectors
Embedding	Photoreactive Polymers
Ferrites	Photovoltaics Cells
Ferroelectrics	Polymers, Conductive
Fiber Optics	Semiconductors
Film Deposition Techniques	Silver and Silver Alloys
Gallium and Gallium Compounds	Superconducting Materials
Germanium and Germanium Com-	Tellurium and Tellurium Compounds
pounds	Thermoelectric Energy Conversion
Glassy Metals	

[a] M. Grayson, Ed., *Encyclopedia of Semiconductor Technology*, Wiley, New York, 1984.

edition. In addition, processes, products, and concepts once believed obsolete may again become interesting because of changing economics, raw materials shortages, or environmental reasons. This is a principal reason why older editions of encyclopedias such as *KO* can maintain a surprisingly long useful life span even though large parts are out of date. Furthermore, the third edition of *KO* contains numerous references to the earlier editions.

2. **Ullmans** *(3).* The German language counterpart of *KO* is *Ullmanns Encyklopaedie der Technischen Chemie.* Publication of the fourth edition of this important work began in 1972 and was completed in 1984. There are 25 volumes.

According to the publisher,

> Volumes 1–6 (*Thematic Section—General Principles & Methodology*) constitute a complete, up-to-date account of general principles and methodology (e.g., thermodynamics, process engineering, and development); physical-chemical analysis methods; and measurement and control, pollution abatement, and works safety.

The following 18 volumes consist of an alphabetical compilation of articles on key subjects. Every third volume of the 18 alphabetical volumes contains a cumulative subject index. Volume 25 is a cumulative subject index to the complete work.

TABLE 12.2. Contents of *Kirk–Othmer Encyclopedia of Chemical Technology*

210

TABLE 12.2. Continued

Computers
Contact Lenses
Contraceptive Drugs
Coordination Compounds
Copolymers
Copper

Volume 7

Copper Alloys
Copper Compounds
Cork
Corrosion and Corrosion Inhibitors
Cosmetics
Cotton
Coumarin
Crotonaldehyde
Crotonic Acid
Cryogenics
Crystallization
Cumene
Cyanamides
Cyanides
Cyanine Dyes
Cyanocarbons
Cyanoethylation
Cyanohydrins
Cyanuric and Isocyanuric Acids
Cyclohexanol and Cyclohexanone
Cyclopentadiene and
  Dicyclopentadiene
Defoamers
Deformation Recording Media
Dental Materials
Dentifrices
Design of Experiments
Deuterium and Tritium
Dialysis
Diamines and Higher Amines
Diatomite
Dicarboxylic Acids
Dietary Fiber
Diffusion Separation Methods
Digital Displays
Dimensional Analysis
Dimer Acids
Diphenyl and Terphenyls
Disinfectants and Antiseptics
Dispersants
Distillation

Volume 8

Diuretics
Driers and Metallic Soaps
Drycleaning and Laundering
Drying
Drying Agents
Drying Oils
Dye Carriers
Dyes and Dye Intermediates
Dyes
Economic Evaluation
Eggs
Elastomers, Synthetic
Electrical Connectors
Electrochemical Processing
Electrodecantation
Electrodialysis

Electroless Plating
Electrolytic Machining Methods
Electromigration
Electrophotography
Electroplating
Electrostatic Sealing
Embedding
Emulsions

Volume 9

Enamels, Porcelain or Vitreous
Energy Management
Engineering and Chemical Data
  Correlation
Engineering Plastics
Enzyme Detergents
Enzymes
Epinephrine and Norepinephrine
Epoxidation
Epoxy Resins
Esterification
Esters
Ethanol
Ethers
Ethylene
Ethylene Oxide
Evaporation
Exhaust Control
Expectorants, Antitussives, and
  Related Agents
Explosives and Propellants
Extraction
Extractive Metallurgy
Fans and Blowers
Fats and Fatty Oils
Feedstocks
Felts
Ferrites

Volume 10

Ferroelectrics
Fertilizers
Fiber Optics
Fibers
Fillers
Film and Sheeting Materials
Film Deposition Techniques
Filters, Optical
Filtration
Fine Chemicals
Flame Retardants
Flame Retardants for Textiles
Flavor Characterization
Flavors and Spices
Flocculating Agents
Flotation
Fluidization
Fluid Mechanics
Fluorine
Fluorine Compounds, Inorganic
Fluorine Compounds, Organic

Volume 11

Fluorine Compounds, Organic
Foamed Plastics
Foams

Food Additives
Food Processing
Foods, Nonconventional
Forensic Chemistry
Formaldehyde
Formic Acid and Derivatives
Friedel-Crafts Reactions
Fruit Juices
Fuels
Fuels from Biomass
Fuels from Waste
Fuels, Synthetic
Fungicides (Agricultural)
Furan Derivatives
Furnaces, Electric
Furnaces, Fuel-Fired
Furs, Synthetic
Fusion Energy
Gallium and Gallium Compounds
Gas Cleaning
Gas, Natural
Gasoline and Other Motor Fuels
Gastrointestinal Agents
Gelatin
Gems, Synthetic
Genetic Engineering
Geothermal Energy
Germanium and Germanium
  Compounds
Gilsonite
Glass
Glass-Ceramics
Glassy Metals
Glue
Glycerol
Glycols
Gold and Gold Compounds

Volume 12

Gravity Concentration
Grignard Reaction
Gums
Hafnium and Hafnium Compounds
Hair Preparations
Hardness
Heat Exchange Technology
Heat-Resistant Polymers
Heat Stabilizers
Helium-Group Gases
Helium-Group Gases, Compounds
Herbicides
High Pressure Technology
High Temperature Alloys
High Temperature Composites
Histamine and Antihistamine
  Agents
Hollow-Fiber Membranes
Holography
Hormones
Hydantoin
Hydraulic Fluids
Hydrides
Hydroboration
Hydrocarbon Oxidation
Hydrocarbon Resins
Hydrocarbons
Hydrochloric Acid

211

TABLE 12.2. Continued

Hydrogen
Hydrogen Energy
Hydrogen-Ion Activity
Hydrogen Peroxide
Hydroquinone, Resorcinol, and
  Pyrocatechol
Hydroxybenzaldehydes
Hydroxy Carboxylic Acids

Volume 13

Hydrogen-Ion Activity
Hydrogen Peroxide
Hydroquinone, Resorcinol, and
  Catechol
Hydroxybenzaldehydes
Hydroxy Carboxylic Acids
Hydroxy Dicarboxylic Acids
Hypnotics, Sedatives,
  Anticonvulsants
Imines, Cyclic
Immunotherapeutic Agents
Incinerators
Indium and Indium Compounds
Indole
Industrial Antimicrobial Agents
Industrial Hygiene and Toxicology
Information Retrieval
Infrared Technology
Initiators
Inks
Inorganic High Polymers
Insect Control Technology
Instrumentation and Control
Insulation, Acoustic
Insulation, Electric
Insulation, Electric
Insulation, Thermal
Insulin and Other Anti-Diabetic
  Agents
Integrated Circuits
Iodine and Iodine Compounds
Ion Exchange
Ion Implantation
Ion-Selective Electrodes
Iron
Iron by Direct Reduction
Iron Compounds
Isocyanates, Organic
Isoprene
Isotopes
Itaconic Acid and Derivatives
Ketenes and Related Substances
Ketones
Laminated and Reinforced Metals
Laminated and Reinforced Plastics
Laminated Materials, Glass

Volume 14

Laminated Wood-Based
  Composites
Lasers
Latex Technology
Lead

Lead Alloys
Lead Compounds, Lead Salts
Lead Compounds, Organolead
  Compounds
Lead Compounds, Industrial
  Toxicology
Leather
Leatherlike Materials
Lecithin
Light-Emitting Diodes and
  Semiconductor Lasers
Lignin
Lignite and Brown Coal
Lime and Limestone
Liquefied Petroleum Gas
Liquid Crystals
Liquid-Level Measurement
Lithium and Lithium Compounds
Lubrication and Lubricants
Luminescent Materials, Phosphors
Luminescent Materials, Fluorescent
  Pigments (Daylight)
Magnesium and Magnesium Alloys
Magnesium Compounds
Magnetic Materials, Bulk
Magnetic Materials, Thin Film
Magnetic Separation
Magnetic Tape
Maintenance
Maleic Anhydride, Maleic Acid, and
  Fumaric Acid
Malonic Acid and Derivatives
Malts and Malting
Manganese and Manganese Alloys
Manganese Compounds
Market and Marketing Research
Mass Transfer

Volume 15

Matches
Materials Reliability
Materials Standards and
  Specifications
Meat Products
Medical Diagnostic Reagents
Membrane Technology
Memory Enhancing Agents and
  Antiaging Drugs
Mercury
Mercury Compounds
Metal Anodes
Metal-Containing Polymers
Metal Fibers
Metallic Coatings, Survey
Metallic Coatings, Explosively Clad
  Metals
Metal Surface Treatments,
  Cleaning, Pickling, and Related
  Processes
Metal Surface Treatments,
  Chemical and Electrochemical
  Conversion Treatments
Metal Surface Treatments, Case
  Hardening
Metal Treatments

Methacrylic Acid and Derivatives
Methacrylic Polymers
Methanol
Micas, Natural and Synthetic
Microbial Polysaccharides
Microbial Transformations
Microencapsulation
Microwave Technology
Milk and Milk Products
Mineral Nutrients
Mixing and Blending
Molecular Sieves
Molybdenum and Molybdenum
  Alloys
Molybdenum Compounds
Naphthalene
Naphthalene Derivatives
Naphthenic Acids
Neuroregulators
Nickel and Nickel Alloys
Nickel Compounds
Niobium and Niobium
  Compounds
Nitration
Nitric Acid
Nitrides
Nitriles
Nitroalcohols
Nitrobenzene and Nitrotoluenes
Nitrogen
Nitrogen Fixation
Nitroparaffins
N-Nitrosamines

Volume 16

Novoloid Fibers
Nuclear Reactors
Nuts
Ocean Raw Materials
Odor Modification
Oils, Essential
Oil Shale
Olefin Fibers
Olefin Polymers
Olefins, Higher
Operations Planning
Optical Filters
Organometallics, ⁿ - Bonded Alkyls
  and Aryls
Organometallics, Metal ⁿ
  Complexes
Oxalic Acid
Oxo Process
Oxygen
Oxygen-Generating Systems
Ozone
Packaging Materials, Industrial
  Products
Packing Materials
Paint
Paint and Varnish Removers
Paper
Papermaking Additives
Particle-Track Etching
Patents
Perfumes

212

TABLE 12.2. Continued

Volume 17

Peroxides and Peroxy Compounds, Inorganic
Peroxides and Peroxy Compounds, Organic
Pet and Livestock Feeds
Petroleum
Pharmaceuticals
Pharmaceuticals, Controlled Release
Pharmaceuticals, Optically Active
Pharmacodynamics
Phenol
Phenolic Resins
Phosgene
Phosphoric Acid and the Phosphates
Phosphorus and the Phosphides
Phosphorus Compounds
Photochemical Technology
Photodetectors
Photography
Photomultiplier Tubes
Photoreactive Polymers
Photovoltaic Cells
Phthalic Acid and Other Benzenepolycarboxylic Acids
Phthalocyanine Compounds
Pigments
Pilot Plants and Microplants
Pipelines
Piping Systems

Volume 18

Plant-Growth Substances (or Regulators)
Plant Layout
Plant Location
Plant Safety
Plasma Technology
Plastic Building Products
Plasticizers
Plastics Processing
Plastics Testing
Platinum Group Metals
Platinum Group Metals, Compounds
Plutonium and Plutonium Compounds
Poisons, Economic
Polishes
Polyamides
Poly(Bicycloheptene) and Poly(Bicycloheptadiene)
Polyblends
Polycarbonates
Polyelectrolytes
Polyester Fibers
Polyesters, Thermoplastic
Polyesters, Unsaturated
Polyethers
Polyhydroxybenzenes
Polyimides
Polymerization Mechanisms and Processes

Polymers
Polymers, Conductive
Polymers Containing Sulfur
Polymethine Dyes
(Polymethyl)Benzenes
Polypeptides
Potassium
Potassium Compounds

Volume 19

Powder Coatings
Powder Metallurgy
Power Generation
Pressure Measurement
Printing Processes
Process Research & Development
Programmable Pocket Calculators
Propyl Alcohols
Propylene
Propylene Oxide
Prosthetic & Biomedical Devices
Proteins
Psychopharmacological Agents
Pulp
Pulp, Synthetic
Pyrazoles, Pyrazolines & Pyrazolones
Pyridine & Pyridine Derivatives
Pyrotechnics
Pyrrole & Pyrrole Derivatives
Quaternary Ammonium Compounds
Quinolines & Isoquinolines
Quinones
Radiation Curing
Radioactive Drugs
Radioactive Tracers
Radioactivity, Natural
Radioisotopes
Radiopaques
Radioprotective Agents
Rare Earth Elements
Rayon
Reactor Technology
Recording Disks
Recreational Surfaces
Recycling

Volume 20

Refractories
Refractory Coatings
Refractory Fibers
Refrigeration
Regulatory Agencies
Reprography
Research Management
Resins, Natural
Resins, Water-Soluble
Reverse Osmosis
Rhenium & Rhenium Compounds
Rheological Measurements
Roofing Materials

Rubber Chemicals
Rubber Compounding
Rubber, Natural
Rubidium & Rubidium Compounds
Salicylic Acid & Related Compounds
Sampling
Sealants
Sedimentation
Selenium & Selenium Compounds
Semiconductors, Theory & Application
Semiconductors, Fabrication & Characterization
Semiconductors, Amorphous
Semiconductors, Organic
Separation Systems Synthesis
Shape-Memory Alloys
Shellac
Silica
Silicon & Silicon Alloys
Silicon Compounds
Silk

Volume 21

Silver & Silver Alloys
Silver Compounds
Simulation & Process Design
Simultaneous Heat & Mass Transfer
Size Enlargement
Size Measurement of Particles
Size Reduction
Soap
Sodium & Sodium Alloys
Sodium Compounds
Soil Chemistry of Pesticides
Solar Energy
Solders and Brazing Alloys
Solvent Recovery
Solvents, Industrial
Sorbic Acid
Soybeans & Other Oil-Seed Proteins
Space Chemistry
Sprays
Stains, Industrial
Starch
Steam
Steel
Sterilization Techniques
Steroids
Stilbene Derivatives
Stilbene Dyes
Stimulants
Strontium & Strontium Compounds
Styrene
Styrene Plastics
Succinic Acid & Succinic Anhydride
Sugar
Sulfamic Acid & Sulfamates
Sulfolanes & Sulfones

TABLE 12.2. Continued

214

As compared to *KO, Ullmanns* has both advantages and disadvantages (4). Although coverage of chemistry is about the same, *Ullmanns* seems to cover process technology in more detail.

There is full and accurate treatment of important topics of interest. The literature citations accompanying each article are of special value, with appropriate emphasis given to pertinent sources, especially patents. Coverage of developments in Europe, and especially West Germany, is especially good, as might be anticipated.

The principal disadvantage of *Ullmanns,* as far as English speaking and reading chemists or engineers are concerned, is that the text has been in German, at least up through and including the fourth edition. An attempt has been made to bridge the language barrier by providing both German- and English-language tables of contents and indexes. Nevertheless, some lessening of speed and comprehension for chemists who are not thoroughly familiar with German chemical terminology and technology is inevitable. In addition, and equally important, practices and conventions described may not correspond with those of the United States and other countries in some cases.

It is important to be aware that the fifth edition of *Ullmanns* is to be published entirely in English under the title *Ullmanns Encyclopedia of Industrial Chemistry.* Publication of this 36 volume edition began in 1984 and is to be complete in 1996. Depending on demand and interest, the text may be made available online.

3. **McKetta** *(5).* Volumes of this comprehensive encyclopedia, targeted at engineers, were first published in 1977 and are still being issued. As of early 1986, some 24 volumes had appeared. The full title is *Encyclopedia of Chemical Processing and Design,* and the editor is John J. McKetta.* The purpose of this publication is to detail current developments in the field of chemical technology and related industries. According to the publisher,

> the encyclopedia rigorously covers chemical processes, methods, practices and standards used in the chemical industry. Although most articles are written by American chemical engineers, scientists and technologists, no important foreign source has been neglected in making the encyclopedia truly international in scope.

As the title of this work indicates, emphasis is placed on chemical processing, equipment, operations, and design aspects. The full potential value of *McKetta* is diminished somewhat by the fact that it is still incomplete at this writing and appears to have a slow publication schedule.

4. **Mark** *(6).* For the many chemists or engineers concerned primarily with polymer chemistry, the major encyclopedic source in lieu of or in addition to those previously mentioned, is the *Encyclopedia of Polymer Science and Technology* of which the first edition was completed in 1976. A new edition was initiated in 1985 under the title *Encyclopedia of Polymer Science and Engineering* (J. I. Kroschwitz, Ed.) Online availability was announced in 1986.

5. Beever *(7)*. A major new reference work, *Encyclopedia of Materials Science and Engineering,* was published by MIT Press in 1986 under the editorship of Michael B. Beever. This 8-volume work should be of considerable interest to chemists and engineers.

B.   Which Encyclopedia to Use

In their use of major chemical encyclopedias, chemists and engineers should consider *all* of those mentioned as appropriate, because each can provide a different perspective, may have different or additional information, and because topics not adequately covered in one encyclopedia may be more fully and adequately covered in another. The staggered and overlapping time frames within which the encyclopedias are issued is another reason for looking at all of those listed. However, *KO* is by far the most important chemical encyclopedia at this time and should be looked at first.

12.3.   THE *HANDBUCH* CONCEPT

The *Handbuch* is a valuable type of publication, significantly different from most handbooks published in the United States. It is a multivolume work, far more extensive in scope, and provides more in-depth coverage, than the typical U.S. handbook.

Þ*Handbuch* coverage usually goes back to the "beginnings" of chemistry—long before *CA* began in 1907—and brings that coverage up to a reasonably current time, as is discussed later in this chapter. Ongoing updating is another feature. For the chemist, this can mean that any needed search of later literature can begin where *Handbuch* coverage ceases.

Although consultation of the *Handbuch* can in many cases obviate the need for further literature searching (assuming the *Handbuch* has the desired information), it would be a mistake for the chemist to conclude that *Handbuch* coverage is 100% complete and without some errors or omissions.

The two most important examples of the German Handbuch in chemistry are *Beilstein* and *Gmelin*. Both are edited at the Carl Bosch Haus in Frankfurt-am-Main, Germany.

The *Handbuch* is an old tradition in German chemistry. In 1817, Leopold Gmelin published his first *Handbuch,* which was a quick success. At the start of the fifth edition (1852), the field of organic chemistry was split off and has been carried on separately as *Beilstein.*

Gmelin is now in its eighth edition, and *Beilstein* is in its fourth. It is, unfortunately, all too easy (and incorrect) to regard *Handbuch* volumes as (a) obsolete; (b) sometimes incomprehensible, because large parts are written in languages other than English; or (c) otherwise difficult to use because of large size (many volumes) and "complex" organization. This view, although prevalent among many chemists, can be counterproductive and should be carefully reconsidered, especially because of recent developments in *Handbuch* publication.

A. The Question of Obsolescense

It is true that some volumes in major *Handbuch* series go back many years. But chemists can ignore or downplay older literature only at the risk of unnecessary repetition of older work that may still be valid. Moreover, some newer *Handbuch* volumes bring literature coverage up to much more recent years. In some cases, this is within about six months of date of publication.

B. On Potential Language Problems

Although most of the older volumes are in German, and will stay that way, this is not a insurmountable hurdle, as indicated later in this chapter in the discussion of *Beilstein*. In addition, all new volumes are now being published in English for at least two of the major *Handbuch* series, that is, *Gmelin* and *Beilstein*.

C. On Potential Complexity of Use

Handbuch volumes occupy many linear feet of shelf space, and this alone may intimidate some at first. There are, however, helpful self-study type aids available without charge from *Handbuch* publishers. Trained librarians or information specialists can assist by introducing use of the volumes to those less familiar. In any event, it is often helpful to motivate oneself to take any extra effort required to use a *Handbuch* on the basis that a *Handbuch* has the potential to save much needless duplication of work. The complexity is often more imaginary than real.

Prices of *Handbuch* sets, like that of almost all scientific literature, have been escalating sharply. This difficult situation is more severe with *Handbuch* volumes because their prices have always been relatively high. Some organizations have been forced by budget constraints to curtail or suspend purchase of one or more of the *Handbuch* sets. In such cases, volumes already on the shelves should, of course, be retained because much older data continues to be valid as previously noted. It is a well accepted principle of chemical literature use that older volumes and editions frequently have data omitted in later volumes or editions.

12.4.   ORGANIC CHEMISTRY—SOME IMPORTANT REFERENCE WORKS

A.   *Beilsteins Handbuch der Organischen Chemie (Handbook of Organic Chemistry)*

Beilstein (8) is the basic reference work on organic compounds. It takes its name from Friedrich Konrad Beilstein, who published the first edition in 1881–1882. Many readers of this book will already have used, seen, or heard about this essential tool for organic chemists. It is available (or should be) in every medium-sized or large chemistry library. Furthermore, there is a recent guide and other aids (9) that describe *Beilstein* and how to use it in detail. Accordingly, it would be redundant to present an in-depth description of *Beilstein* here and more appropriate to concentrate on key features. Sample pages from *Beilstein* are shown in Figure 12.1.

  Beilstein celebrated its 100th anniversary in May, 1981. Ever since its origins, its value to chemists throughout the world has continued to increase. This is a major undertaking, with more than 300 (in 1986) volumes occupying more than 48 linear feet.

  The Beilstein Institute is located in the Carl Bosch Haus in Frankfurt-am-Main*, several hundred yards from the Frankfurt fairgrounds and exhibition halls and near the University of Frankfurt. Also located in the same large and modern building are other important chemical information organizations, including the Gmelin Institute (see Section 12.5.A), and, until recently, the Internationale Dokumentationsgesellschaft für Chemie. The latter is an industrial consortium that provides sophisticated chemical structure searching and other information services to members (see Section 10.11.B), and they have since moved to Sulzbath, West Germany.

  Beilstein has a staff of 160, of whom 110 have a doctorate in chemistry or equivalent. In addition, there have been more than 500 outside contributors.

  Since 1978, the Managing Director* of the Beilstein Institute has been Reiner Luckenbach who came to *Beilstein* from the University of Mainz where he continues to instruct students on special topics in organic chemistry. Table 12.3 lists the editors of the Beilstein Handbook.

  One reason why *Beilstein* is so highly regarded by chemists is that it covers organic chemistry from the beginning up to within about 20 years of the present; the most recent Supplementary Series covers 1960–1980. It includes in one place (actually several adjacent volumes including supplements) much of the most pertinent data (primarily physical and chemical properties) about a very large number of organic compounds. *Beilstein* editors critically review the data to be included, and the procedures used for this review are well described in an important paper by *Beilstein's* Managing Director (10).

  For the individual laboratory chemist to attempt to locate in other ways all of the data included in *Beilstein* about specific compounds of interest is

*In 1985, Luckenbach was named President, Scientific Content/Printed Media Division and Clemens Jochum was named President, Electronic Data Processing Department.

2,3-Epoxy-bornane, 1,8,8-Trimethyl-3-oxa-tricyclo[3.2.1.0²·⁴]octane $C_{10}H_{16}O$.

a) **(1R)-2endo,3endo-Epoxy-bornane,** formula XI (E III/IV 331).

Prep. From (1R)-2endo-(toluene-4-sulfonyloxy)-bornan-3endo-ol [KOBuᵗ/BuᵗOH] (*Cole et al.,* Aust. J. Chem. **25** [1972] 361, 363). — mp: 175—176°; [α]_D: +16° [cyclohexane; c = 6]; ¹H-NMR.

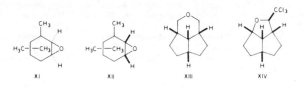

b) **(1S)-2endo,3endo-Epoxy-bornane,** formula XI [mirror image].

Confirmation of configurative purity (*Chabudziński, Sędzik-Hibner,* Bull. Acad. Pol. Sci. Ser. Sci. Chim. **17** [1969] 343).

Prep. From (1R)-born-2-ene and 4-NO₂-C₆H₄CO₃H (*Borowiecki, Chrétien-Bessière,* Bull. Soc. Chim. Fr. **1967** 2364, 2368).

mp: 175—176°; [α]²⁰ from 589 nm (−6.5°) to 365 nm (−19.04°) [CHCl₃; c = 3.4]; ¹H-NMR (*Bo., Ch.-Be.*).

c) **(1R)-2exo,3exo-Epoxy-bornane,** formula XII.

Prep. From (1R)-bornane-2exo,3exo-diol, by sequential treatment with (i) TsCl/Py, and (ii) KOBuᵗ/BuᵗOH (*Cole et al.*). — Waxy solid; [α]_D: +5° [cyclohexane; c = 4.2]; ¹H-NMR.

(3ar,5ac,7ac,7bc)-Decahydro-pentaleno[1,6-cd]pyran $C_{10}H_{16}O$, formula XIII.

Prep. From 1t,6t-bis-hydroxymethyl-(3ar,6ac)-octahydro-pentalene, by sequential treatment with (i) TsCl/Py, and (ii) NaH/THF (*Paquette et al.,* J. Am. Chem. Soc. **97** [1975] 6124, 6131). From 2a,7b-dihydro-5H,7H-pentaleno[1′,6′:1,3,2]cyclopropa[1,2-c]furan, by hydrogenation on Rh (*Pa. et al.*). — mp: 71—72°; ¹H-NMR; IR.

2c-Trichloromethyl-(2ar)-octahydro-pentaleno[1,6-bc]furan $C_{10}H_{13}Cl_3O$, formula XIV + mirror image.

Prep. From cycloocta-1c,5c-diene and chloral [AlCl₃; CS₂] (*Fritz et al.,* Helv. Chim. Acta **58** [1975] 1345, 1350, 1352, 1356). — bp_{0.04}: 80.5—83°; ¹H-NMR; ¹³C-NMR; IR.

(3S)-3,6t-Dimethyl-octahydro-3r,5c-cyclo-benzofuran, (1R,2R,4R)-4,8-Epoxy-pinane $C_{10}H_{16}O$, formula I.

Configuration inferred from the genetic relationship with (+)-nopinone [E IV 7 176] (*P. Magnus,* personal communication, 1983).

Prep. From (−)-neoverbanol ((1R,2R,4R)-pinan-4-ol), by treatment with HgO/Br₂ [tungsten lamp] (*Hobbs, Magnus,* J. Chem. Soc. Perkin Trans. 1 **1973** 2879). — bp₅: 62°; [α]_D^{21.5}: +62.7° [CHCl₃; c = 4]; ¹H-NMR; IR.

(3S)-3,7a-Dimethyl-octahydro-3,5-cyclo-benzofuran, (1R,2R)-2,8-Epoxy-pinane $C_{10}H_{16}O$, formula II (X = H).

Prep. From (1R,2R)-pinan-2-ol, by treatment with HgO/Br₂ (*Grison, Bessière-Chretien,* Bull. Soc. Chim. Fr. **1972** 4570, 4574; *Gibson, Erman,* J. Am. Chem. Soc. **91** [1969] 4771, 4774). — bp_{10.5}: 64°; n_D^{26}: 1.4702; [α]_{546}: +51.7° [EtOH; c = 3]; ¹H-NMR; IR (*Gi., Er.*).

Reaction with BF₃/Et₂O [Ac₂O]: product distribution (*Gr., Be.-Ch.*). Reaction with Ac₂O/Py· HCl: product distribution (*Bessière-Chrétien, Grison,* Bull. Soc. Chim. Fr. **1975** 2499).

(3S)-7a-Chloromethyl-3-methyl-octahydro-3,5-cyclo-benzofuran $C_{10}H_{15}ClO$, formula II (X = Cl). *Prep.* From (+)-nopinone (E IV 7 176) [several steps] (*Gi., Er.*). — bp_{1.3}: 65°; n_D^{26}: 1.4969; [α]_{546}: +46.2° [EtOH; c = 2.17]; ¹H-NMR.

Figure 12.1. Representative pages from *Beilstein Handbook of Organic Chemistry*. Volume 17/1 Fifth Supplementary Series. This copyrighted material is reprinted with permission of the Director of the Handbook and the publisher, Springer-Verlag, Heidelberg, W. Germany, and reduced in size from original publication.

et al., Helv. Chim. Acta **54** [1971] 1845, 1851, 1861). − $bp_{0.001}$: $67-69°$; d_4^{20}: 0.9924; n_D^{20}: 1.4990; ¹H-NMR; ¹³C-NMR.

6,7-Epoxy-3-isopropyl-6,8a-dimethyl-1,2,4,5,6,7,8,8a-octahydro-azulene $C_{15}H_{24}O$.

a) **(8a*R*)-6*c*,7*c*-Epoxy-3-isopropyl-6*t*,8a-dimethyl-(8a*r*)-1,2,4,5,6,7,8,8a-octahydro-azulene,** formula II.

Prep. From (3a*S*)-6*c*,7*c*-epoxy-3*c*-isopropyl-6*t*,8a-methyl-(3a*r*,8a*c*)-decahydro-azulen-3a-ol and SOCl₂/Py (*de Broissia et al.,* Bull. Soc. Chim. Fr. **1972** 4314, 4318). − $[\alpha]_D$: +9° [CHCl₃; c = 2.2]; ¹H-NMR.

b) **(8a*R*)-6*t*,7*t*-Epoxy-3-isopropyl-6*c*,8a-dimethyl-(8a*r*)-1,2,4,5,6,7,8,8a-octahydro-azulene.**

Prep. From (8a*R*)-3-isopropyl-6*c*,8a-dimethyl-(8a*r*)-1,2,4,5,6,7,8,8a-octahydro-azulene-6*t*,7*c*-diol, by sequential treatment with (i) TsCl/Py, and (ii) KOH/MeOH (*Gonzalez et al.,* Phyto≠chemistry **16** [1977] 265). − ¹H-NMR; MS.

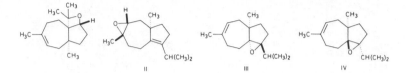

3,3a-Epoxy-3-isopropyl-6,8a-dimethyl-1,2,3,3a,4,5,8,8a-octahydro-azulene $C_{15}H_{24}O$.

a) **(3a*R*)-3*c*,3a-Epoxy-3*t*-isopropyl-6,8a-dimethyl-(3a*r*,8a*c*)-1,2,3,3a,4,5,8,8a-octahydro-azu≠lene,** formula III.

Prep. [Together with b)] from (+)-daucene [(8a*R*)-3-isopropyl-6,8a-dimethyl-1,2,4,5,8,8a-hexahydro-azulene] and monoperoxyphthalic acid (*de Broissia et al.,* Bull. Soc. Chim. Fr. **1972** 4314, 4317). − $[\alpha]_D$: +54° [CHCl₃; c = 1.2]; ¹H-NMR.

b) **(3a*S*)-3*c*,3a-Epoxy-3*t*-isopropyl-6,8a-dimethyl-(3a*r*,8a*t*)-1,2,3,3a,4,5,8,8a-octahydro-azu≠lene,** formula IV.

$[\alpha]_D$: +29° [CHCl₃; c = 1.3]; ¹H-NMR.

***ent*-(11Ξ)-8β,12-Epoxy-(10β*H*)-cadin-4-ene, (3a*S*)-1ξ,5*c*,8-Trimethyl-(3a*r*,5a*c*,9a*t*,9b*c*)-1,2,3a,4,5,5a,6,7,9a,9b-decahydro-naphtho[2,1-*b*]furan,** Brachyl oxide $C_{15}H_{24}O$, formula V.

Isolation from Brachylaena hutchinsii (*Klein, Schmidt,* J. Agric. Food Chem. **19** [1971] 1115). − d_4^{20}: 0.9983; n_D^{20}: 1.4992; $[\alpha]_D^{20}$: +28°; ¹H-NMR spectrum; IR.

3ξ,4-Epoxy-(4ξ*H*)-cadin-1(10)-ene, (4a*S*)-2ξ,3-Epoxy-5*c*-isopropyl-3ξ,8-dimethyl-(4a*r*)-1,2,3,4,4a,5,6,7-octahydro-naphthalene $C_{15}H_{24}O$, formula VI.

Prep. From cadina-1(10),3-diene and monoperoxyphthalic acid (*Vlahov et al.,* Collect. Czech. Chem. Commun. **32** [1967] 822, 826). − mp: $46-48°$.

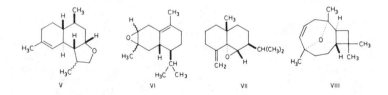

5,6α-Epoxy-eudesm-4(15)-ene $C_{15}H_{24}O$, formula VII.

Prep. From sibirene [eudesma-4(15),5-diene] and peroxybenzoic acid (*Dubovenko et al.,* Sib. Chem. J. (Engl. Transl.) **1969** 478; orig. **1969** (4) 108; CA **72** [1970] 67127). − $[\alpha]_D^{24}$: +54.2° [CHCl₃; c = 3.1]; ¹H-NMR; IR.

Figure 12.1. Continued.

by heating to 100° (*Ishii et al.*, Chem. Pharm. Bull. **20** [1972] 203). — Liquid; ¹H-NMR.

**2,2,4,8-Tetramethyl-(3ac,9ac)-2,3,3a,4,5,8,9,9a-octahydro-1H-4r,8c-epioxido-cyclopenta‐
cyclooctene** $C_{15}H_{24}O$, formula XV + mirror image.

Prep. [Together with other products] from humulene (E IV **5** 1171), by sequential treatment
with (i) $Hg(OAc)_2$ [aq. THF], and (ii) $NaBH_4$ (*Misumi et al.*, Tetrahedron Lett. **1976** 2865).
— ¹³C-NMR (*Misumi et al.*, Tetrahedron Lett. **1979** 35). ¹H-NMR; IR (*Mi. et al.* 2685).

***ent*-5,11-Epoxy-(7βH)-eudesm-3-ene, α-Agarofuran** $C_{15}H_{24}O$, formula XVI.

Configuration: see *Barrett, Büchi*, J. Am. Chem. Soc. **89** [1967] 5665; *Marshall, Pike*, J.
Org. Chem. **33** [1968] 435.

Isolation from fungus infected agar wood [Aquillaria agallocha] (*Maheshwari et al.*, Tetrahe‐
dron **19** [1963] 1079, 1087), from Amyris balsamifera (*Rohmer et al.*, Phytochemistry **16** [1977]
773), from Asarum europaeum (*Biering, Jork*, Planta Med. **37** [1979] 137, 140).

Prep. From 10α-eudesma-4,11-dien-3-one [several steps] (*Ba., Bü.; Ma., Pike*). From 5-hy‐
droxy-(10α)-eudesm-11-en-3-one [several steps] (*Asselin et al.*, Can. J. Chem. **46** [1968] 2817;
Huffman et al., J. Org. Chem. **41** [1976] 3705, 3708).

bp_6: 134° [bath temp.]; n_D^{30}: 1.5061; $[\alpha]_D^{30}$: +37° [$CHCl_3$; c = 6.12] (*Ma. et al.*). $bp_{2.3}$:
120°; n_D^{22}: 1.5087; $[\alpha]_D^{24}$: +41.5° [$CHCl_3$; c = 3.6] (*Ba., Bü.*). ¹H-NMR spectrum; IR spectrum
(*Ma. et al.* 1080, 1083). ¹³C-NMR spectrum (*Bi., Jork* 139).

Photo-initiated Wagner-Meerwein rearrangement (*Thomas, Ozainne*, Helv. Chim. Acta **59**
[1976] 1243).

5,11-Epoxy-(4ξ,5β,10α)-eudesm-2-ene $C_{15}H_{24}O$, formula I. mp: 30−32°, $[\alpha]_D^{24}$: +18.7°
[$CHCl_3$; c = 7]; ¹H-NMR; IR (*Ba., Bü.*).

***ent*-5,11-Epoxy-(7βH)-eudesm-4(15)-ene, β-Agarofuran** $C_{15}H_{24}O$, formula II.

Configuration: see *Marshall, Pike*, J. Org. Chem. **33** [1968] 435.

Isolation from fungus infected agar wood [Aquillaria agallocha] (*Maheshwari et al.*, Tetrahe‐
dron **19** [1963] 1079, 1087).

Prep. From α-agarofuran [see above], by UV irradiation (*Ma., Pike*). From (1R)-5-chloro-
4exo,7,7-trimethyl-4endo-pent-4-ynyl-6-oxa-bicyclo[3.2.1]octane, by sequential treatment with
(i) EtMgBr, (ii) Me_3SiCl, (iii) Bu_3^nSnH [α,α'-diazenediyl-di-isobutyronitrile], and (iv) toluene-4-
sulfinic acid (*Büchi, Wüest*, J. Org. Chem. **44** [1979] 546).

bp_8: 130° [bath temp.]; d_4^{30}: 0.9646; n_D^{28}: 1.4973; $[\alpha]_D^{30}$: −127.1° [$CHCl_3$; c = 8] (*Ma. et al.*).
$bp_{0.1}$: 55°; $[\alpha]_D^{25}$: −151° [$CHCl_3$; c = 3.8] (*Bü., Wü.*). ¹H-NMR spectrum; IR spectrum (*Ma.
et al.* 1080, 1083).

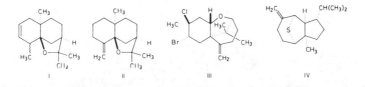

**(2S)-7c-Bromo-8t-chloro-8c,10,10-trimethyl-5-methylene-(9at)-octahydro-2r,5ac-methano-
benzo[b]oxepine, Nidifocene** $C_{15}H_{22}BrClO$, formula III.

Position of Cl and Br: see *Waraszkiewicz et al.*, Tetrahedron Lett. **1977** 2311.

Isolation from Laurencia nidifica (*Waraszkiewicz, Erickson*, Tetrahedron Lett. **1976** 1443).

mp: 79−81° [from hexane]; crystal structure determination (*Wa. et al.*). ¹H-NMR (*Wa.,
Er.*).

(4R)-1c-Isopropyl-3a-methyl-7-methylene-(3ac,8ac)-decahydro-4r,8c-episulfano-azulene, M i n t
s u l f i d e $C_{15}H_{24}S$, formula IV.

Isolation from Mentha piperita (*Yoshida et al.*, J. Chem. Soc. Chem. Commun. **1979** 512).

Prep. From germacrene-D [(8S,[1,2,3,4]M,[6,7,8,9]P)-8-isopropyl-1-methyl-5-methylene-

Figure 12.1. Continued.

TABLE 12.3. Editors of the *Beilstein Handbook*

Years	Editor	Years	Editor
1881–1896	F. K. Beilstein	1933–1961	F. Richter
1896–1918	P. Jacobson	1961–1978	H. G. Boit
1918–1923	P. Jacobson and B. Prager	1978–	R. Luckenbach*
1923–1933	B. Prager and F. Richter		

*In 1986, Luckenbach was named President, Scientific Content/Printed Media Division and Clemens Jochum was named President, Electronic Data Processing Department.

often prohibitively time consuming, and all of the journals and other sources that would need to be looked at to accomplish such a task are frequently not readily available. *Beilstein* is thus an excellent starting point, and its use frequently permits the chemist to concentrate on obtaining more current and pertinent data as needed.

Industrial chemists, in particular, will frequently need to augment their use of *Beilstein* with additional examination of the literature. *Beilstein* is not a prime source of commercial applications or use data, even though patent literature is included. It is, however, an extremely important source to *all* chemists, industrial and other. Table 12.4 is a convenient guide to the several series in which *Beilstein* appears.

TABLE 12.4. *BEILSTEIN* Series Coverage

Series	Abbreviations	Literature Years Covered	Spine Label Color	Status
Basic Series	H	Up to 1909	Green	Complete
Supplementary Series I	EI	1910–1919	Dark red (on brown cover)	Complete
Supplementary Series II	EII	1920–1929	White	Complete
Supplementary Series III	EIII	1930–1949	Blue	Complete
Supplementary Series III/IV[a]	EIII/IV	1930–1959	Blue/black	To be complete in 1986
Supplementary Series IV	EIV	1950–1959	Black	To be complete in 1987
Supplementary Series V[b] (in English)	EV	1960–1979	Red (on blue cover)	Begins in 1984

[a] Volumes 17–27 of Supplementary Series III and IV, covering heterocyclic compounds, are combined in a joint issue.
[b] The first volumes to be published in this series relate to heterocyclic compounds (Vol. 17–27), based on a survey of user requirements. Other volumes are to follow thereafter.

For the time period until 1960, approximately 1.5 million compounds are included in *Beilstein*. This compares with the present (1986) total of more than 7 million compounds in the CAS Registry System. These two sources define "compound" differently. *Beilstein* includes those species whose constitution is fully and undoubtedly established. Although the CAS Registry System includes many more compounds than *Beilstein,* at least one recent unpublished study indicates that a significant number of compounds are in *Beilstein* but not yet registered by CAS.

As noted earlier, *Beilstein's* coverage goes back to the beginnings of organic chemistry. Ongoing supplementing brings the data up to more recent years as shown in the compilation of data about the *Beilstein* series in Table 12.4. For each chemical compound included, selected information is presented on, for example:

Structure of the molecule

Natural occurrence and isolation from natural products

Preparation and manufacture

Physical properties alone and in mixture with other compounds

Chemical properties

Analytical methods

Salts and addition compounds

Input is based on a critical review of the published literature, including journals, patents, reports, and other publications (10). Each piece of data is accompanied by reference to the publication that is the source of the data. See also Section 15.14.

Beilstein editors present material in a highly condensed, telegraphic form and make extensive use of abbreviations as shown in Table 12.5. A list of abbreviations and symbols is provided in the front of each volume.

To further assist in use of the text and of abbreviations in those *Beilstein* volumes in German, the publisher makes available a free German–English pocketbook dictionary (11). This small book, aimed specifically at *Beilstein* users, can be obtained by writing the publisher, Springer Verlag,* as can other user aids (9).

Any additional language problems are expected to become virtually nonexistent as the volumes of Supplementary Series V, totally in English, become widely available.

Chemists can find compounds in *Beilstein* in three ways: (a) molecular formula index, (b) subject index, and (c) *System Number* based on a kind of hierarchy or classification unique to *Beilstein*. Each volume of *Beilstein* contains a subject (actually compound) and formula index for that volume. In addition, there are cumulative or collective indexes that appear periodi-

TABLE 12.5. Some Abbreviations Frequently Used in *Beilstein*

Abbreviation	Meaning	Abbreviation	Meaning
A.	Ethanol	opt.-inakt.	Optically inactive
Acn.	Acetone	PAe.	Petroleum ether
Ae.	Diethyl ether	Py	Pyridine
alkal.	Alkaline	RRI	*Ring Index* (2nd
Anm.	Annotation, footnote		ed., 1960)
B.	Formation, synthesis	RIS	*Ring Index* Supplement
Bd.	Volume	S.	Page
Bzl.	Benzene	s.	See
Bzn.	Petroleum ether	s	Second
bzw.	Or, respectively	s.a.	See also
Diss.	Dissertation	s.o.	See above
E	Supplemental volume	sog.	So-called
	in *Beilstein* series	Spl.	Supplement
E	Freezing point	. . . stdg.	For . . . hours
E.	Ethyl acetate	s.u.	See below
Eg.	Acetic acid	Syst. Nr.	System Number in
engl.	English edition		*Beilstein*
Ausgabe		T1.	Part
Gew-%	Percentage by weight	U.S.P.	U.S. Patent
H	Main volume in	unkorr.	Uncorrected
	Beilstein series	unverd.	Undiluted
h	Hour(s)	verd.	Diluted
konz.	Concentrated	vgl.	Compare
korr.	Corrected	w.	Water
Kp	Boiling point	wss.	Aqueous
Me.	Methanol	z.B.	For example
min	Minute	Zers.	Decomposition

cally. Work on the next edition of an updated general subject and formula index, covering all 27 volume numbers, is to begin soon. This will include all compounds in *Beilstein* through the end of the fourth supplement (1959).

Of the several approaches to locating a compound in *Beilstein,* the formula index is the quickest and most reliable. Use of the subject index is a second choice because of the potential vagaries of chemical nomenclature. *Beilstein* editors are, however, aware of these vagaries and have attempted a compromise by providing more than one name (but by no means all) for a single compound in the subject index.

Once a compound is located in the general index, it should be a relatively simple matter to locate the same compound in later volumes that may not yet be in a general index. The chemist need merely use the same *Beilstein*

System Number, and the search procedure is further aided by the *coordinating references* found at the top of each odd-numbered page. But all of these steps may still not locate the desired compound quickly, especially if the desired compound is described in *Beilstein* for the first time in a newly published volume, and is not yet covered in a general index. In this case, the chemist can use such additional aids as indexes and tables of contents in individual volumes.

Finally, as a last resort—short of contacting *Beilstein* editors for advice—the chemist can look for the compound of interest using the *Beilstein* System rules. A letter to the publisher can obtain free literature on use of the System (and other features) of *Beilstein*. This literature goes into considerable detail on how to use the System. The first principle is that the *Beilstein* volumes (actually groups of volumes) are divided as follows: acyclic compounds, Volumes 1–4; isocyclics, Volumes 5–16; and heterocyclics, Volumes 17–27. Further determination on placement of compounds is then based primarily on types and numbers of functional groups according to a series of rules. For chemists who want to learn how to use the System efficiently, a few hours of study and practice are strongly recommended.

Plans call for 1986 availability of a floppy disk for personal computers that will help chemists find the correct volumes in *Beilstein* based on electronic input (drawing) of chemical structures. This software package has been given the name *SANDRA* (Structure and Reference Analyzer). Online availability of data from the *Beilstein* volumes is currently scheduled for 1987 in cooperation with one or more of the major online systems. The most recent of the published volumes have been printed using electronic typesetting and are therefore computer-readable. However, the online file is to include all of the compounds and most of the data and references in *Beilstein*. In addition, *Beilstein* plans to make its computerized system available for in-house use so that organizations can mount the system on their own computers. Finally, subsets are to be available in disk form for personal computers, and, in yet another product, users will be able to record their proprietary structures using personal computers.

Sets of *Beilstein* are available to chemists in hundreds of major university and industrial libraries throughout the United States, and many other sets are located throughout the world. The cost of maintaining an up-to-date set of *Beilstein* is about $10,000 per year.* This expenditure would permit purchase of all published volumes each year. This figure is mentioned here because, in many organizations, availability of funds, or lack of such, could be a major factor in the hoped for continuing widespread availability of *Beilstein*. A review of a recent volume of *Beilstein* by Herbert O. House is a lucid discussion of the cost situation and presents cogent arguments for continued subscription to both *Beilstein* and *Chemical Abstracts* (12). When the plan for computerized access is fully implemented, this should be an important step to making the content of *Beilstein* more readily and quickly available to working chemists.

*In 1986, cost rose to $40,000 per year.

B.   *Dictionary of Organic Compounds*

The *Dictionary of Organic Compounds* (13) is a seven volume set (plus supplements) that has concise data on many common compounds. Completion of the new fifth edition of this work was first announced at a Symposium of the American Chemical Society on September 16, 1982 in Kansas City, MO.

Some older chemists may still prefer to identify these volumes as *Heilbron,* after the late Ian Heilbron, Professor of Chemistry at the University of Manchester, who edited the first three editions beginning in 1934. His name remains associated with the publication, but the more correct designation is now *DOC,* an acronym taken from the words in the title. The Executive Editor of the fifth edition is John Buckingham.

A comparison of *DOC* with *Beilstein* is helpful in providing perspective. *DOC* covers many chemicals, but far fewer than *Beilstein* (about 150,000 in the main volumes versus approximately 1.5 million in *Beilstein*). As an example of an omission that the editors of *DOC* plan to correct, certain chemicals of commerce such as toluene diisocyanate are not listed in the first seven volumes. In addition, data provided are much less complete than those in *Beilstein*.

Although *DOC* is not intended, nor should it be used, as a direct replacement for *Beilstein,* it does have a number of attractive features. One is its relatively low original cost ($1950), which may make it more readily available in chemistry libraries than the much more expensive *Beilstein*. Chemists should find *DOC* handy for quick look-up of basic data such as chemical constitution, some selected physical and chemical properties, references to some preparative methods, types of uses, and hazards. Figure 12.2 shows a typical entry in *DOC*.

In addition, some of the data in *DOC* may be more recent than those in *Beilstein*. The publishers claim a closing date of 1980 for the original volumes, with many references as recent as 1981 and 1982. Currency is further enhanced by the planned issuance of supplements on an annual basis, beginning in 1983, and by the use of computer-assisted typesetting. Because *Beilstein* editors are also striving to be more current in the 1980s, a number of years may go by before a completely objective comparison on this point can be made.

Finally, many chemists find *DOC* easy to use. The main volumes have a straightforward alphabetical arrangement, and there are indexes by chemical name (including systematic, trivial, and trade names) molecular formula, heteroatom, and CAS Registry Number.

The text of *DOC* was made available online in 1986 on *DIALOG*® Information Services; another file planned to be introduced later is to include the structures. In addition, a series of paperback subsets of *DOC* was initiated in 1985. These deal with discrete topics such as amino acids and peptides, carbohydrates, and others. The volumes are part of the *Chapman and Hall*

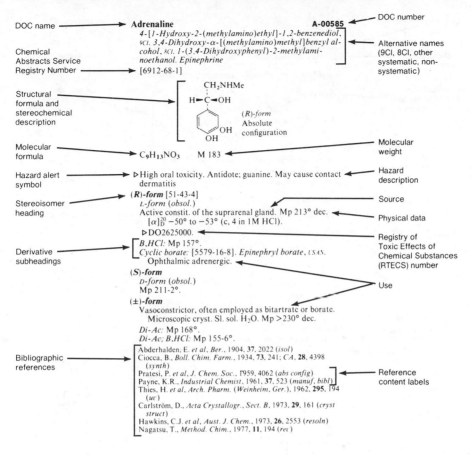

Figure 12.2. Sample entry from *Dictionary of Organic Chemistry*, fifth ed. Reprinted with permission of Methuen, Inc., Chapman & Hall.

Chemistry Sourcebooks series and also contain some information not found in *DOC*.

C. Some Other Important Organic Chemistry Reference Works

1. A major English-language reference work in organic chemistry is *Rodd's Chemistry of Carbon Compounds* (14), the second edition of which began publication in 1964 under the editorship of S. Coffey or M. F. Ansell. This is an outstanding work of broad scope. Although not as comprehensive as *Beilstein,* it is, in many fields, more current, and provides an excellent starting point.

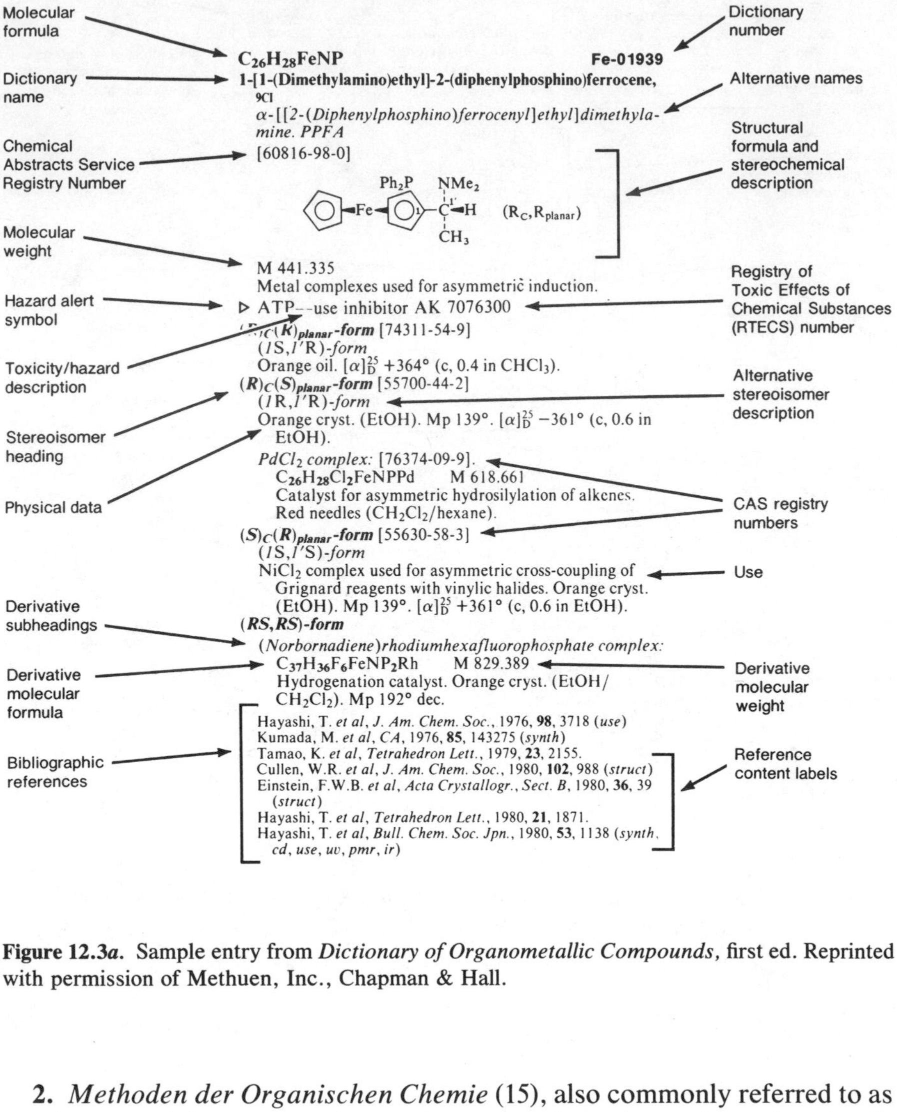

Molecular formula

Dictionary name

Chemical Abstracts Service Registry Number

Molecular weight

Hazard alert symbol

Toxicity/hazard description

Stereoisomer heading

Physical data

Derivative subheadings

Derivative molecular formula

Bibliographic references

Dictionary number

Alternative names

Structural formula and stereochemical description

Registry of Toxic Effects of Chemical Substances (RTECS) number

Alternative stereoisomer description

CAS registry numbers

Use

Derivative molecular weight

Reference content labels

$C_{26}H_{28}FeNP$ Fe-01939
1-[1-(Dimethylamino)ethyl]-2-(diphenylphosphino)ferrocene,
9Cl
α-[[2-(Diphenylphosphino)ferrocenyl]ethyl]dimethyla-
mine. *PPFA*
[60816-98-0]

Ph$_2$P   NMe$_2$

Fe — C⟵H   (R_C,R_{planar})
CH$_3$

M 441.335
Metal complexes used for asymmetric induction.
▷ ATP--use inhibitor AK 7076300
$(S)_C(R)_{planar}$-*form* [74311-54-9]
$(1S,1'R)$-*form*
Orange oil. $[\alpha]_D^{25}$ +364° (c, 0.4 in CHCl$_3$).
$(R)_C(S)_{planar}$-*form* [55700-44-2]
$(1R,1'R)$-*form*
Orange cryst. (EtOH). Mp 139°. $[\alpha]_D^{25}$ −361° (c, 0.6 in
EtOH).
PdCl$_2$ complex: [76374-09-9].
$C_{26}H_{28}Cl_2FeNPPd$ M 618.661
Catalyst for asymmetric hydrosilylation of alkenes.
Red needles (CH$_2$Cl$_2$/hexane).
$(S)_C(R)_{planar}$-*form* [55630-58-3]
$(1S,1'S)$-*form*
NiCl$_2$ complex used for asymmetric cross-coupling of
Grignard reagents with vinylic halides. Orange cryst.
(EtOH). Mp 139°. $[\alpha]_D^{25}$ +361° (c, 0.6 in EtOH).
(RS,RS)-form
(*Norbornadiene*)*rhodiumhexafluorophosphate complex:*
$C_{37}H_{36}F_6FeNP_2Rh$ M 829.389
Hydrogenation catalyst. Orange cryst. (EtOH/
CH$_2$Cl$_2$). Mp 192° dec.
Hayashi, T. *et al, J. Am. Chem. Soc.*, 1976, **98**, 3718 (*use*)
Kumada, M. *et al, CA*, 1976, **85**, 143275 (*synth*)
Tamao, K. *et al, Tetrahedron Lett.*, 1979, **23**, 2155.
Cullen, W.R. *et al, J. Am. Chem. Soc.*, 1980, **102**, 988 (*struct*)
Einstein, F.W.B. *et al, Acta Crystallogr.*, Sect. B, 1980, **36**, 39
(*struct*)
Hayashi, T. *et al, Tetrahedron Lett.*, 1980, **21**, 1871.
Hayashi, T. *et al, Bull. Chem. Soc. Jpn.*, 1980, **53**, 1138 (*synth.
cd, use, uv, pmr, ir*)

Figure 12.3a. Sample entry from *Dictionary of Organometallic Compounds,* first ed. Reprinted with permission of Methuen, Inc., Chapman & Hall.

2. *Methoden der Organischen Chemie* (15), also commonly referred to as *Houben–Weyl,* is an extensive set (over 60 volumes) now in its fourth edition, which began in 1952. It gives selected methods for preparation of many organic compounds, in considerable detail, so that consultation of the original reference is frequently not necessary. Publication of supplementary and additional volumes is to be complete in 1991. To date, the volumes are entirely in German, an important barrier for many.

3. A six-volume *Comprehensive Organic Chemistry* (16), edited by Derek Barton, was published in 1983.

4. *Organic Syntheses* (17) is an annual compilation of "satisfactory" and

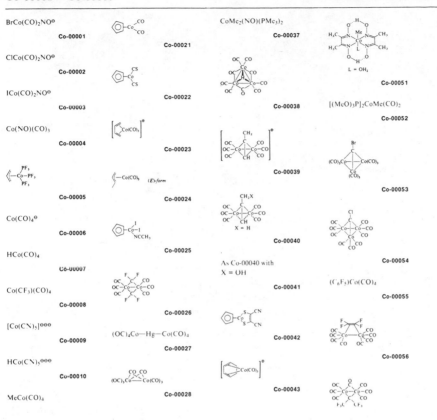

Figure 12.3b. Structure Index page from the *Dictionary of Organometallic Compounds,* first ed. Reprinted with permission of Methuen, Inc., Chapman & Hall.

checked laboratory methods for preparing organic compounds in one-half to five-pound lots. There is a collective volume every 10 years that revises and updates the annual volumes where necessary. Collective Volume 6 (1983) includes cumulative indexes for Collective Volumes 1–5. An effort by Molecular Design, Limited, has made much of the content of the series computer accessible (see Section 10.11.A.4).

Recent volumes, beginning with Volume 41 (1961), emphasize

widely applicable, model procedures that illustrate important types of reactions; . . . many of the procedures selected . . . have major significance in the synthetic method, rather than in the product that results. However, preparations of reagents and products of special interest are also included, as in previous volumes (18).

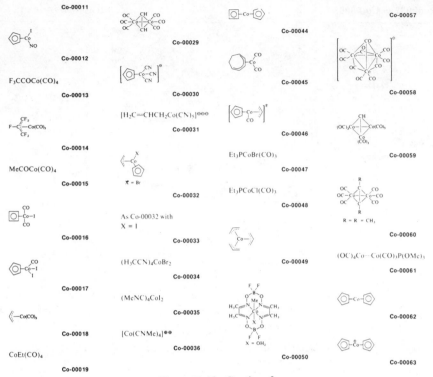

Figure 12.3b. Continued

The change in policy is easily explained. When the series was founded by Roger Adams at the end of World War I, there was an urgent need for preparing useful compounds in reasonable quantities, because supplies of organic chemicals from abroad were no longer available. Today, a wide variety of fine organics is commercially available, and reactions can be studied with decreasing starting material quantities because of improved techniques.

5. *Organic Reactions* (19) volumes, edited by William G. Dauben, (University of California at Berkeley), are "collections of chapters each devoted to a single reaction, or a definite phase of a reaction, of wide applicability." Emphasis is on the preparative aspects; several detailed procedures are given for each method. Many examples and extensive bibliographies are included. Beginning with Volume 22 (1975), a program of updating earlier reviews was initiated. Volume 30 (1984) includes author, chapter, and topic indexes for Volumes 1–30 and a subject index for Volume 30.

6. *Survey of Organic Syntheses* (20) originally edited by Buehler and Pearson, is an example of several similar works that "bring together the principal methods for synthesizing the main types of organic compounds . . . How the functional group is created from other functional groups is the main concern of [this] work." The third volume of this series is edited by Robert Williams of Colorado State University.

7. One of the most complete collections of chemical reactions in a single place is *Synthetic Methods of Organic Chemistry* (21). This is often referred to simply by the last name of the original editor, William Theilheimer, but it is now edited by A. F. Finch, Derwent Publications, Ltd. The set began in 1948 (Vol. 1 covers literature from 1942–1944), and the 40 volumes published or scheduled to date (1986) contain approximately 35,000 chemical reactions.

8. A related source is the *Journal of Synthetic Methods* for which Theilheimer serves as Consulting Editor. This monthly publication of Derwent Publications, Ltd. provides informative abstracts of approximately 6000 reactions per year and is intended as a direct supplement to the previously published books.

9. *Chemical Reactions Documentation Service,* another Derwent product, provides computer-based access to over 50,000 reactions in Volumes 1–30 (1942–1974) of the original Theilheimer work and its continuation in the *Journal of Synthetic Methods* from 1975 to date. Online searching is via SDC®'s *ORBIT®* (see Section 10.4) for subscribers to the *Journal,* or the magnetic tapes can be purchased for in-house use. Additional computer access to Theilheimer's volumes from 1942 is being developed by Molecular Design, Limited (see Section 10.11.A.4).

10. *Compendium of Organic Synthetic Methods* (22), originally edited by Harrison and Harrison, and now by Leroy G. Wade (Colorado State University), is a compilation of published organic function group transformations. Synthetic methods are presented in the form of reactions with references.

11. A nine-volume *Comprehensive Organometallic Chemistry,* edited by Geoffrey Wilkinson, was published in 1982 (23).

12. Chapman and Hall, the publishers of *DOC*, have announced a three-volume *Dictionary of Organometallic Compounds* (24) under the editorship of John Buckingham. Contents are sequenced by molecular formula rather than compound name, and overall arrangement is element by element, with a short introduction to each by a specialist in the field. Figure 12.3*a* shows a typical entry. A unique structure index (see Figure 12.3*b*) follows each section, and there are indexes by chemical name, molecular formula, and CAS Registry Number. The text of this publication is searchable online via *DIALOG®*. Plans call for the publication of annual supplements and also a series of paperback volumes, each covering individual families of the main work. These paperbacks are part of the series *Chapman and Hall Chemistry Sourcebooks,* and organometallic compounds covered in 1985 are in the following groupings:

Aluminum, gallium, indium and thallium
Boron

Cobalt, rhodium and iridium

Germanium, tin and lead

Iron

Lanthanides, actinides and early transition metals

Nickel, palladium, platinum, copper, silver and gold

Ruthenium and osmium

Silicon

Zinc, cadmium and mercury

A new *Dictionary of Organophosphorous Compounds* is planned by Chapman and Hall for publication in 1986 or 1987.

13. Within *Gmelin* (see Section 12.5.A), a supplementary work attempts to build a bridge between classical organic and inorganic chemistry. Thus, such volumes on metalloorganic chemistry as the following examples are available:

Chromium—organic bonds

Cobalt—organic bonds

Hafnium–organic bonds

Iron–organic bonds

Nickel–organic bonds

Vanadium–organic bonds

Zirconium–organic bonds

14. Another important work is *Fieser and Fieser's Reagents for Organic Synthesis* (25). This well-known multivolume series covers the reagent literature. As of 1986, 12 volumes had appeared covering the literature from 1967 through 1982. A collective index covering these volumes is in preparation.

15. John Wiley & Sons, Inc., New York, have been publishing a series of monographs on the chemistry of heterocyclic compounds (26), originally under the editorship of the late Arnold Weissberger, who was for many years with Eastman Kodak Co., Rochester, NY. This extensive series goes back a number of years and is still in progress.

16. An eight-volume work on heterocyclic chemistry, edited by Alan R. Katritzky and Charles W. Rees, was published in 1984 (27) by Pergamon Press.

D.   Organic Reactions Sources

This chapter is a convenient place to summarize the principal sources and systems that provide comprehensive access to large numbers of organic reactions:

1. Theilheimer's *Synthetic Methods of Organic Chemistry*.
2. Derwent's *Journal of Synthetic Methods*.
3. Derwent's *Chemical Reactions Documentation Service*.
4. *Current Chemical Reactions®*, a product of *ISI®*.
5. Molecular Design's *REACCS* system.
6. Beckman Instruments* has announced a synthesis library routine designated as SYNLIB. This computer software, which is available for running on an organization's own computers, is intended to help improve literature search efficiency for organic synthesis reactions. The database contains over 26,000 reactions as reported in the literature and is to be updated on a regular basis. Structures of interest may be drawn graphically on the screen using a light pen. Such parameters as reaction conditions and yield, for example, may also be specified. A new and separate company, Distributed Chemical Graphics, has been formed to market this product.

    Both Distributed Chemical Graphics and Molecular Design, Limited plan to make available systems that can be economically operated on personal computers.
7. The Royal Society of Chemistry now offers a monthly current awareness bulletin in the field of synthetic chemistry, *Methods in Organic Syntheses*. This provides coverage of novel and unusual reaction schemes as reported in worldwide journals.
8. Chemical Abstracts Service has announced plans to offer reactions searching capability in addition to what is already available in *CA*.
9. Compendia such as that by Wade cited earlier in this Section.

12.5. INORGANIC CHEMISTRY—SOME IMPORTANT REFERENCE WORKS

A. *Gmelin Handbook of Inorganic Chemistry*

Chemists working on elements and inorganic chemistry and compounds, including organometallic compounds, should usually consult *Gmelin* (28) as the next step after studying any available chemical encyclopedia article on the topic of interest.

As compared to the other classic source of information on inorganic chemistry (*Mellor,* discussed in Section 12.5.B), *Gmelin* is by far the source of first choice. *Gmelin* is better organized, more complete, much more current, and easier to use, and it is now published totally in English. Most of the older *Gmelin* volumes are, however, in German.

The Gmelin Institute is part of the Max Planck Society for the Advancement of Science. It is staffed by about 120 full-time employees, of whom about 80 have doctorates.

Gmelin is located in Frankfurt in the same building as *Beilstein*. The Director (since 1979) is Ekkehard Fluck, who is also a Professor at the University of Stuttgart and an eminent inorganic chemist. Walter Lippert is Deputy Director.

Gmelin presently adds about 30,000 new compounds per year to its base of about 400,000 compounds. In contrast, *Beilstein* has a base of about 1.5 million compounds.

Whereas all *Beilstein* editorial work is in house, some *Gmelin* contributions have been by outside experts, including several from the United States, such as Therald Moeller, the late Earl Muetterties, and Glenn Seaborg.

The *reporting period* for *Gmelin,* like that of the *Landolt–Börnstein Handbook* (see Section 15.9), begins with the literature closing date of the previous volume and extends up to about six months before publication date of the new volume. On the other hand, *Beilstein* divides its reporting period into decades.

There are no plans for a ninth edition of *Gmelin*. Rather, the current (eighth) edition, begun in 1922, will continue to be updated by more recent volumes. This edition now contains well over 400 volumes and more than 130,000 pages. About 100 of the volumes are in English.

Thus, *Gmelin* consists of:

- Main Series (Hauptband) volumes—began in 1922.
- Supplement (Ergänzungsband) volumes—began in 1954. These are now the principal vehicle for the updating of *Gmelin*.
- New Supplement Series volumes—began in 1970. This was the first of the *Gmelin* series to introduce text in English. These volumes should now be found shelved among the regular Supplement volumes of the corresponding elements.

Supplements thus update the coverage of the Main volumes and are frequently much more extensive. For the latest, and possibly the *best,* data, users of *Gmelin* should consult Supplements first, and then, only if needed, Main volumes.

As an aid to using *Gmelin,* the chemist will find that the reverse of the title page for each volume contains the latest date through which literature for that system volume is evaluated. This means that the chemist can elect to begin searching of other sources, if needed, issued subsequent to that date and thereby save considerable time.

Gmelin provides key information in concise form. The original source of the data is referred to, as, in many cases, are abstracts of these sources.

One of the most significant features of *Gmelin* is inclusion of many valuable tables of numerical data, curves, and other graphic material, including diagrams of apparatus. Another significant feature is extensive coverage of the more applied or "practical" aspects. Considerable attention is paid to commercial manufacturing practices. The staff covers patents liberally, but selectively. About 20% of patents considered are included.

Gmelin staff has kept pace with major developments affecting chemists by increasing scope of coverage. Thus, since about 1975–1977, more attention is paid to toxicology than in the past because of the great surge of interest in environmental matters, including occupational health and safety. In addition, there is more emphasis on uses and applications than in the past because much research, especially in industry, is more focused in this direction.

Beginning in 1982, *Gmelin* volumes are published totally in English, although some volumes prior to then are also in English, or may have tables of contents, marginal captions, and indexes in English. Most American and other English speaking chemists find even those older *Gmelin* volumes in German easy to use because of the high degree of organization throughout the treatise. For example, chemists using *Gmelin* will find its indexes invariably helpful. This is especially so for elements that are covered in more than one volume, such as sodium and its compounds. To help in use of volumes on sodium, the editors have provided this series with its own formula and subject indexes. Another example of the strength of *Gmelin* indexes is in helping chemists locate data on multicomponent systems. There is also an English-language *General Formula Index* that covers all elements and defined compounds in the eighth edition for volumes of the main work that appeared before the end of 1974, as well as for Volumes 1–12 of the New Supplement series. A supplement to this index for 1974–1979 began appearing in 1983 and is to be completed in 1986; this includes all main volumes and supplement volumes. The formula indexes through 1979 may be online in 1987. Another supplement covering 1980–1986 is already in preparation.

There are plans to add CAS Registry Numbers (see Chapters 6, 7) to the *Formula Index* and to include these Numbers in the text of the volumes. There are, however, no plans at this time to provide a separate index by Registry Number. Inclusion of these features in the future would be helpful to chemists using the volumes.

Gmelin staff are working on making the text searchable through an online computer system, perhaps as early as 1988 or 1989.

As a further aid to helping chemists use *Gmelin* more easily, subscribers should now shelve the volumes as recommended by *Gmelin* staff, for example, by atomic symbol, rather than by System Number (see Figures 12.4 and 12.5), as had been done in the past. This outmodes previous distinctions among the several types of Supplemental volumes. Special dividers are supplied by *Gmelin* to facilitate this new, much more appropriate arrangement. Finally, helpful descriptive booklets about *Gmelin* are available without charge from the publisher, Springer Verlag.* Sample pages from *Gmelin* are shown in Figure 12.6.

Key to the Gmelin System of Elements and Compounds

System Number	Symbol	Element
1		Noble Gases
2	H	Hydrogen
3	O	Oxygen
4	N	Nitrogen
5	F	Fluorine
6	**Cl**	**Chlorine**
7	Br	Bromine
8	I	Iodine
	At	Astatine
9	S	Sulfur
10	Se	Selenium
11	Te	Tellurium
12	Po	Polonium
13	B	Boron
14	C	Carbon
15	Si	Silicon
16	P	Phosphorus
17	As	Arsenic
18	Sb	Antimony
19	Bi	Bismuth
20	Li	Lithium
21	Na	Sodium
22	K	Potassium
23	NH_4	Ammonium
24	Rb	Rubidium
25	Cs	Caesium
	Fr	Francium
26	Be	Beryllium
27	Mg	Magnesium
28	Ca	Calcium
29	Sr	Strontium
30	Ba	Barium
31	Ra	Radium
32	**Zn**	**Zinc**
33	Cd	Cadmium
34	Hg	Mercury
35	Al	Aluminium
36	Ga	Gallium

System Number	Symbol	Element
37	In	Indium
38	Tl	Thallium
39	Sc, Y	Rare Earth
	La—Lu	Elements
40	Ac	Actinium
41	Ti	Titanium
42	Zr	Zirconium
43	Hf	Hafnium
44	Th	Thorium
45	Ge	Germanium
46	Sn	Tin
47	Pb	Lead
48	V	Vanadium
49	Nb	Niobium
50	Ta	Tantalum
51	Pa	Protactinium
52	**Cr**	**Chromium**
53	Mo	Molybdenum
54	W	Tungsten
55	U	Uranium
56	Mn	Manganese
57	Ni	Nickel
58	Co	Cobalt
59	Fe	Iron
60	Cu	Copper
61	Ag	Silver
62	Au	Gold
63	Ru	Ruthenium
64	Rh	Rhodium
65	Pd	Palladium
66	Os	Osmium
67	Ir	Iridium
68	Pt	Platinum
69	Tc	Technetium[1]
70	Re	Rhenium
71	Np,Pu . . .	Transuranium Elements

HCl

CrCl$_2$

ZnCrO$_4$

ZnCl$_2$

Material presented under each Gmelin System Number includes all information concerning the element(s) listed for that number plus the compounds with elements of lower System Number.

For example, zinc (System Number 32) as well as all zinc compounds with elements numbered from 1 to 31 are classified under number 32.

[1] A Gmelin volume titled "Masurium" was published with this System Number in 1941.

Figure 12.4. Key to the *Gmelin* System of Elements and Compounds. Reprinted with permission of Gmelin Institute, Frankfurt/M., West Germany.

Figure 12.5. Periodic Table of the Elements with the *Gmelin* System Numbers circled. Reprinted with permission of Gmelin Institute, Frankfurt/M., West Germany.

Main table (Gmelin System Numbers in parentheses):

1 H (2)																	2 He (1)
3 Li (20)	4 Be (26)											5 B (13)	6 C (14)	7 N (4)	8 O (3)	9 F (5)	10 Ne (1)
11 Na (21)	12 Mg (27)											13 Al (35)	14 Si (15)	15 P (16)	16 S (9)	17 Cl (6)	18 Ar (1)
19 K (22)	20 Ca (28)	21 Sc (39)	22 Ti (41)	23 V (48)	24 Cr (52)	25 Mn (56)	26 Fe (59)	27 Co (58)	28 Ni (57)	29 Cu (60)	30 Zn (32)	31 Ga (36)	32 Ge (45)	33 As (17)	34 Se (10)	35 Br (7)	36 Kr (1)
37 Rb (24)	38 Sr (29)	39 Y (39)	40 Zr (42)	41 Nb (49)	42 Mo (53)	43 Tc (69)	44 Ru (63)	45 Rh (64)	46 Pd (65)	47 Ag (61)	48 Cd (33)	49 In (37)	50 Sn (46)	51 Sb (18)	52 Te (11)	53 I (8)	54 Xe (1)
55 Cs (25)	56 Ba (30)	57** La (39)	72 Hf (43)	73 Ta (50)	74 W (54)	75 Re (70)	76 Os (70)	77 Ir (67)	78 Pt (68)	79 Au (62)	80 Hg (34)	81 Tl (38)	82 Pb (47)	83 Bi (19)	84 Po (12)	85 At	86 Rn (1)
87 Fr	88 Ra (31)	89*** Ac (40)	104 (71)	105 (71)													

* NH₄ (23)

**Lanthanides (39):

58 Ce	59 Pr (51)	60 Nd (55)	61 Pm	62 Sm	63 Eu	64 Gd	65 Tb	66 Dy	67 Ho	68 Er	69 Tm	70 Yb	71 Lu

***Actinides:

90 Th (44)	91 Pa (51)	92 U (55)	93 Np (71)	94 Pu (71)	95 Am (71)	96 Cm (71)	97 Bk (71)	98 Cf (71)	99 Es (71)	100 Fm (71)	101 Md (71)	102 No (71)	103 Lr (71)

237

$S_3N_3Cl_3 + 9H_2O \rightarrow 3NH_4Cl + 3H_2SO_3$ [7]. When $S_3N_3Cl_3$ is treated with 10 to 30% alkali sulfite solutions, the following reactions occur simultaneously:

$$S_3N_3Cl_3 + 9OH^- \rightarrow 3SO_3^{2-} + 3Cl^- + 3NH_3 \tag{1}$$

$$S_3N_3Cl_3 + 3SO_3^{2-} + 9H_2O \rightarrow 3S(OH)_2 + 3SO_4^{2-} + 3Cl^- + 3NH_4^+ \tag{2}$$

With decreasing pH-value the importance of reaction (2) increases. At the same time the rate of decomposition continuously decreases. The intermediate sulfoxylic acid, $S(OH)_2$, decomposes to give $S_2O_3^{2-}$ and $S_3O_6^{2-}$ [8]. When $S_3N_3Cl_3$ is treated with ice-cold water in a moderate acidic medium, SO_2, NH_4^+, SO_4^{2-}, and a dark blue compound of composition $H_4N_2S_2O$ form, probably via $4S_3N_3Cl_3 + 27H_2O \rightarrow 3OS_2(NH_2)_2 + 6(NH_4)HSO_4 + 12HCl$ [6, 9], see also "Schwefel" B 3, 1963, p. 1850.

Reactions with Nitrogen Compounds

NH₃. The ammonolysis of $S_3N_3Cl_3$ in liquid NH_3 ($-78°C$) is said to produce the unstable compounds $(NSNH_2)_3$ and $S(NH)_2$. The metal derivative of $S(NH)_2$, $Hg[N_2S] \cdot NH_3$, can be isolated if a mixture of $S_3N_3Cl_3$ and NH_3 is treated with HgI_2 [10]. According to recent investigations the reaction of $S_3N_3Cl_3$ with NH_3 at $-78°C$ produces an orange-red solid, $NH_4^+ S_4N_5^-$, probably via the intermediate compounds $(NSNH_2)_3$ or $S(NH)_2$ [11]:

$$S_3N_3Cl_3 \xrightarrow[-3\,NH_4Cl]{+6\,NH_3} \{(NSNH_2)_3 \text{ or } 3\,S(NH)_2\} \rightarrow \tfrac{3}{4}NH_4^+S_4N_5^- + \tfrac{1}{4}N_2 + NH_3$$

C₆H₅S–NH–C₆H₅. C₆H₅S–NH–SC₆F₅. C₆F₅S–NH–SC₆F₅. C₆F₅S–NH–Si(CH₃)₃. The reaction of $S_3N_3Cl_3$ with sulfenyl amines produces sulfur diimides in the presence of an HCl trapping base B $(B = C_5H_5N$ or $(C_2H_5)_3N)$ per $R-NH-R' + \tfrac{1}{3}S_3N_3Cl_3 \xrightarrow{B} R-N=S=N-R' + B \cdot HCl$, with $R = C_6H_5$, $R' = C_6H_5$ (in CCl_4/petroleum ether, at 0°C), $R = C_6H_5S$, $R' = C_6F_5S$ (CCl_4/ether, 50°C, 1 h reflux) [12], $R = R' = C_6F_5S$ (CCl_4, 60°C) [13], or $R = C_6F_5S$, $R' = Si(CH_3)_3$ (CCl_4, 25°C, 2 h) [14].

[(CH₃)₃Si]₂NH. $S_3N_3Cl_3$ reacts with $[(CH_3)_3Si]_2NH$ and C_5H_5N (mole ratio $\sim 1:4:4$) in CCl_4 at 25 to 60°C in 12 h to produce $(CH_3)_3SiN=S=NSi(CH_3)_3$ along with S_4N_4 and $(CH_3)_3SiN=S=N-S-N=S=NSi(CH_3)_3$ as side products [14, 15].

(CH₃)₃SiN₃. The reaction of $S_3N_3Cl_3$ with trimethylsilylazide, $(CH_3)_3SiN_3$, leads to a variety of products depending on the reaction conditions: temperatures -15 to $+40°C$, reaction time 0.5 to 48 h, mole ratios $\sim 1:4$ to $\sim 1:15$, solvents CH_3CN, tetrahydrofuran (THF), and CH_2Cl_2. Main products have been found to be up to 67% S_4N_4 and up to 72% $(SN)_x$ in CH_3CN. No $(SN)_x$ is obtained in THF and CH_2Cl_2; instead $S_4N_3^+Cl^-$ and $[S_3N_2Cl]_2$, respectively, form in addition to S_4N_4. Other products are $(CH_3)_3SiCl$ and NH_4Cl [16]. In an earlier paper the reaction of $S_3N_3Cl_3$ with $(CH_3)_3SiN_3$ (mole ratio $\sim 1:4$) in CH_3CN was found to produce 40 to 50% $(SN)_x$, along with $(CH_3)_3SiCl$, NH_4Cl, S_4N_4, N_2 and sulfur [17].

NO. NO₂. Treating a suspension of $S_3N_3Cl_3$ in CH_3NO_2 with gaseous NO gives the green five-membered ring compound $[S_3N_2Cl]_2$: $8S_3N_3Cl_3 + 24NO \rightarrow 3[S_3N_2Cl]_2 + 12NOCl + 3S_2Cl_2 + 12N_2O$ [6]. $S_3N_3Cl_3$ is oxidized by NO_2 and the ring completely destroyed. The reaction proceeds similarly to that of NO_2 with S_4N_4 (see "Schwefel" B 3, 1963, p. 1548), forming the compound $(NO)_2S_2O_7$ [6, 9].

(CF₃)₂NO. $S_3N_3Cl_3$ reacts with excess $(CF_3)_2NO$ to form $S_4N_4[ON(CF_3)_2]_4$ [18, 19]. Reaction of $S_3N_3Cl_3$ with iodine and a large excess of $(CF_3)_2NO$ at room temperature for 12 h also yields $S_4N_4[ON(CF_3)_2]_4$ [19].

(CF₃)₂CN₂. When $S_3N_3Cl_3$ is reacted with bis(trifluoromethyl)diazomethane, $(CF_3)_2CN_2$, in a 1:3 mole ratio in a glass bomb at 40°C, ring cleavage occurs: $S_3N_3Cl_3 + 3(CF_3)_2CN_2 \rightarrow 3(CF_3)_2C=N-SCl + 3N_2$ [20].

Figure 12.6a,b. Representative pages from *Gmelin Handbook of Inorganic Chemistry*, 8th ed., Fe: Organoiron Compounds, Part B8; Mononuclear Compounds 8. This copyrighted material is reprinted with permission of the Director of the Gmelin Institute for Inorganic Chemistry of the Max Planck Society, Frankfurt, W. Germany, and the publisher, Springer-Verlag, Heidelberg, W. Germany, and reduced in size from original publication.

[2-CH₃O(5-C₃H₄N₂)C₆H₆Fe(CO)₃]BF₄ (Table 9, No. 21) is quantitatively prepared in situ by mixing equimolar amounts of [2-CH₃OC₆H₆Fe(CO)₃]BF₄ and imidazole in acetone-d₆ [81].

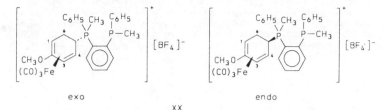

exo   endo

XX

[2-CH₃O(5-(2-CH₃(C₆H₅)PC₆H₄P(C₆H₅)CH₃)C₆H₆Fe(CO)₃]BF₄ (Table 9, No. 26) is prepared by attack of racemic [2-CH₃OC₆H₆Fe(CO)₃]BF₄ by (−)₅₈₉(S,S)-CH₃(C₆H₅)PC₆H₄P(C₆H₅)CH₃-2. This reaction exhibits considerable kinetic diastereoselectivity in CH₃CN and OC(CH₃)₂. The corresponding reaction in CH₂Cl₂ is of lower selectivity [78]. The phosphine preferentially selects the (−)₅₈₉(R)-enantiomer of the salt [72]. Thus, recovery of unreacted dienyl salt from reactions between racemic [2-CH₃OC₆H₆Fe(CO)₃]BF₄ and the (−)₅₈₉(S,S)-phosphine (1:0.5 molar ratio) provides a convenient method for preparing optically active [2-CH₃OC₆H₆-Fe(CO)₃]BF₄ [78] (6 to 11% enantiomeric excess [78, 79]). Unexpectedly, the 50/50 mixture of diastereoisomers XX, obtained immediately upon reacting both compounds in a 1:1 molar ratio, equilibrates over 3 d in CH₃CN to a 60:40 mixture [78]. However, this equilibration is not confirmed by [79] for no change in the CD or ¹H NMR spectra is observed over a period of a week at room temperature [79]. Treatment at room temperature in acetone gives, upon immediate evaporation, a high yield (>80%) of the yellow salt XX. The ¹H NMR spectrum (acetone-d₆) confirms a 50:50 mixture of the two diastereoisomers XX by the presence of δ=3.48 and 3.68 (OCH₃), and 5.40 and 5.65 (H-3) ppm for the endo and exo isomer. The IR spectrum (acetone) shows bands at 1990 and 2060 cm⁻¹. Preliminary attempts to separate the diastereoisomers XX by crystallization from various solvents are unsuccessful [78], cf. Nos. 27 and 28.

CH₃
(C₆H₅)₂P
CH(CH₃)₂

XXI

H₃C   CH₃
H--C–C—H
(C₆H₅)₂P   P(C₆H₅)₂

XXII

[2-CH₃O(5-R''(C₆H₅)₂P)C₆H₆Fe(CO)₃]BF₄ (Table 9, Nos. 27 and 28 with R'' = cyclo-C₆H₉(C(CH₃)₂H-6)CH₃-3 and CH(CH₃)CH(CH₃)P(C₆H₅)₂) are prepared by the reaction of racemic [2-CH₃OC₆H₆Fe(CO)₃]BF₄ with equimolar amounts of (−)₅₈₉-XXI or (−)₅₈₉(S,S)-XXII in high yield. In contrast, a 1:0.5 molar ratio mixture of racemic [2-CH₃OC₆H₆Fe(CO)₃]BF₄ and (−)₅₈₉(S,S)-XXII gives a quantitative yield of XXIII. Each compound, No. 27 and 28, is expected to be a 50:50 mixture of diastereoisomers (see No. 26). However, in the ¹H NMR spectra in CH₃CN, distinct OCH₃ or H-3 resonances are not observed for the two diastereoisomers in each case even in the presence of Eu(fod) (for fod see No. 6). This coincidence of diastereoisomer proton resonances prevents quantitative in situ determination of the chiral discrimination. However, the unreacted dienyl salt recovered from the

References on pp. 228/30

Figure 12.6b. This copyrighted material is reprinted with permission of the Director of the Gmelin Institute for Inorganic Chemistry of the Max Planck Society, Frankfurt, W. Germany, and the publisher, Springer-Verlag, Heidelberg, W. Germany, and reduced in size from original publication.

reaction of [2-CH₃OC₆H₆Fe(CO)₃]BF₄ with XXI (1:0.5 molar ratio) in acetone is found to contain an 11% enantiomeric excess of [(+)₅₈₉(S)-2-CH₃OC₆H₆Fe(CO)₃]BF₄. Thus (−)₅₈₉-XXI preferentially selects the [(−)₅₈₉(R)-2-CH₃OC₆H₆Fe(CO)₃]BF₄. Similarly a 5% enantiomeric excess of [(−)₅₈₉(R)-2-CH₃OC₆H₆Fe(CO)₃]BF₄ is found for the reaction with (−)₅₈₉(S,S)-XXII (molar ratio 1.5:0.5). However, the relative low enantiomeric excess of recovered [(−)₅₈₉(R)-2-CH₃OC₆H₆Fe(CO)₃]BF₄ from such a reaction in acetone suggests that (−)₅₈₉(S,S)-XXII is less diastereoselective than either (−)₅₈₉(S,S)-CH₃(C₆H₅)PC₆H₄P(C₆H₅)CH₃-2 (see No. 26) or (−)₅₈₉-XXI [79].

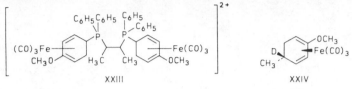

XXIII   XXIV

2-CH₃O(5-CH₃)C₆H₆Fe(CO)₃ (Table 9, No. 30) can also be prepared by reduction of [2-CH₃O(5-CH₃)C₆H₅Fe(CO)₃]BF₄ with NaBH₄ in H₂O. The reaction mixture is extracted with ether and the organic layer is washed with H₂O, dried (MgSO₄), and evaporated. The residue is dissolved in pentane, and filtered through silica. Evaporation gives a mixture of 1-CH₃O-(4-CH₃)C₆H₆Fe(CO)₃ and No. 30 (ratio 2:3) in 90% yield. Chromatography has shown compound No. 30 to be a 1:1 mixture of 5-exo and 5-endo isomers. This mixture of the exo and endo isomers of No. 30 reacts with [C(C₆H₅)₃]BF₄ in CH₂Cl₂ to give [2-CH₃O-(5-CH₃)C₆H₅Fe(CO)₃]BF₄ in 50% yield. Chromatography of the neutral washings gives pure 2-CH₃O(5-endo-CH₃)C₆H₆Fe(CO)₃ [11]. Similar reduction of [2-CH₃O(5-CH₃)C₆H₅Fe(CO)₃]BF₄ with NaBD₄ gives XXIV [19]. Optical isomers of compound No. 30 are prepared as follows: a cooled solution of [(−)(2R)-2-CH₃OC₆H₆Fe(CO)₃]PF₆ ([α]²⁰_D = −107°, c = 9 mg/100 mL, CH₃CN) in CH₃CN is added to a solution of LiCu(CH₃)₂ at −45 °C. After 2 min the reaction is quenched by iced 5% aqueous HCl and extracted with ether. The extract is washed with H₂O, brine, dried (MgSO₄), and filtered through silica. Evaporation gives a yellow oil of (+)(2R, 5R)-2-CH₃O(5-CH₃)C₆H₆Fe(CO)₃ ([α]²⁰_D = +11.7°, c = 4 mg/100 mL, CHCl₃) in 75% yield [62]. 36 mg of fully resolved [(−)(2R, 5S)-2-CH₃O(5-CH₃)C₆H₅Fe(CO)₃]PF₆ (see Formula XXVI, [α]²⁰_D = −138°, c = 4 mg/100 mL, CH₃CN) is mixed with 324 mg of racemic [2-CH₃O-(5-CH₃)C₆H₅Fe(CO)₃]PF₆ to obtain a sample with [α]²⁰_D = −13.2°. NaBH₄ is added to this sample in CH₃CN at 5 °C. After 10 min the mixture is filtered and evaporated. The residue is taken up in pentane and filtered through basic alumina. Evaporation affords a pale brown gum (84% crude yield). Chromatography on silica with C₆H₆/hexane (1:1) gives the more polar (−)(1R, 4S)-1-CH₃O(4-CH₃)C₆H₆Fe(CO)₃ (see Formula XXVII, [α]²⁰_D = −12°, c = 1 mg/100 mL, CHCl₃) in 7% and (+)(2R, 5S)-2-CH₃O(5-CH₃)C₆H₆Fe(CO)₃ (see Formula XVIII, [α]²⁰_D = +13.1°, c = 2 mg/100 mL, CHCl₃) in 31% yield and in an enantiomeric excess of 9.5% [65].

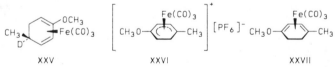

XXV   XXVI   XXVII

The (+)(2R, 5S) isomer (see Formula XXVIII) shows in the ¹H NMR in CDCl₃ chemical shifts at δ = 0.86 (d, CH₃; J = 7 Hz), 1.20 (ddd, H-6 exo; J = 14 Hz, J = 5 Hz), 1.63 (m, H-5 exo),

Gmelin Handbook
Fe-Org. Comp. B8   References on pp. 228/30

Figure 12.6c. Representative page from *Gmelin Handbook of Inorganic Chemistry*, 8th ed., S: Sulfur-Nitrogen Compounds, Part 2; Compounds with Sulfur of Oxidation Number IV. This copyrighted material is reprinted with permission of the Director of the Gmelin Institute for Inorganic Chemistry of the Max Planck Society, Frankfurt, W. Germany, and the publisher, Springer-Verlag, Heidelberg, W. Germany, and reduced in size from original publication.

240

It is difficult to find fault with so thorough and competent a work as *Gmelin*. If there is any failing, it is lack of complete up to dateness. Even that is understandable in a publication so ambitious in scope, and major efforts are made to be as current as possible, especially in fields of high interest. For much of the material included, coverage is of recent vintage.

B. *Mellor*

As compared to *Gmelin*, *Mellor* (29) has the advantages of being completely in English (at least this is an advantage for chemists who read only English) and of a full (except for supplements) cumulative index, which *Gmelin* lacks at this writing.

The original volumes of *Mellor* were published over a time period beginning in 1922 and ending in 1937. In addition, the following supplements have been published:

Volume 2, Supplement 1: *The Halogens* (1956)
Volume 2, Supplement 2: *The Alkali Metals, Part 1* (1961)
Volume 2, Supplement 3: *The Alkali Metals, Part 2* (1963)
Volume 8, Supplement 2: *Nitrogen, Part 2* (1967)
Volume 8, Supplement 3: *Phosphorus* (1972)

The vintage of *Mellor* makes it increasingly less useful than *Gmelin*, which is continuously updated.

C. Some Other Inorganic Chemistry Reference Works

In addition to such major treatises as *Gmelin* and *Mellor*, works intended to be of lesser scope have been issued. In the field of inorganic chemistry, a good example is the four volume treatise by Bailar (30).

Inorganic Syntheses (31) contains checked synthetic procedures for inorganic compounds of "more than routine interest." New volumes issue every year or two.

Pergamon Press has announced plans for publishing a six-volume *Comprehensive Coordination Chemistry*, to be edited by Geoffrey Wilkinson (32).

12.6. KEEPING UP WITH AND IDENTIFYING BOOKS

Several important sources help the chemist keep up with the many new books, which appear with almost alarming frequency. The most complete source is *CA*. Here, new books are cited in a special place in the sections in each current issue. The indexes to *CA* signify that an entry is to a book or encyclopedia by use of the designation B. In addition, one of the *CA SELECTS* series, *New Books in Chemistry,* provides information on new books cited in *CA.*

Current journals provide lists of books and/or book reviews. For example, *Chemistry and Industry, Chemistry in Britain, Chemical Engineering,* and *Journal of the American Chemical Society* contain excellent reviews. Extensive listings of new books appear in *Chemical & Engineering News.* The *Journal of Chemical Education* has a useful, comprehensive listing, categorized by broad subject, in its September issue each year.

The librarians who work with chemists have available a number of tools that can help locate books in all fields of science and technology. For example, *Books in Print* (33) lists most English-language books; it is conveniently arranged by title, broad subject, and author, and it can be searched in either printed or online form. *ISI®* has announced a new service that is a guide to chapters in multiauthor scientific books.

REFERENCES

1. M. Grayson, Ed., *Kirk–Othmer Encyclopedia of Chemical Technology,* 3rd ed., Wiley, New York, 1978–1984. Also available online.

2. M. Grayson, Ed., *Encyclopedia of Semiconductor Technology,* Wiley, New York, 1984.

3. *Ullmans Encyclopedia of Industrial Chemistry,* 5th ed. (in English), Verlag Chemie, Weinheim, Germany and Deerfield Beach, FL, 1984 to date; *Ullmans Encyclopädie der Technischen Chemie,* 4th ed., 1972–1984.

4. J. Matley, *Ullman's Encyclopedia of Industrial Chemistry,* in *Chem. Eng,* **93**(8), 95–97 (1986). This review apparently is misleading regarding comparative costs.

5. J. J. McKetta and W. A. Cunningham, Eds., *Encyclopedia of Chemical Processing and Design,* Marcel Dekker, New York, 1976–to date.

6. J. Kroshwitz, Ed., 2nd ed. *Encyclopedia of Polymer Science and Engineering,* Wiley, New York, 1984 to date. First edition completed in 1976 and commonly known as *Mark.* Also available online.

7. M. B. Beever, *Encyclopedia of Materials Science and Engineering,* MIT Press, Cambridge, MA, 1986.

8. R. Luckenbach, Ed., *Beilsteins Handbuch der Organischen Chemie,* 4th ed., 4th and 5th Supplements, Springer-Verlag, Berlin, 1972 to date. Now in English.

9. *How to Use Beilstein* (pamphlet). See also: *Shortcut to Locating a Compound* (chart); *Stereochemical Conventions* (pamphlet); *What Is Beilstein* (pamphlet); *Outline of Beilstein,* 1984–1985 (chart).

10. R. Luckenbach, R. Ecker, and J. Sunkel, "The Critical Screening and Assessment of Scientific Results Without Loss of Information—Possible or Not", *Angew. Chem. Int. Ed. Engl.* **20**, 841–849 (1981).

11. *Beilstein Dictionary German–English,* Springer-Verlag, New York, 1979.

12. H. O. House, *Review of Beilstein Handbook of Organic Chemistry*, (5th Suppl., Vol. 17, Part 1), *J. Chem. Inf. Comp. Sci.*, **24**(4), 277, (1984).

13. J. Buckingham, Ed., *Dictionary of Organic Compounds*, 5th ed., Chapman and Hall, London, 1982, and supplements thereafter, as well as online and paperback versions.

14. S. Coffey or M. F. Ansell, Ed., *Rodd's Chemistry of Carbon Compounds*, 2nd ed., Elsevier, Amsterdam, 1964 to date.

15. E. Müller, O. Bayer, H. Meerwein, and K. Ziegler, Eds., *Methoden der Organischen Chemie (Houben-Weyl)*, 4th ed. (H. Kropf, Ed.), Georg Thieme Verlag, Stuttgart, 1952 to date.

16. D. Barton, Ed., *Comprehensive Organic Chemistry*, Pergamon, London, 1983.

17. *Organic Syntheses*, Wiley, New York, 1941 to date.

18. From *Organic Syntheses*, Vol. 55, p. IX, reprinted with permission of John Wiley, & Sons, Inc. New York.

19. W. Dauben, Ed., *Organic Reactions*, Wiley, New York, 1942 to date.

20. C. A. Buehler and D. A. Pearson, Eds., *Survey of Organic Syntheses*, Vols. 1 and 2, Wiley, New York, 1970 and 1977. Vol. 3, Robert Williams, Ed., in preparation (1987).

21. A. Finch or W. Theilheimer, Ed., *Synthetic Methods of Organic Chemistry*, S. Karger, A. G. Basel, New York, 1946 to date (approximately annual). Also available online.

22. L. G. Wade, Ed., *Compendium of Organic Synthetic Methods*, Wiley, New York, Vol. 4, 1980, Vol. 5, 1984.

23. G. Wilkinson, Ed., *Comprehensive Organometallic Chemistry*, Pergamon, London, 1982.

24. J. Buckingham, Ed., *Dictionary of Organometallic Compounds*, Pergamon, London, 1984.

25. M. Fieser, Ed., *Fieser and Fieser's Reagents for Organic Synthesis*, Wiley, New York, 1967 to date.

26. Various authors, *The Chemistry of Heterocyclic Compounds—A Series of Monographs*, Wiley, New York, still being published.

27. A. R. Katritzky and C. W. Rees, Eds., *Comprehensive Organometallic Chemistry*, Pergamon, London, 1984.

28. E. Fluck, Ed., *Gmelins Handbuch der Anorganischen Chemie*, 8th ed., Springer-Verlag, Berlin, 1924 to date.

29. J. W. Mellor, Ed., *A Comprehensive Treatise on Inorganic and Theoretical Chemistry*, Wiley, New York, 1922–1937 (main volumes), 1956–1972 (supplements).

30. J. C. Bailar, H. J. Emeleus, R. Nyholm, and A. F. Trotman Dickinson, Eds., *Comprehensive Inorganic Chemistry*, Pergamon, New York, 1973.

31. *Inorganic Syntheses*, Wiley, New York, 1939 to date.

32. G. Wilkinson, Ed., *Comprehensive Coordination Chemistry*, Pergamon, London, to begin publication in 1987.

33. *Books in Print*, R. R. Bowker, New York, published annually. Also available online.

LIST OF ADDRESSES

Beckman Instruments
2500 Harbor Blvd.
Fullerton, CA 92634

Beilstein Institute
Varrentrapstrasse 40/42
D-6000 Frankfurt (Main) 90
West Germany

John J. McKetta, Professor
Department of Chemical Engineering
University of Texas
Austin, TX 78712

Springer Verlag
175 Fifth Avenue
New York, N.Y. 10010

13 PATENTS

13.1. INTRODUCTION

Patents are probably the most significant forms of chemical literature. In the United States, granting of a patent indicates that the inventor has the right to *exclude others* from making, using, or selling his or her invention in the United States, its territories and possessions, for 17 years after the patent is issued by the Patent Office.

A patent may be granted to the inventor of any *new* and *useful* process, machine, manufacture (includes all manufactured articles), or composition of matter (may include new chemical compounds or mixtures). For a more complete legal definition and explanation, the reader should consult publications (1) of the Patent Office, or a patent attorney or agent registered to practice before that office.

Other countries have similar provisions. One major difference among the countries is duration of the monopoly granted. In many countries, (such as the United Kingdom), the life of a patent has been increased from the previous 15 or 17 years to 20 years from date of filing. Another difference is speed of publication of the patent after filing the application. A third difference is ultimate validity, especially for some countries in which applications are not examined before initial publication (see Section 13.8).

13.2. TYPES OF PATENT DOCUMENTS

The two principal types of published patent documents discussed in this chapter are published patent applications (referred to in this chapter as patent applications or quick-issue patents) and issued or granted patents. Published patent applications are documents filed by inventors with a national or multinational patent office and subsequently published (made available to the public or open to inspection) prior to the official examination process that may be necessary to determine whether a patent meets requirements to be granted and issued. Granted or issued patents are published only after successful completion of the examination process in countries that require this; only these patents permit the patent owner (inventor or assignee) to exclude others as previously described. For ease of expression, this chapter usually refers to all patent publications as "patents," unless otherwise specified.

13.3.   WHY PATENTS ARE IMPORTANT

Chemists ought to display a keen interest in patents for two principal reasons: (a) many chemists discover new products, processes, or uses that can be patented and can lead to substantial financial and other rewards for the inventor or his or her organization, and (b) patents are a rich and unique source of chemical information. More specifically, the benefits of the patent as an information tool include these:

1. It permits the chemist to evaluate an idea rapidly and accurately.
2. It tells whether the idea belongs to anyone else.
3. It gives a feel for directions in which others are working, thus making it possible to evaluate one's own competitive situation.
4. It gives a trading position, for several inventors might have parts of an idea and might need each other to make an economical operation.
5. It is a good source of ideas, and especially component parts, to build a total system.

13.4.   PATENT STRUCTURE

The parts of a typical patent include:

1. The front page, which contains basic bibliographic information and an abstract.
2. The drawings (if any).
3. The so-called specification, which constitutes the main body of the patent (in terms of size) and contains a description of the invention.
4. The claims, which always appear at the end of the patent and delineate the scope of the monopoly granted under law.

If chemists read the abstract and the first few paragraphs of the specification, they can identify the general purpose and nature of the invention. They should look for sections captioned *Background of the Invention* and/or *Summary of the Invention*.

The specification usually contains in the first few paragraphs indication of the uniqueness (that which is new) and of the advantages of the invention over "prior art" (that which is already known). In most chemical patents, the specification includes *examples,* which are so labeled. These examples are useful to the chemist because they give specific experimental details.

Claims express the legal essence of the patent and are of special interest to chemists in industrial organizations, particularly to patent attorneys, who are trained to interpret what is, and is not, covered by the patent from a legal standpoint. The first claims of most patents tend to be broad (generic); succeeding claims tend to become more and more specific. Almost all claims are, quite properly, highly legalistic in phrasing.

Official patent titles, particularly those of several years ago and many issued today, may be too general to be a meaningful guide to the technical content of a patent.

Good patent information tools, such as most of those mentioned in this chapter, try to provide the chemist adequate and reliable clues to patent content in several ways:

1. Patent titles are enriched with words that help make the context of the patent more understandable to the chemist. For example, in the title "Compound XYZ, *A Solvent, Manufacture by Process ABC Using Catalyst C,*" the italicized words represent hypothetical enrichment of a hypothetical title.

2. The content of patents is indicated by indexing of key features, especially examples and novel subject matter covered by the invention.

13.5. THE ROLE OF PATENTS IN IDEA GENERATION AND CREATIVITY

Patents are invaluable in helping guide chemists in evaluation and development of ideas. Although the patent system is available to chemists looking for information, it is frequently not used to the extent that it should be. Effective use of patents can lead to more productive research and development in chemistry. Such use often requires the cooperative efforts of the research chemist, the patent attorney, and the literature chemist or information specialist.

Since the beginning of time man has had an urge to create. To *create* means to carry a new idea to the public. This creation may take the form of a book, a painting, a musical arrangement, a floral design, a structure such as a house or bridge, or in the case of chemistry, a new chemical compound, process, or use.

Those who create things usually want to be identified with them. All men and women recognize that progress is made by these creative people. The government has seen the need to develop systems to encourage them to create. The three most important ways used by the government to encourage creativity are (a) the copyright for creative works, as writing and music; (b) trademarks; and (c) patents.

In the patent system: (a) the inventor describes the invention; (b) files the detailed description, including claims, with government patent offices in one or more countries; (c) has it examined by others (officially designated patent examiners in the United States and many other countries) to determine if it is really new, unobvious, and useful; and (d) if it is, to grant the right to exclude others from using the claimed invention, for 17 years in the case of patents granted by the U.S. Patent Office.

The purpose of granting patents is to encourage people to create. It is anticipated by the government that these concepts will be used or that the dissemination of the information contained therein will promote the advance of the technology.

The patent system was developed to tell chemists and others what has been going on and what areas of technology have been staked out by others. So the patent system was, and is, not only a *legal* tool or system, but an *information* tool. A limited monopoly is granted by the government in exchange for disclosure of details of the invention.

Not to use the patent system is a mistake, because if an idea is old, the sooner the chemist knows it the better. Even if the idea is old and patented, the chemist might still be able to use it. For example, the first person to conceive a specific chemical idea could have been 20 years ahead of his or her time. If this is so, then present-day chemists could have the concept *free*. Also, it may be that the basic concept was conceived, but something was missing to make it satisfactorily useable. This may permit a modern-day chemist to modify an older invention to make it more useful. This is building on the technical foundation of the past and is what the creators of the patent system wanted to happen.

13.6. PATENTS AS INFORMATION TOOLS

How can the chemist use the patent system as an information tool? Searching the literature, reading, and thinking are cheaper and faster than experimenting. So, a chemist's approach to a problem could be as follows:

1. Think and guess what he or she would like to accomplish.
2. Make a quick survey of patent and other scientific literature.
3. Study.
4. Think and guess again.
5. Make a more detailed study of patent and other scientific literature.
6. Read, study, think.
7. Experiment.

To do the literature searching noted, the laboratory chemist needs expert help—people who know their way around the patent system and other parts of the scientific literature. The patent system has become too complex for most laboratory chemists to "go it alone." That is why it is important for the laboratory chemist to confer with the information chemist, tell that person what is being thought and guessed about, and ask for help in pulling together the works of others for study.

In this type of interactive relationship, the laboratory person tells the information person the type of search needed and also suggests how much time is worth spending on the search. Time to be spent is based on such factors as potential profit or other impact and priority considerations.

Next, the laboratory chemist studies the results of the search, does some additional searching as needed, and makes decisions based on these studies.

The decision could be one of these:

1. Dig a little deeper.
2. Modify the original idea—shift to the left or right.
3. It's old stuff—forget it.
4. It's old stuff, but by adding A or B it can be used.

The benefits of this approach include learning, using appropriately the skills of others, and determining as quickly and economically as possible if an idea is good and novel. The next, or perhaps concurrent, step could be to work in close cooperation with a patent attorney or agent toward the ultimate filing of a patent application if that is perceived as a goal.

13.7. PATENTS COMPARED TO OTHER INFORMATION SOURCES

Patents are a major information source about new chemical products, processes, and uses. The 29 leading patent-granting offices publish approximately 12,000 patents each week according to some estimates; most of these are patent applications and the figure does not include many Japanese patent publications.

One way in which patents are unique sources of chemical information is that they are primarily legal documents. Some chemists find patents more difficult to read and understand than books and journal articles because patents are usually written in legal phraseology by patent attorneys or agents who act for inventors. But this should not deter the chemist from taking full advantage of the unique and rapid access to the enormously broad range of chemical information that patents provide.

Similarly, patents should not deter chemists from using journals, books, and other forms of chemical information. All forms are important, and all have certain advantages that vary with the situation and intended use. Although journals and books are more familiar to most chemists, patents are at least on a par with books and journals in value of information provided with regard to chemical technology.

Some people do not look on patents as a source of reliable information. It has, however, been the experience of many chemists that it is as easy to replicate an example in a patent as it is to duplicate an experimental description in a scientific journal. In both cases, the inventor and the scientist usually report the highest yield obtained, but in both cases this yield was not always obtained in exactly the way the process was described. Nevertheless, with a little ingenuity and a knowledge of the field, the chemist can easily duplicate the majority of examples in patents.

Because chemists do not usually publish anything proprietary in journals and books, at least until there is fully adequate patent coverage, patents are usually good clues to what the competition is thinking. It is believed that much, although by no means all, of what appears in patents cannot be found published in journals or in other forms. It has been claimed that about 70% of the material in patents is not duplicated in journal literature,[*] and, although this has never been proven definitively, the lion's share of patent information never appears elsewhere, except perhaps in publications which aim to summarize patent documents (Section 13.21).

Patent documents frequently contain numerous detailed examples, discussion of variables, and drawings. The amount of highly specific experimental detail given in patent examples can be outstanding. These details can include not only methods of synthesis or manufacture, but also results of physical tests and application studies. In contrast, journals may be constrained as to length, because of budget and other limitations, and they are also very often limited in their coverage of the most modern commercial methods of synthesis, manufacture, and testing. Patents are not subject to conventional editorial or refereeing processes, so their content can be broad. In addition, because generic scope is frequently sought to provide the broadest possible patent coverage, the number and variety of examples given can be considerable and the detailed discussion of variables can be useful.

Another unique and important advantage of patents is the invaluable perspective and background they can provide on technological history and development. For example, perusal of patents in individual subclasses of the U.S. Patent Office usually provides chemists with an excellent picture of how a technology has unfolded and progressed. Reading of the initial pages of patents, prior to the examples, can provide insights into the state of the art. These pages usually provide background and help identify what is important. Furthermore, references cited by U.S. patent examiners, as noted on the cover pages of U.S. patents, can provide access to other relevant literature. All of this can be very helpful in creating and planning new research and development programs.

---

[*] For example, see U.S. Patent Office Technology Assessment and Forecast Report No. 8, 1977, pp. 23–34.

The exercise just described can be carried out in the Public Search Room of the U.S. Patent Office where chemists can find all U.S. patents conveniently arranged in the appropriate subclasses and can also obtain advice as to which subclasses to examine. This exercise can sometimes be carried out elsewhere, including at one's own desk, particularly if a subclass is small enough such that all patents in that subclass can be economically purchased. As discussed later in this chapter, however, use of the subclass approach requires considerable expertise.

An alternative approach is to scan Derwent manual code classes (see Section 13.11.A); this is much more complete for the years covered but does not go back nearly as far in time as U.S. Patent Office files.

Patent literature has also replaced conventional trade and scientific journals as the most up-to-date source of information on technological progress in numerous fields. This is partly because many countries have revised their laws so that patent applications are published without examination a few months after filing. Such countries are called *quick-issue* or *quick publication* countries, and most are listed in Section 13.11.A.

The examination system used in the United States requires patent examiners to evaluate all applications on the basis of such criteria as unobviousness, novelty, and usefulness. These criteria are more specifically defined by law.

The examination process, to be done well, takes time. When the first edition of this book was written, the average time from filing of U.S. patent applications to issuance or abandonment (called the pendancy period) was 19 months. Subsequently, this increased to over two years (Fiscal Years 1982–1985).

The official stated goal of the U.S. Patent Office is to reduce pendancy to 18 months by 1988. This task will not be easy in view of the flood of 20,000 documents received daily for incorporation into the files, in addition to material already on hand. If the target of 18 months pendancy can be achieved, this would put the accessibility of an *average* issued U.S. patent within the time range of patent applications published by some so-called quick-issue countries. The word *average* is emphasized because if the state of the art in a field is *close* or *crowded,* issue of the U.S. patent could take considerably longer than 18 months, whereas in other fields, issue could be much faster. (For example, inventions relating to energy or to environment as well as in other fields, could issue in six to eight months after filing, if expedited examination is requested at the time of filing.)

13.8. QUICK-ISSUE PATENT PUBLICATIONS

By screening quick-issue patent publications (patent applications) from the more rapidly publishing countries, chemists can gain advance knowledge of inventions whose counterparts (equivalents) may be issued later in countries that publish more slowly, and as mentioned, this knowledge is usually ob-

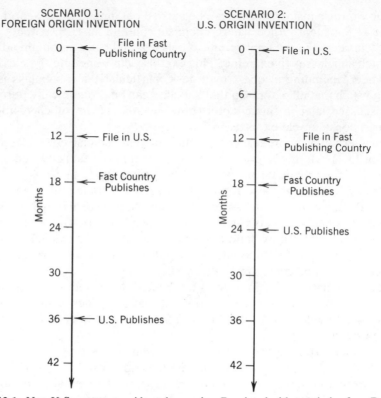

tained much before it appears in conventional trade and professional journals (see Figure 13.1).

The essence of quick-issue or quick-publication systems, such as those of West Germany, Japan, Belgium, the Netherlands, and the European Patent Convention (EPC) (which is described further in Section 13.20), is that patent applications are *published* within 18 months after filing. Examinations as to patentability may be deferred for several years after filing, and allowed patents are subject to an *opposition period*.

This plan was first adopted by the Netherlands in 1964, by West Germany on October 1, 1968, and by Japan on January 1, 1971, for the following purposes:

1. To eliminate backlogs of unexamined patent documents.
2. To make the contents of these documents available more quickly.
3. To defray legal costs until inventions are more fully developed.
4. To examine only the more worthy inventions.

The principal benefits to the public have been, as mentioned, early access to competitor activities and to new technologies, as well as minimizing surprises and duplication of R & D effort.

On the negative side, this arrangement is said to shift the burden of scope and validity to the public, and there is a certain amount of uncertainty as to what will actually be patented. Practicing chemists and other patent users have also been flooded with large volumes of *published* unexamined patent applications, especially from Japan, and keeping current with all of these has proven to be virtually impossible.

As examples of quick-issue access, consider the following:

1. Belgian patent specifications are commonly *laid open to public inspection* in Brussels *without examination*. About half the documents are published within 2 to 3 months of filing and the rest within 6 months. This is usually before their counterparts anywhere else.

2. West German patent applications are published weekly under the new law *without examination,* about 18 months after the earliest priority date—which for convention applications (see Section 13.20) is normally about 6 months after the filing date. Some chemists believe that published West German applications may be among the most useful of the quick-issue patents, because: copies can be obtained quickly and easily; if the technology and chemistry are important, an application will probably be filed in West Germany; the language is more familiar to chemists and translators than any other language with the exception of English; and, in addition to outstanding coverage by Derwent, West German patent applications are thoughtfully abstracted and thoroughly indexed by *CA* if they are first issued in that country as noted in Section 7.12. *Examined* West German applications may be published up to 10 years later. This is an illustration of the importance of quick-issue (unexamined) patent applications.

In recent years, European Patent Office patent applications (see Section 13.20), have become increasingly important. A large percentage are in English, and the documents sometimes come complete with search reports. British applications are also very important for similar reasons.

Because the current track record (or performance) of countries such as Japan and West Germany is impressive, any chemist who wants to keep up with the latest technology is advised to follow information sources covering these quick-issue countries and, in addition, those for other major quick-issue locations, especially from the European and British Patent Offices. Close watch of foreign patent activity is particularly important in industry, but can be of equal value in academic and other organizations.

Despite all that has been said about some of the advantages of quick-issue foreign (non-United States) patent applications, issued U.S. patents are frequently of great importance and deserve the careful attention of chemists in the United States and in other countries. Significant foreign inventions are usually patented in the United States if possible, because it is such an important market. In 1985, 45 percent of all U.S. patents were issued to foreign applicants.

13.9.   OFFICIAL GOVERNMENT SOURCES OF PATENT INFORMATION

An important way for chemists to learn about new patents is to read the patent gazettes usually published weekly by patent offices of most major industrialized nations. Here a chemist will find the title, and sometimes principal claim or abstract, depending on the country, of new patents.

A chemist can use the gazettes to keep up to date with patents; this is usually the quickest way for any single country. But this can be difficult for a variety of reasons. One is that key patents may turn up in material issued by unexpected countries and in unfamiliar languages. Furthermore, many chemists find that information given in the gazettes is much less useful than the patent abstracts that appear in such services as *Derwent* and *CA* described elsewhere in this chapter. The gazettes, however, provide some information useful to laboratory chemists, for example, in helping decide whether to try to locate an abstract that would provide more information or to obtain a copy of the full patent.

13.10.   OBTAINING INFORMATION ABOUT UNITED STATES PATENTS

The *Official Gazette,* compiled since 1872 by the U.S. Patent Office, provides a listing of claims (usually only the main claim) for all U.S. patents issued each week. Copies of the *Gazette* can be found in many university and public libraries and in almost all industrial research organizations.

Patents listed in the *Gazette* are arranged in three major categories: (1) general and mechanical, (2) chemical, and (3) electrical. Note that there may be some overlap among these categories. A patent of chemical interest may be found in one of the other sections. Within each category, patent claims are grouped in numerical sequence according to a classification system established and maintained by the Patent Office.

Other categories in the *Gazette* are Defensive Publications and Reissue and Reexamination patents. Defensive publications are unexamined patent applications published at the applicant's request to prevent another inventor from patenting the concept (see Section 13.22). In May 1985, the Defensive Publication series was replaced by the Statutory Invention Registration series which now appears as a separate category. A reissue patent is one that is granted when a substantive error is discovered in a previously issued patent. Approximately 500 reissue patents are published annually.

Patent applications are currently (1986), being filed with the U.S. Patent Office at the rate of approximately 125,000 applications per year. Patent Office examiners presently allow about two-thirds of all applications received. In 1985, a total of 75,302 utility, design, and plant patents were issued.

The U.S. Patent and Trademark Office recently established a new procedure for review of granted patents as a way to reduce the amount of patent related litigation presented to the courts. Under the new rules, on payment of a large fee, anyone may apply for reexamination of any U.S. patent. If the reexamination procedure results in changes to the patent, the patent is assigned a reexamination number, and the affected portions of the patent are published in the *Official Gazette*.

To use the *Gazette* effectively, the chemist must identify pertinent classes and subclasses that reflect the interests of the chemist or his or her organization. This identification can best be achieved with the aid of a patent attorney or by direct communication with the U.S. Patent Office, since the classification system is complex and is revised at intervals. This approach to locating patents (scanning of the *Gazette*) is at best very risky and can result in significant information loss.

An obvious weakness of the *Gazette* is lack of cross-references to related patents. This means that important subject matter will often appear in subclasses other than that in which a patent of interest appears in the *Gazette*.

The annual indexes to the *Gazette* have been appearing many months or even several years after completion of the calendar year. Beginning in 1976, a commercial service (IFI/Plenum Data Co.) started publishing quarterly and annual indexes to help bridge the gap (2) (see Section 13.13).

Furthermore, the annual index to the *Official Gazette* does not contain a subject index, although all U.S. patents issued that year are classified by class and subclass. This feature is useful to the patent attorney or agent. It can be a difficult means of access for the working chemist because of the complexities of the classification, as noted previously, but use becomes much easier if the chemist studies the classification and knows the field.

If a specific inventor or organization is of interest, access by this key is relatively easy. It involves scanning the indexes to the *Gazette* that appear weekly and are cumulated annually.

When looking for organizations in patent indexes, the searcher must consider corporate name changes and other variations. Companies merge, acquire partially or totally owned subsidiaries, disinvest or spin off, or change names to reflect new activities and policy changes. For example, a historical search over a period of years for patents issued to one company might include such variations as:

Mathieson Alkali Works
Mathieson Chemical Corp.

Olin Corp. (this is the current name)
Olin Industries, Inc.
Olin Mathieson Chemical Corp.

Alternatively, some patents exhibit organization name variations for other reasons, including simple error.

Several organizations that provide online patent databases, as described later, have attempted to facilitate searching for alternative organization names by standardizing these names whenever possible. These include Derwent, IFI, and *INPADOC*. Even so, patents issued to subsidiaries or affiliated companies may be missed, as may patents that are not initially assigned.

Nomenclature is even more highly variable than organization names and can also lead to loss of information, unless the searcher is on the lookout for all possible variations. Use of CAS Registry Numbers in online databases can sometimes eliminate or minimize problems, and another approach is use of one of the systems that permits graphic searching for structures (see Section 10.11). However, patent searching often requires a highly generic approach and online systems have not yet perfected this approach.

Another way to search for and identify U.S. patents is to visit the Public Search Room of the Patent Office located at Crystal Plaza,* near Washington National Airport. This is the only place where all U.S. patents are arranged by subject matter according to a *Manual of Classification* (3), which helps the user decide which of over 300 main categories or *classes* to search. These classes are further broken down into more than 90,000 subclasses. The searcher removes the selected subclass bundles of patents from their open-stack arrangement, takes them to a desk, and inspects or reads them in their chronological order.

If a patent discloses subject matter classified in two or more subclasses, a copy is placed in each subclass. To distinguish these copies from each other, the copy placed in the subclass selected as the principal basis of classification is called an original, and all others are called cross-references. Both original and cross-reference patents are contained in classification bundles in the search room.

It is important to be aware that patents are constantly being reclassified by the U.S. Patent Office, especially in the most active technologies. For this and other reasons, the class and subclass groupings need to be used with both care and skill.

Note that almost all U.S. patent applications being examined for possible issue are *not* available to the public. (Exceptions to this rule include all patent applications developed by the U.S. Government that meet certain criteria; these applications are published by the U.S. National Technical Information Service.)*

Although most U.S. patent applications are confidential, the quick-issue (quick-publication) systems used in some other countries can often provide important clues to developments for which applications might be filed in the United States. Unfortunately, the public search room at the U.S. Patent Office is geographically inconvenient for regular use by most chemists, except those who live in the Washington, D.C. area, or whose business brings them to that location. All are, however, always welcome to use this facility.

For some years now, there has been talk in government and private circles about establishing one or more counterparts of the existing search room in other locations. This concept is still "alive," but no decisions have been made on whether the idea will be implemented. Chemists would do well to follow any discussions on the concept, because the result could be more ready access to categorized patent files for many.

Although the search room is available to all, it is designed primarily for use by patent attorneys, patent agents, and other trained patent searchers. Laboratory chemists and other inventors desiring to use the search room for the first time should seek assistance from the staff of the room, or ideally be accompanied and guided by a patent attorney or agent from their own organization who knows all Patent Office facilities and how to use them efficiently. Chemists and other search room users can make good use of Patent Office booklets on how to use the search room and associated facilities. In addition to the patents in the public search room, unofficial subclasses and foreign patents and literature can be found in examiner search areas and may be searched with the permission of appropriate officials in these areas.

A Public Search Center is operated by Derwent, Inc. at the Patent Office on a fee basis. The Center can respond to mail or telephone inquiries.

The scientific library of the Patent Office includes official journals of foreign (non-United States) patent offices and millions of foreign patents, as well as many other reference sources.

As a convenience for those who cannot visit its facilities in Arlington, the Patent Office will supply, for a fee, lists (patent numbers) of original or cross-reference patents contained in the subclasses comprising the *field of search*. The same information may be obtained on a visit to the Patent Office or to any of the Patent Depository Libraries (see Table 5.1) through the Patent Office's Classification and Search Support Information System (CASSIS). In addition to containing the original and cross-reference classification numbers for all U.S. patents, this online system searching of: (a) patent abstracts for the 3 most recent years (e.g., in 1985 for 1982–84)*; (b) titles in the official Manual of Classification by word or number; (c) Index to U.S. Patent Classification; and (d) manual of patent examination procedures. There is no charge for use of CASSIS by Patent Depository Libraries. However, it is not nearly as powerful a chemical patent search tool as Derwent, *CA,* or IFI *Claims*™ for the more recent years. Original and cross-reference information may also be obtained through use of the online systems offered by Derwent, IFI, Pergamon, and others (see Table 13.1). The

---

*However, actual number of years covered varies with budget.

TABLE 13.1. Some Representative Online Databases That Have Chemical Patent Information[a−e]

Vendor	Name	Primary Country Coverage	Approximate Start of Coverage
Derwent (SDC®, DIALOG®, Questel)	WPI, WPIL	International	Mostly 1970, some back to 1963
Derwent (SDC® only at present)	USP70 and USP77; USPA	United States	August 1970
CAS	CAS ONLINE (STN^SM)	International; includes other types of documents in addition to patents	Primarily 1967
IFI/Plenum (DIALOG®, SDC®, STN^SM)	Claims™/U.S. Patents and Claims™/ UNITERM; Claims™ Comprehensive Data Base; Claims™/ Citation	United States	1950; 1947 for citation files
BRS® Information Technologies	PATDATA®	United States	1975
Pergamon InfoLine®	PATSEARCH®	United States; Patent Cooperation Treaty	1970; 1978
Pergamon InfoLine®	INPADOC; IN-PANEW	International equivalents	Late 1960s (varies by country)
Questel	INPI[c]	France	1969
	INPI[d]	European & Patent Cooperation Treaty	1978
	INPI[e]	International (European Patent Office)	Some as early as 1964, but varies with country
Mead Data Central	LEXPAT	United States	1975

[a] The files listed in this table are not equivalent. Some provide considerably more information and are more fully indexed than others. The most significant databases are those of Derwent (WPI and WPIL), IFI/Plenum (Claims™), and CAS.

[b] Most of the files that cover U.S. patents exclusively also provide patent citation searching capability.

[c] See World Patent Inf., 7(1/2) (1985). This issue is dedicated to online patent information.

[d] Time period is only approximate and there may be gaps.

[e] There are a number of other pertinent databases. For good lists, see E. S. Simmons and F. C. Rosenthal, "Patent Databases," World Patent Inf., 7, 33–67 (1/2) (1985) and C. Culp, "Patent Databases," Database, 7, 56–72 (1984).

chemist should seek the help of a patent attorney or of other appropriate experts in defining the pertinent subclasses.

Copies of patents can be purchased, or they can be inspected, at a library hopefully located within reasonable travel distance (for a list of these libraries see Table 5.1), which has a numerically arranged set of patents. Additional libraries are added to this list as state or local funds become available. Unfortunately, 12 states still have no Patent Depository Libraries (as of 1986). These are Connecticut,* Hawaii, Iowa, Kansas, Kentucky, Maine, Mississippi, North Dakota, South Dakota, Vermont, West Virginia, and Wyoming. See Section 5.6 on obtaining patent copies.

Although the United States Patent and Trademark Office is an impressive operation, with an annual budget of about $200 million, in the 1970's, the volume of prior art (older patents and articles) and the influx of new materials became so great that the staff of approximately 800 examiners was unable to keep up. The average pendancy period (time from patent filing to issuance or abandonment) increased to 25.3 months in Fiscal Year 1984. Another problem has been that an unacceptably large percentage (estimated at about 7%) of Public Search Room files has been incomplete because of misfiling or for other unexplained reasons. To combat this situation, several important steps were initiated by the since-resigned Commissioner of Patents and Trademarks, Gerald J. Mossinghoff, in 1982–1983. The present (1986) Commissioner is Donald J. Quigg.

The corps of examiners is being increased to about 1500 persons. In addition, extensive automation plans are in progress, with completion scheduled for 1990. Almost all U.S. patents from August 1970 to date are already machine readable; however, approximately 128,000 patents, largely chemical, have still not been keyed into the system by the Patent Office except claims and bibliographic data.

Patent Office plans call for all post-1970 (and for selected earlier patents) to be full-text searchable and for the corresponding drawings to be computer processable. The entire patent back file is to be searchable by subclass. In addition, foreign references and technical literature are to be computer retrievable or searchable, systems such as those originally developed by CAS are to be employed to facilitate location of chemical information in patents, and CAS is to handle automation software under contract with the Planning Research Corporation.

When fully implemented, these improvements should materially help Patent Office examiners as well as chemists and other users—both those who come to the Search Room in person and those who use it remotely by means of data terminals in Patent Depository Libraries throughout the United States (Table 5.1).

Computerized online searching of U.S. and foreign patents—using databases such as those of Derwent, IFI (United States only), Pergamon, *CA,*

*A patent depository is to be established at Science Park, New Haven, CT in 1987.

and Mead (United States only) as described in this chapter, and those that the U.S. Patent Office is developing—is a very powerful technique. Nevertheless, some patent attorneys and others believe that there is no substitute for first-hand, on-site seaching through the files of the Patent Office in Crystal Plaza.

The following sections describe major patent information tools or sources of interest to chemists. Almost all of these are available in both printed and online forms. Online searching has become an especially powerful and efficient patent searching procedure. In addition to the discussion in the text, Table 13.1 summarizes some representative databases that have patent data.

13.11. DERWENT PATENT INFORMATION SERVICES

The leader in coverage of worldwide chemical patent information is Derwent Publications, Ltd., the main office of which is located in England.* Derwent was founded in 1951 by Montagu Hyams, their principal executive for many years. It is now one of the operating units within Thomson Data Ltd. There is a full-time staff of 450 and, in addition, there are some 600 part-time abstractors and 100 data input operators, mostly working from their homes.

A major enhancement of Derwent for chemists in the United States and Canada is that Derwent now has a full-time staff in the Washington, D.C.* area in addition to the staff at the home offices in London. The staff can be of expert help in answering questions about how to use the online files and the other services that are available.

The first Derwent product was an abstracting service covering new British patents. This was followed by a series of similar bulletins covering German, Belgian, French, Dutch, Soviet, and Japanese patents. In the 1960s, Derwent initiated a number of discipline-based current awareness and retrospective services to supplement the previously existing country-based current awareness services. The new subject service included pharmaceutical patents (*Farmdoc,* founded in 1963), agricultural chemical patents (*Agdoc,* 1965), and polymer patents (*Plasdoc,* 1966). These three services were consolidated, and chemical patents coverage was expanded, to produce the *Chemical Patents Index,* which began in 1970 as the *Central Patents Index.*

A. Derwent *Chemical Patents Index*

Derwent *Chemical Patents Index* (*CPI*) is a major, highly significant abstracting and retrieval service that deals with chemical and chemically related patents worldwide. In *CPI* and other services, Derwent covers about 5500 newly described inventions each week that it terms "basic patents." The other (approximately 6500 per week) patents covered by Derwent are called "equivalents." Usually, equivalents are simply patent applications (or issued patents) corresponding to the basics but filed in other countries.

All patents for a single invention are grouped into a "family," and each family is assigned a unique number by Derwent. This procedure greatly simplifies efforts of chemists and others to cope with the voluminous and sometimes confusing multilingual and multinational patent literature. Patent equivalents and families are further discussed in Section 13.17.

Each week, Derwent publishes Alerting Bulletins with English-language abstracts of the latest inventions arranged by country, and a week later, by Derwent class. The following week, a bulletin giving more detailed and coded abstracts of "first disclosure" and "basic" patents appears.

Reading of Derwent classified alerting bulletins is the single best way for laboratory chemists to conveniently keep up with new advances in patented chemical technology on a worldwide basis. Sections A (*Plasdoc*) and E (*Chemdoc*) are especially recommended. A complete list of all Derwent sections is given later in this section.

The 12 *major* countries covered from the start of *CPI* in 1970 are Belgium, East Germany, France, Japan, the Netherlands, South Africa, United Kingdom, and West Germany, all of which publish patent applications just a few months after filing and are known as *quick-issue* countries; and the slower publishing countries of Canada, the Soviet Union, Switzerland, and the United States. In addition, European Patent Office and Patent Cooperation Treaty (PCT) patents (see Section 13.20) have been covered beginning in 1979. Sweden became a major country in the same year. Coverage of all articles from the publication *Research Disclosure* (see Section 13.16) was introduced early in 1978, and of articles from the publication *International Technology Disclosures* in 1984. Australia has been covered as a *major* country from the start of 1983.

The 14 *minor* countries introduced since 1975 are the rapidly publishing (quick-issue) countries of Brazil, Denmark, Finland, Luxembourg (in 1984), Norway, Portugal, Spain (from July 1983), and Sweden (now major); and slower publishing Austria, Czechoslovakia, Hungary, Israel, Italy, and Romania.

TABLE 13.2. Derwent Printed and Microfilm Indexes

Alerting Bulletin Indexes	Microfilm Indexes
Patentee	Patentee
Patent number	International Patent Classification
Derwent classification	Accession number
Accession number	Patent number family
Basic Journal Indexes	Priority concordance
Patentee	
Patent number	
Derwent number	
Manual code	

Chemists can use and retrieve Derwent patent information using one or more of several approaches:

1. Conventional (visual) scanning of the printed product. Of the several approaches listed here, this is by far the most common and easiest way of keeping up with new patents covered by Derwent each week.
2. Consulting of issue and cumulative indexes. These are available by patent number, patentee, Derwent accession number, patent number family, manual *CPI* class, manual code, priority, and International Patent Classification. There is no subject or molecular formula index, however. Not all of the different types of indexes are available in printed or cumulative forms. See Table 13.2.
3. Use of Derwent manual code cards.
4. Use of magnetic tapes within one's own organization.
5. Use of online service (see Section 13.11.C).
6. Use of Derwent *Bureau Service* under which Derwent personnel will search their files on either a current or retrospective basis. Bureau Service is available to anyone on a fee basis whether or not a person is a subscriber to Derwent.

Abstracts and complete specifications of all basic patents covered are available from Derwent in microfilm form, and full paper copies of all specifications may also be obtained from Derwent.

In *CPI,* patent documents are divided about equally between those that relate to entirely new inventions or basics, and those for which corresponding patents have already been published in another country or equivalents.

For the basics, Derwent editors prepare a meaningful descriptive title in English. Almost all patents are accompanied by English-language abstracts of considerable length. An example and/or drawing is frequently included.

Chemical patents account for about 45% of total Derwent coverage. They are selected for inclusion in one or more of the 12 sections:

A Plastics and Polymers (*Plasdoc*)—covers polymers, fabrication, additives, uses, and specified monomers; began in 1966.
B Pharmaceuticals (*Farmdoc*)—patents of pharmaceutical, veterinary, and related interest; began in 1963.
C Agricultural Chemicals (*Agdoc*)—compounds of agricultural and specified veterinary interest; began in 1965.
D Food, Detergents—includes disinfectants; began in 1970.
E General Chemicals (*Chemdoc*)—general organic and inorganic compounds and dyestuffs; began in 1970.

F Textiles, Paper—began in 1970.

G Printing, Coating, Photographic—began in 1970.

H Petroleum—began in 1970.

J Chemical Engineering—began in 1970.

K Nucleonics, Explosives, Protection—began in 1970.

L Glass, Refractories, Electrochemistry—began in 1970.

M Metallurgy—began in 1970.

To meet the needs of chemists who have highly specific areas of interest, a number of *basic* patent abstracts booklets covering quite narrow fields are issued as *Profile Booklets*. An outline of topics covered in all sections and booklets is given in brochures readily available without charge from Derwent.

All Derwent patent data are available for online searching by subscribers. At a higher rate, nonsubscribers may also search the online files except for certain coding categories.

B. Derwent *World Patents Index* and *World Patents Abstracts*

The first announcement of a patent in any of the Derwent services is in *World Patents Index* gazettes (*WPI*)—a feature that makes this a unique and important publication. This is a collection of indexes giving details of all inventions covered in the countries previously mentioned and classified according to patentee, subject matter, patent family, patent number, and priorities claimed.

WPI gazettes are published in four editions—general, mechanical, electrical, and chemical—based on the *International Patent Classification* (*IPC*) and patentee. Abstracts are *not* included in *WPI*.

The weekly *WPI* gazettes are useful for watching for patents by patentee, for determining whenever a given invention of interest appears or reappears in another country, and for following by subject matter using the *IPC* that provides a generic form of access.

World Patents Abstracts (*WPA*) is a collection of abstracts of inventions reported one week earlier in *WPI*. On a current basis, *WPA* takes the form of 12 different weekly publications, each devoted to a particular country or treaty, and 7 weekly journals, each dealing with a different area of technology on a multicountry basis, as shown in Figure 13.2.

Note, however, that these *WPA* journals arranged by subject cover only nonchemical subjects. Abstracts of chemical patents are provided by Derwent's *CPI,* which is more suited to the needs of chemists than either *WPI* or *WPA*.

Some of the newer Derwent products and services are summarized in Table 13.3.

WPA — PRINTED PRODUCTS BY COUNTRY

COMPLETE SUBJECT MATTER COVERAGE

BELGIAN PATENTS ABSTRACTS
Each week covers about 50 unexamined Belgian specifications OPI within 3-6 months of filing. Also gives a numerical index of granted European patents designating Belgium.
Complete Edition only.

BRITISH PATENTS ABSTRACTS
About 400 unexamined British, including PCT applications designating Great Britain, are covered each week, together with the main claim of the 400 granted patents published each week, along with a drawing. Also gives a numerical index of granted European patents designating Great Britain.
Three Sectional Editions – Chemical, General/ Mechanical and Electrical.
Two complete Editions one week later in numerical (and main GB class) order – Unexamined and Granted (including PCT applications designating GB).

EUROPEAN PATENTS REPORT
Detailed abstracts plus drawings of some 650 unexamined EP applications (including PCT documents designating EP) and 350 granted patents per week.
Three Sectional Editions – Chemical, General/ Mechanical and Electrical.
Two complete Editions one week later in numerical order – Unexamined (including quarterly delayed search report Edition) and Granted.

GERMAN PATENTS ABSTRACTS
Covers about 400 examined applications each week and includes a numerical index of granted European patents designating West Germany.
Chemical Edition and combined General, Mechanical, Electrical Edition.

GERMAN PATENTS GAZETTE
Unexamined applications and PCT applications designating West Germany totalling about 700 patents per week are covered.
Three Editions published. Chemical, General/Mechanical and Electrical.

PCT PATENTS REPORT
Issued every four weeks and covering about 400 unexamined applications per issue.
Complete Edition only.

SOVIET INVENTIONS ILLUSTRATED
Includes abstracts of patents and authors certificates based upon the complete documents. Also covers PCT applications examined by the Soviet Patent Office. Total coverage 1560 documents per week.
Three Editions published. Chemical, General/Mechanical and Electrical.

UNITED STATES PATENTS ABSTRACTS
About 1350 documents per week are covered, including PCT applications passed to the United States Patent Office for examination and US National Technical Information Service documents.
Three Editions published. Chemical, General/Mechanical and Electrical.

CHEMICAL COVERAGE ONLY

FRENCH PATENTS ABSTRACTS
Contains about 110 unexamined applications each week and 350 South African abstracts monthly. A numerical index of granted European patents designating France is also provided.

JAPANESE PATENTS GAZETTE
Covers about 1600 unexamined 'Kokai' documents each week, including PCT applications passed to the Japanese Patent Office for examination.

JAPANESE PATENTS REPORT
Covers about 400 examined applications each week.

NETHERLANDS PATENTS REPORT
Four issues per month: two each containing about 70 abstracts from OPI documents, with PCT applications passed to the Netherlands Patent Office for examination; the other two issues each containing abstracts from around 60 examined specifications. All issues have a numerical index of granted European patents designating Netherlands.

Figure 13.2. Derwent WPA printed products by country.

C.   Online Access to Derwent *WPI/WPIL*

Chemists and others can use Derwent through online files designated as *WPI* and *WPIL* (*World Patents Index Latest*). The latter covers patent documents from 1981 to date, while the former covers earlier patents.

Online access to Derwent was first offered beginning in 1976 through the facilities of System Development Corporation (*SDC*®), Santa Monica, CA.

TABLE 13.3. New Derwent Products and Services for 1985

European Patents Report (unexamined)—overall numerical order
European Patents Report (granted)—overall numerical order
British Patents Abstracts (unexamined)—overall numerical order
British Patents (granted)—overall numerical order
Primary–secondary accession number index
New online hosts—*Questel, DIALOG®* Information Services
Statistical analysis—microcomputer based
   This is intended to help provide statistics relating to time, country, company,
   and technology. Other parameters include patent priority, size of patent
   families, designated states (countries), and cited patents. Results can be used
   in detection and analysis of trends.
Chemical code menu

Derwent online files consist of more than 2 million patent families and about 5 million patents in all and are increasing at the estimated rate of approximately 225,000 patents per year. This online database is now also loaded on the SDC® computer in Tokyo.

An important development regarding Derwent *WPI* and *WPIL* patent files is that they are also available through vendors and systems beyond availability through SDC®. The new systems through which Derwent is available are *Questel* and *DIALOG®* Information Services (see Table 13.1 and Chapter 10).

In *Questel,* Derwent files are to be *cross-file* searchable with other files, especially those related to chemical structure, and the French Patent Office file. Cross-file means that when pertinent references are located in one file they may be saved and transferred for further use in other related files. Graphic chemical substructure access is planned through *DARC* (see Section 10.11).

WPI coverage online begins in 1963 for pharmaceuticals, in 1965 for agricultural chemicals, in 1966 for polymer chemistry, in 1970 for all other fields of chemistry, and in 1974 for other fields of technology.

Derwent time period coverage online compares favorably with the start of *CA* files online (these begin in 1967), but less favorably with that of the IFI *Claims*™ file of U.S. patents, which begins in 1950 (see Section 13.13D).

Patents covered by Derwent can be searched for online in several ways, the simplest of which is by title. Highly informative titles are assigned by Derwent staff to all basic patents covered. Words in the title may be accessed through a thesaurus that lists preferred forms of words to which certain listed variants have been converted. In addition, title words can also be searched in their free language form.

Full texts of Derwent abstracts are online, and words in abstracts are also searchable in their free language forms. If desired, abstracts may also be printed online.

Other approaches to searching Derwent online are listed here. These can be used alone or in any combination, including title words and abstract words when available.

- Derwent accession number.
- Derwent accession number year.
- Derwent company code (this is the Derwent code for the company to which the inventor has assigned the patent; the company so designated is known as the assignee).
- Derwent classes (these are broad categories that provide a degree of subject access).
- International Patent Classification (IPC) numbers (these are more extensive categories than Derwent classes and provide additional access by subject category but not nearly in as much detail as the indexes of *CA*.
- Derwent Manual and Fragmentation Codes (see Section 13.11.D)
- Priority year and priority country (this information can help identify related patents and families and can help limit searches to specific countries).
- Patent number and patent country (this information can be useful in locating equivalents or patent families—thus, if one has a Japanese patent number and wants to determine if there is an English-language equivalent, it can be done in this way).

Derwent online files are continually updated with additions to patent family lines, *IPC,* and inventors' names.

D.   Derwent Manual and Fragmentation Codes

Manual codes are alphanumeric index entries intended to provide another means of access to the patents covered. Some organizations subscribe to the Derwent manual code cards. A complete file of these permits chemists to scan manually, but rapidly, through large numbers of patents in specific categories of interest. These codes may also be searched online.

Fragmentation codes provide the indexing needed to assist in searching for the structural concepts and generic aspects so typical of chemical patents. These codes are searchable only online. Use of manual and fragmentation codes online is restricted to those Derwent subscribers whose subscription fees qualify them for use of these features.

E.   Efficient Use of Derwent

Searching efficiently takes special knowledge, best acquired by starting with attendance at training classes that Derwent offers at intervals, followed by

GENERAL REFERENCE

1A   Derwent Patents Manual – General
Describes in detail the Derwent patents products, countries covered including special points relating to each, classification systems and descriptors used (April 1986).

1C – Online Instruction Manual
Explains the implementation of the Derwent WPI/L patents database on SDC, DIALOG and QUESTEL. Describes the available search parameters, print commands and retrieval techniques for successful patents searching. (June 1985).

1E – List of Standard Thesaurus Terms
Contains the authority list of some 43,000 terms used to edit the Derwent assigned title words, incorporates all previous amendments made since the first edition and includes a further 300 new terms. (January 1983; Suppl. January 1984).

2A – Company Code Manual
Lists all 13,700 standard companies and their codes, including related companies and cross-reference entries. Gives rules for establishing non-standard codes with numerous examples, and a glossary of preferred word abbreviations. Two volumes, name order (Part 1) and code order (Part 2) (April 1985).

SPECIAL CODING

2B – Manual Code Manual – CPI
Gives all Manual Codes applicable to each section of CPI, including section N (Catalysts), each code having a brief description of its meaning with scope notes where appropriate. Contains an alphabetical index of concepts for all CPI sections with their corresponding Manual Codes. (January 1983).

2C – Manual Code Manual – EPI
Lists all Manual Codes applicable to each section of EPI, each code having a description of its meaning. Contains an alphabetical index of concepts for all EPI sections with their corresponding Manual Codes. (January 1983).

2E – Derwent CPI Registry Compounds
This manual contains two separate alphabetical lists of commonly occurring compounds and their corresponding Derwent registry numbers. One list (750 compounds) is for CPI Section A and the other (2,000 compounds) for Sections B to M (April 1984).

3A – Chemical Punch Code Manual
This is a detailed description of the Fragmentation Code used for the coding and retrieval of chemical compounds with their uses, applications and production, in CPI Sections B, C and E.

3D – Chemical Code Dictionary
Simple-to-use guide to the Chemical Code. After a brief explanation of the code, gives a detailed alphabetical list of various concepts which are retrievable for sections B, C and E. The list includes terms for uses, applications and processes as well as those for chemical fragments. For each term the online search logic is provided. The manual also contains an alphabetical list of the 2,000 Derwent registry compounds. (January 1983).

3E – Chemical Code Menu Program Manual
Contains as Part 1 full instructions on the use of the menu program, and provides as Part 2 a list of associated formulae, definitions and notes.

Also included, as Part 3, is a printed version of the menu frames which can be used, as an alternative to the program itself, by anyone who does not have a suitable microcomputer. (August 1984).

4A – Combined Plasdoc Punch Code and Key Terms
Contains a description of the multipunch code, and instruction for retrieval, in both subject and term order with definitions alongside. Also included are specific serial numbers (KS) used to retrieve Plasdoc coded material from the A series onwards and a list of Plasdoc registry compounds.

4B – Numerical Lists of Plasdoc Codes and Key Terms
Contains separate numerical lists of: (a) Individual punch positions and concepts associated therewith and (b) Key Term serial numbers. (September 1984).

4F – Plasdoc Code Dictionary
Simple-to-use, detailed alphabetical list of the various concepts which are retrievable using the Derwent Plasdoc Code. The list includes terms for polymers and their associated concepts. For each term the online search logic is provided. (September 1984).

The manuals are separately bound but are readily convertible by the user into loose-leaf format. Holes are punched for easy insertion into a variety of binders. Corrections are issued on loose-leaf pages.

Figure 13.3. List of Derwent patent instruction manuals.

ongoing experience. Also, up-to-date user manuals and other user aids are available (see Figure 13.3).

From a current awareness standpoint, however, it takes no special skill for chemists to scan and read the printed Derwent bulletins. Using these bulletins, chemists can keep up with a broad range of new developments in chemistry and chemical technology on virtually a worldwide basis. Although most chemists do not have the time or need to read all Derwent bulletins, they can select those of most interest, and within those bulletins, limit themselves to categories of principal interest.

F. Other Derwent Services

Derwent provides coverage of electrical and electronics patents through its *EPI* (*Electrical Patent Index*) service, which was started in mid-1980. This product is included in the *WPI* and *WPIL* online files described previously.

In addition to its own services, Derwent, Inc. offers U.S. Patent and Trademark Office (USPTO) files online. These files contain data from USPTO computer readable files regarding issued U.S. patents from 1970 to date. Data include patent classification, full claims, abstract, citations, and bibliographic details. Users may execute cross-file searches between the Derwent USPTO file and *WPI/WPIL*. This means that results of a search in one database can be held and entered as a search statement in another without having to keyboard it in. This has such advantages as simplifying searches for equivalents.

Derwent also provides specialized journal literature information services for the pharmaceutical industry and a chemical reactions documentation service (see Section 12.4.D), both available online separately. Other fields covered by Derwent include pest control, biotechnology, and veterinary science.

G.   Advantages and Disadvantages of Derwent

The principal advantages of Derwent services are these:

1. Coverage is comprehensive and virtually worldwide. Derwent editors interpret the definition of what constitutes a chemical patent broadly, so that more material is likely to be found in Derwent than any other comparable service.
2. Abstracts are provided for both equivalents and basics. Almost all patents included are given abstracts.
3. Abstracts are lengthy and detailed, often including an example and/or a drawing. All abstracts are in English.
4. Speed of coverage is excellent.
5. Indexes to assignees are helpful in locating patents assigned to a given organization.
6. Weekly and cumulative indexes that show related families of patents (or equivalents) are thorough and especially useful.
7. The Derwent Patents Copy Service saves time and minimizes or eliminates the red tape and frustration often associated with obtaining patent copies.
8. Abstracts contain patent-filing details and other related legalities not ordinarily found in any other comparable service. This is of special value to patent-oriented chemists and patent attorneys.
9. Online searching is quick, flexible, and highly efficient.

Some disadvantages of Derwent include:

1. There are no conventional printed substance or subject indexes with standardized and systematic nomenclature such as those of *CA*. Instead, title term and abstract word access is available online, and Derwent codes and classification systems may be used.

2. If the chemist's organization subscribes to all parts of Derwent, the service can be an expensive addition. This may mean that only organizations with excellent budgets can afford to get all parts of this important service. Subscribers must pay for a certain minimum number of subscription *units* to have complete access to all data online.

3. Time coverage of the service is relatively limited. In the case of general chemistry, the beginning of truly comprehensive coverage goes back only to about 1970.

4. Many chemists find overall use of Derwent and its components somewhat difficult, despite the availability of printed manuals and other aids and of training classes on appropriate use. This difficulty can be overcome by careful ongoing study by the chemist or information specialist who intends to become a regular user of Derwent and to use it to maximum efficiency.

5. Some believe that Derwent should provide coverage of additional countries, and that present coverage of the large volume of unexamined Japanese patent applications could be improved, even though these improvements would be expensive and would increase subscription costs.

6. As of early 1986, graphic structure searching of Derwent files is not possible. Ongoing discussions with *Questel–DARC* are expected to resolve this matter soon.

13.12. COVERAGE OF PATENTS BY *CHEMICAL ABSTRACTS*

Chemical Abstracts (*CA*) is a major source of chemical patent information. Some aspects of the treatment of patents by CAS are touched on in Chapters 6 and 7, and elsewhere in this chapter. From 1981 through 1985, patents constituted about 31% of all documents cited by *CA*. The largest fraction of these citations is to unexamined Japanese patent applications, which began an explosive growth trend in 1972.

CAS claims to cover patents of chemical and chemical engineering interest for certain countries, as shown in Table 6.7.

Published West German patent applications (Offenlegungschriften) are designated by *CA* as *Ger. Offen.* Granted patents (Patentschriften) are designated simply as *Ger.* The previous Auslegeschriften series was discontinued by the German Patent Office on January 1, 1981.

CA abbreviates the unexamined Japanese patent application as *Jpn.*

Kokai Tokkyo Koho. (Some chemists refer to a Japanese patent application simply as a *Kokai.*) The examined Japanese patent application is designated as *Jpn. Tokkyo Koho.*

CA handling of West German and Japanese patent information is especially significant. For both countries CAS provides extensive coverage and thorough indexing of unexamined patent applications. The significance of these two countries is based on their technological strength and on their quick-issue policies noted earlier in this chapter.

Compared to some other patent information sources and tools, *CA* is characterized by such features as the following:

1. *CA* is probably more widely available and used in all types of organizations and libraries than any other comparable tool.

2. Indexing and abstracting of patents is excellent.

3. For *CA* to cover a chemical substance reported in a patent *specification,* there must be sufficient scientific or technical evidence; so-called *paper chemistry* is not enough. This evidence can consist of such data as: examples describing synthesis or manufacture; yield; boiling or melting point; or end use test results. (Note that Derwent does not require this kind of evidence but rather merely that a chemical substance be disclosed or exemplified.) However, *CA* now indexes chemical substances specifically mentioned in patent *claims,* even when unsupported by any examples or other evidence; this is the result of a gradually implemented policy change first started in 1979, and that was in place for all *CA* sections by 1982.

4. More emphasis is placed on the chemical aspects than the legal aspects, with emphasis on details found in examples.

5. Although speed of coverage is good, the primary thrust appears to be the aspects noted previously and accuracy.

Disadvantages of *CA* are discussed in Chapters 6 and 7.

CA has recently announced plans for a new online patent information service that will have expanded abstracts, more through indexing, and prompter coverage.

13.13.   IFI/PLENUM DATA COMPANY PATENT SERVICES

For the chemist interested in U.S. patents, and their equivalents in certain other countries, a key source of information is the IFI/Plenum Data Company.*

A.   *IFI Comprehensive Data Base to U.S. Chemical Patents*

At the beginning of 1986, this service contained over 600,000 chemical and chemically related patents dating back from 1950 to within a few months of the most recently issued patents. Access points (searchable) include: IFI accession number, U.S. patent number, assignee, inventor Patent Office classes and subclasses, chemical compounds, chemical fragments, and so-called general terms. The file is distinguished by such features as depth of indexing (the current average is reportedly 50 descriptors per patent), relative ease and power of structure searching, and broad span of time covered. There are several other benefits that IFI officials can point out to potential users. The complete database is best searched by computer.

Access is possible in several ways:

1. IFI operates a service bureau that will conduct searches for a fee. Contact can be made by letter or wire, but a telephone call is also desirable, because search scope and strategy can best be developed this way.

2. In some large research organizations chemists can have access through their organization's computer center, which can conduct searches as needed. Because this database is expensive, relatively few organizations have this valuable file in house at this writing.

3. This database is available online (*DIALOG*® Information Services *SDC*® Information Services, and *STN*^SM), but only to subscribers to the service.

B.   *IFI Uniterm Index to U.S. Chemical Patents*

This is a modified version of the *Comprehensive Data Base* described previously. One feature is an arrangement for manual (visual) searching—a computer is not required, although online searching capability is available and is the preferred method of use. Also, the cost is considerably less than the *Comprehensive Data Base;* this is a tool affordable by more organizations. One disadvantage is that it includes only about one half the depth of indexing of the previously mentioned IFI tool (although even at that relatively lower level, the indexing is respectable). Use of the printed (manual) version involves frequent visual comparison of long columns of numbers; this difficulty can be eliminated by using the online version. As compared to the IFI *Claims*® files mentioned in Section 13.13.D, there is considerably more in-depth indexing for chemical and chemically related patents.

C.   *IFI Assignee Index to U.S. Patents*

This service provides access to U.S. patents alphabetically either by inventor or assignee from 1970 forward. Within company or organization (assignee), patents are listed in ascending order by class and subclass. Along with the patent number, the data given include the expanded title and a letter denoting type of claim (e.g., composition (C), process (P), and machine and device or article of manufacture (M).

As compared to the index to the *Official Gazette,* this service is much more rapid: it is issued quarterly and cumulated annually. It provides im-

mediate (one-step) access to patent title and number—a process that requires two steps in the *Gazette* index. There is a printed version, and online
access is available through the *Claims*™ service mentioned below.

D.   *IFI Claims*™/*U.S. Patents*

Online access to parts of the IFI system is available in a file known on
DIALOG® as *Claims*™/*U.S. Patents.* A segment of this file covers the same
patents listed in the *Uniterm Index* and the *Comprehensive Data Base,* but
depth of indexing is much more shallow. Nevertheless, those features that
are indexed and searchable, especially when coupled with coverage back
through 1950, make this a handy tool. Searchable features include: inventors; abstract (since 1971); most comprehensive claim (for all patents since
1965); title (official title through 1971, but enriched or expanded thereafter);
assignee; class and subclass; *CA* reference when available; foreign patent
numbers (equivalents) for certain countries; U.S. patent number; and other
information. The ability to rapidly compile a list of U.S. patents to a
specified organization is just one example of many uses to which chemists
can put this file. Note that the same file (with similar or different names) is
available through *SDC*® and *STN*^SM.

A related file, known as *Claims*™/*Class,* is an online index to the *Manual
of Classification* of the U.S. Patent Office.

IFI also offers citation searching capability (see Section 13.19).

E.   Some Advantages and Disadvantages of IFI/Plenum Services

Some advantages include these:

1. Coverage extends back through 1950 for U.S. chemical and chemically
   related patents.
2. The subject indexing is in depth—considerably more so, it is said, than
   for some comparable services—and structure seaching capability of
   the *Comprehensive Data Base* file is very good.
3. *CA* references and CAS Registry Numbers are provided whenever
   possible.
4. Beginning in 1972 the type of patent (composition; process; machine or
   device) is indicated.
5. Electrical and mechanical patents can be searched online beginning in
   1963. Many of these patents are important to chemists.
6. IFI will search their files for a fee.

Some disadvantages include these:

1. Content is limited to U.S. patents. There are, however, cross-references to equivalents from five countries: Belgium, France, West Ger-

many, Great Britain, and the Netherlands. Through 1979, equivalents coverage is for chemical patents only.

2. The text of abstracts is given only since 1971, although, as noted previously, *CA* references are given when available. The printed *Uniterm Index* includes a reproduction of the claim reported in the *Official Gazette*.

3. From 1950 to 1971 the titles are mostly the official (legal) titles of patents. This practice was changed, however, in 1972, when keywords began to be added to help make titles more meaningful. More than 80% of all chemical patent titles are now modified.

4. Some chemists find the keyword indexing more difficult to use than that of a *conventional* index such as *CA*.

5. Speed of issue of some of the publications is sometimes slower than users would like for a service of this importance.

13.14. MEAD *LEXPAT*

A major step forward in availability and searching of patent literature occurred when, in 1983, Mead Corp.* introduced its *LEXPAT* online file of U.S. patents. This significant file covers the full text of all U.S. patents from 1975 to the present. Full text means that all words in the patents are both searchable and printable online.

This system is thus completely independent of potential variabilities of indexing and classification systems that are, of course, constructed by persons other than the patentees. Other patent information tools are "limited" to indexes, keywords, codes, classification systems, abstracts, claims, or first page.

Users of the Mead system can conveniently print the full text of patents included. This same feature also provides the file integrity (completeness) that the Public Search Room of the U.S. Patent Office has sometimes lacked, that is, there are no missing patents in this computer-based file for the years covered.

Some potential disadvantages of *LEXPAT* include the fact that drawings and diagrams are not currently searchable or printable. Use of the system can be expensive, but this may be equivalent to some more conventional routes. In addition, some critics claim that searching using traditional controlled vocabulary indexes may be more efficient than full-text searching. These critics argue that use of a full-text system requires users to select the proper words and phrases for search from a totally uncontrolled base developed randomly by the attorneys who write patent applications. Because there have been no definitive studies on the merits of full-text searching of U.S. patents versus other approaches, the issue remains open.

Depending on success of the initial *LEXPAT* file, Mead reportedly has plans to include earlier years (prior to 1975) as well. This would further enhance utility of the file.

13.15. PERGAMON *InfoLine*®

Pergamon *InfoLine*® offers three important online patent files. These are *PATSEARCH*®, *INPANEW,* and *INPADOC.*

PATSEARCH® contains *front page* information for all U.S. patents issued since 1970 and all published *PCT* (*Patent Cooperation Treaty*) patent applications. *PATSEARCH*® can be searched on the basis of words in abstracts and titles; U.S. and *IPC* (*International Patent Classification*) codes; cited references; bibliographic data, such as patent numbers; and applicants and inventors.

INPADOC and *INPANEW* are databases produced by the International Patent Documentation Center, Vienna, Austria. These contain information on patent documents published by 55 patent offices worldwide. *INPANEW* includes data received by the International Patent Documentation Center during the most recent 15 weeks. *INPADOC* covers from 1968 for most major patent offices and from 1973 for most others. Both databases are searchable on the basis of title words, *IPC* codes, and bibliographic data. The significance of these important files is that they provide comprehensive worldwide patent family (equivalent) information. *INPADOC* files may also be searched by direct links to *INPADOC* computers, but this is not as sastisfactory for most purposes as the Pergamon file. IFI represents *INPADOC* in the United States and Canada and can conduct searches.

As mentioned on p. 193, a special feature of these and other Pergamon files is the valuable capability of ranking output. The most likely applications for this feature include patent classification numbers (patent classes), assignees, and patent citations. The U.S. offices of Pergamon International Information Corporation are in Virginia* and the headquarters of *InfoLine*® are in England.* For further discussion of *InfoLine*®, see Section 10.12.B.

13.16. *RESEARCH DISCLOSURE*

Research Disclosure is a newer information source. It includes descriptions of inventions that organizations have decided not to pursue, for one reason or another, through the costs and other rigors of the patenting process. Publication of the description in this source is intended to block anyone from getting a patent in the area described. The purpose is thus defensive. This publication, which is issued each month by Kenneth Mason Publications, Ltd.*, is indexed and/or abstracted by *CA* Derwent and IFI/Plenum, among other sources. There is at least one other similar publication, *International Technology Disclosure.*

13.17. PATENT EQUIVALENTS OR FAMILIES

Many organizations file patent applications in more than one country, especially for inventions believed to have widespread commercial or other importance that transcends national boundaries. This kind of filing cannot be done lightly, because extensive costs can be involved.

Accordingly, if a patent application has been published or issued in a number of countries, one can be reasonably sure that the organization that filed the patent application may be practicing, or at least strongly considering practicing, what is taught in the patent on a commercial basis.

From the point of view of the chemist interested in finding and using chemical information that represents actual practice, what has just been said is important. It is equally important to note that identification of equivalents (or so-called patent families) can be useful in other ways. For example, Section 5.7 on translations notes the importance of finding English-language equivalents of foreign-language patents, thereby probably obviating the need for a translation. Also, identification of a patent family indicates that the chemist probably does not need to obtain copies of all patents in that family; they are probably, although not necessarily, similar in technical content. Members of a patent family are subject to the laws and practices of several different countries, and these vary as to what is patentable. Accordingly, the content of the members of a family may vary somewhat.

How can the chemist most effectively achieve identification of patent equivalents or patent families? This is best done through use of one of several patent concordances.

One good patent concordance system is that of Derwent. Updates are done weekly. Coverage is comprehensive, and online access is very efficient. Equivalent patents are fully abstracted. A disadvantage of the Derwent patent concordance system is that it does not go back far enough in time—only until about 1970 in many cases.

The patent index to *CA* is issued weekly, but it is a cumulated only semiannually. The concordance feature goes back to 1963 and is probably more readily available to most chemists than any other concordance. Until recently, however, the number of countries covered was very small.

To put the history of CAS reporting of equivalent patents into proper perspective, it is helpful to review the chronology:

1. Since 1981 (Vol. 94) all patents (abstracted and equivalent) are listed in a single index, *CA Patent Index*.

2. This replaced the *CA Numerical Patent Index* and the *CA Patent Concordance*. The latter was introduced in 1963 (Vol. 58) and last appeared in 1980 (Vol. 93).

3. Prior to that, between 1961 (Vol. 55) and 1962 (Vol. 57), cross-references from equivalent patents to patents first abstracted in *CA* were inserted in the *CA Numerical Patent Index*.

4. Prior to that, that is, in 1960 and earlier, title-only patent-abstract cross-references were inserted in *CA* Issues such as:

CA 54: 25941c (1960):
Aqueous dispersions of elastomers. B. F. Goodrich Co. Brit. 840,093, July 6, 1960. See U.S. 2,905,649 (*CA* 54:15990d).

In the *CA Numerical Patent Indexes* for 1960 and earlier, each such equivalent patent was listed with the reference to the title-only abstract without any indication that the *abstract* was in fact merely a cross-reference to another abstract.

5. There is no readily available record of when this title-only practice started, but there are examples as far back as 1930 (Vol. 24).

Several additional comments are appropriate:

a. The original purpose of inserting title-only patent abstract cross-references was simply to eliminate duplicate abstracting of essentially the same document, and not to keep a record of equivalent patents.

b. In those days, manual card files on patents were kept by one or two individuals who often depended on memory in trying to screen equivalent patents from duplicate abstracting.

c. Later efforts in building patent files based on priority records (cards, microfilm) were described by G. O. Platau (4).

d. Some aspects of the basically manual operation of determining patent families at CAS until 1980 were described by P. J. Pollick (5).

e. Current computerized operation based on *INPADOC* tapes was described by P. J. Pollick (6).

f. *CA* patent equivalency data are not searchable in *CA* online files. This is a major shortcoming of these files.

The most complete patent equivalent coverage for a specified time period is widely believed to be that offered by *INPADOC*. In this regard, it competes most directly with Derwent's patent equivalent location capabilities.

INPADOC files are most conveniently utilized in the United States, Canada, and the United Kingdom by means of the corresponding online file on Pergamon's *InfoLine®*, although direct access to *INPADOC* computers is also possible as a second choice and most of its services are now available via IFI.

Searching by conventional subject indexes, as can be done with *CA* or by other means, as with Derwent, is, unfortunately, not possible with *INPADOC*. There is, however, broad category access through the *IPC*, and words in the patent title may be searched. Nevertheless, *INPADOC* is not a recommended source of choice for searching patents by subject. In addition, the time period covered by *INPADOC* is considerably less extensive than that of *CA*. Some of the advantages claimed for the *INPADOC* files can be summarized as follows:

1. More complete than any other equivalents file.
2. Rapid updating—in some cases within approximately one week (versus up to two to three months for some other files).

3. Japanese and Russian titles translated into English.

4. For major national patent offices, coverage as far back as 1968.

IFI's capabilities in searching for equivalents were mentioned earlier.

13.18. FILE WRAPPERS

In addition to reading issued patents themselves, it is often helpful to read
the complete files of those U.S. patents that have previously been identified
as of interest. These are available once patents have been issued and can be
obtained in forms known as "file wrappers." These include all correspondence between the applicants and the Patent Office during prosecution of the
application and the original specifications and claims. The complete files can
often have information not in issued patents and can include original data,
including laboratory notebook pages, and explanations of the rationale behind the invention. File wrappers may be purchased from the U.S. Patent
Office.

13.19. PATENT CITATIONS

Patent citations are references cited by patent examiners in the course of
their examining patent applications to determine whether or not these applications should be issued. Citations may be to journals or books, but most
frequently they are to other patents. Literature cited by examiners is to prior
art related to the applications in hand.

The following of citations to a patent, after it has issued, helps indicate
how later inventions may employ the earlier invention as a foundation. To an
industrial company, citations of a company's patents in other patents can
indicate either potential customers or competitors. In addition, this can help
trace development of a technology. Similarly, chemists can look at patent
citations going back in time, that is, look at citations in a group of patents,
known to be of interest, to identify related art.

The principal utility of patent citations is in the clues given as to identity
of organizations or inventors performing related work of interest. Use of
patent citations is similar to the use of journal citations (see Section 8.3)
except that the reason for the citation is a technical/legal decision by patent
examiners.

Searching of patent citations is ideally adaptable to the use of online
computer systems. Online tools available to facilitate patent citation searching include:

1. IFI *Claims™/Citation* files on *DIALOG®* Information Services system.
Covers U.S. patent citations. Begins with patents issued in 1947. This is by

far the most comprehensive of the patent citation files. It is also the most
expensive to use. The files are produced by Search Check, Inc.*

2. Derwent Ltd. Files *WPI* and *WPIL* on System Development Corp.
ORBIT® system and is now available on other systems such as *Questel®* and
DIALOG®. Covers citations in European Patent Office and Patent Coopera-
tion Treaty patent documents beginning in 1979.

3. Derwent, Inc. U.S. Patent Files on System Development Corp. *OR-
BIT®* system. Covers U.S. patents from late 1970 to date. The citation data
for many patents from 1970–1974 are missing.

4. Pergamon *PATSEARCH®* files on *InfoLine®* system. Covers U.S. patents from late 1970 to date and Patent Cooperation Treaty patents from 1978
to date. A valuable feature of this database permits identification and ranking
of patents most frequently cited and other statistical analyses. In addition,
other parts of the file may be identified and ranked, for example, assignees
for patents on specified topics.

5. *BRS® PATDATA* file. Covers U.S. patents from 1975 to date.

Computer Horizons, Inc.,* offers a tracking service based on U.S. patents and patent citations from 1971 to date. Appropriate use of this service is
intended to help in assessing technological trends and in comparing and
evaluating industrial firms.

13.20. INTERNATIONAL TREATIES AND OTHER DEVELOPMENTS

International patent cooperation dates back to the Paris Convention for the
Protection of Individual Property to which the United States has been a
party since 1887. This Convention, since revised, has made it easier to
protect inventions across national borders. The most important new international patent agreements since then are the European Patent Convention of
1977 and the Patent Cooperation Treaty of 1978.

The European Patent Convention (EPC) came into effect in 1977 and full
operation started in 1978. The first patents were granted on January 9, 1980.
Ultramodern headquarters offices of the European Patent Office are located
on the banks of the Isar River near the famous Deutsches Museum in
Munich. The output, following successful uniform examination and granting, is a "bundle" of national patents, each subject to laws and regulations
of the European member country or countries for which a patent has been
requested by the applicant. This bundle is achieved on the basis of a single
application. Member countries include: Austria, Belgium, Federal Republic
of Germany, France, Italy, Liechtenstein, Luxembourg, the Netherlands,
Sweden, Switzerland, and the United Kingdom.

The EPC (or European) patent is important to chemists and their organizations for several reasons:

1. It provides an economical way to get patents filed in a number of
   countries.

2. It is another major source of quick-issue patent applications.
3. The *European Patent Bulletin* provides an English-language title and in addition, in many cases, the patents are filed in English.

Under the Patent Cooperation Treaty, which is separate from the EPC, on the basis of one international application, U.S. patent applicants can designate as many treaty-member countries as they wish at the time of filing. Such designation has the effect of lodging a national patent application in the patent offices of each of the countries designated. Patentability search reports prepared by the U.S. Patent Office are forwarded to applicants and to the World Intellectual Property Organization in Geneva. Based on these reports, and other factors, inventors can decide whether to pursue the obtaining of patents in the various countries designated. They have 20 months from the filing of an international patent application—if that was the first to be filed—to make their decisions. Use of this treaty is mainly outside Europe.

Good coverage of patents from both of the treaties is provided by both Derwent and *CA*. For European Patent Office patents, Derwent offers, in addition to its classified alerting bulletins, *European Patents Report,* which is in numerical (patent number) order.

An expectation held by some is that a single European patent will ultimately replace all individual country patents in Europe. This simplification, if it happens, would help chemists and other scientists by streamlining document access and eliminating most remaining language barriers.

In October, 1983, representatives of the European, Japanese, and U.S. patent offices agreed to continue and extend cooperation including exchange of specialists, introduction of automation, document classification and indexing, and sharing of PCT search results. For example, the Japanese Patent Office agreed to provide the other two offices with magnetic tapes of English-language abstracts and bibliographic data from *Patent Abstracts of Japan*. Another development relating to Japan is that the Japanese Patent Information Organization has been formed. Its database, which includes English language abstracts of more than 1 million Japanese patents since 1976, is searchable on SDC *Orbit*®.

Increasing use of computer-based patent searching systems by the larger patent offices is clearly a development of great importance. In addition to automation efforts in the U.S. Patent Office, automation plans are being considered by the patent offices of Germany, the United Kingdom, Japan, and elsewhere. The European Patent Office has plans to begin to switch to computerized patent application and printing by the mid-1980s. Officials from the various patent offices are talking with one another to determine whether automation efforts may be coordinated from nation to nation.

Other computer developments can be expected at the user level. For example, Derwent has announced that it is making available a software package that will permit use of microcomputers to statistically analyze patent search results. This should help detect trends by field of technology and by company and country. The results could give leads to emerging technologies, competition activities, acquisition opportunities, and personnel activity. As previously mentioned, the developers of Pergamon's *PATSEARCH®* online files have introduced powerful new capabilities that facilitate obtaining of statistical data such as, for example, most frequently cited patents, ranking of patents by organization (assignee), ranking of inventors, and ranking of U.S. Patent Office Classes on a given subject. This may be done online using Pergamon's main-frame computer.

13.21. OTHER SOURCES OF PATENT INFORMATION

Other sources of patent information include, for example:

- Numerous other indexing and abstracting services, such as the *Abstract Bulletin* of the Institute of Paper Chemistry (see Chapter 8), useful for pinpointing patents in specific fields.
- News magazines, such as *Chemical Week,* which in identifying new technology frequently note pertinent new patents associated with that technology.
- Articles, especially review articles.
- Encyclopedias and treatises, such as those mentioned in Chapter 12— good especially for the older patent literature, which can still be useful.
- Other books, especially those from the Noyes Data Corporation,* which bring together U.S. patent information in specialized fields. These usually excellent volumes provide a quick and relatively inexpensive overview of the technologies covered.
- Programs such as those mentioned in Chapter 17, such as the SRI Process Economics Program, which does an excellent job of technoeconomic interpretation of important patents on manufacture of many major commercial chemicals.

13.22. OTHER REMARKS ON PATENTS*

- Any chemist who undertakes industrial or academic research today and has not first made a study of the state of the art (i.e., a patent search), in addition to looking at the other literature, is wasting time. That has been a theme of this chapter and is worth repeating.
- Many chemists complain that there is too much *secrecy* in information and especially too much secrecy in chemical and other technical processes. That may be because they are not willing to study the patents and determine from them the current state of the art. In addition, patents are an excellent stimulus for new ideas.

*Some of the material in this Section reinforces points made earlier.

- One major goal of research, especially in industry, is to invent new products and patents that can withstand keen competition. Because competition does not stand still, patents and patent systems are a spur to ongoing reseach.

- Issuance of a patent does not necessarily mean that the invention will be commercialized; the patent may be primarily defensive (to exclude competition), or it could represent a potential interest that may never be exploited because of changes in policy, technology, or economics. Many patents are not commercialized immediately, but rather at a later date, perhaps a few years after issue of the patent, when conditions are more favorable for the technology or use.

- The U.S. Patent Office has a disclosure document program under which papers filed with that office in a prescribed manner may be used as evidence of dates of conception of inventions. Such papers are retained in confidence for 2 years and then destroyed unless referred to in a separate letter in a related patent application filed within 2 years. Persons desiring to use this program are advised to consult an attorney or agent registered to practice before the U.S. Patent Office.

- In 1969 the U.S. Patent Office instituted a defensive publication program whereby an individual or corporation may elect to publish an abstract of a patent application in the *Official Gazette* in lieu of examination by the U.S. Patent Office. On publication of the abstract, the applicant also agrees to open the complete application to inspection by the general public. On May 8, 1985, this program was replaced by the Statutory Invention Registration series.

- For details of the legal aspects, the reader should consult a patent attorney or agent—this book is written from neither of those perspectives, but rather from the viewpoint of chemical (not legal) information. Other good sources of information include, as previously mentioned, the U.S. Patent Office and its publications; Derwent user manuals (Figure 13.3); the book by Maynard (7); the American Chemical Society's audio tape course on patents (8); and the articles on patents in Volume 16 of the third edition of *Kirk–Othmer Encyclopedia of Chemical Technology*. For keeping up with patent information tools on a current basis, a good source is *World Patent Information,* published by Pergamon Press. This journal includes a regular column about online databases and techniques by Stuart M. Kaback (Exxon Corp.) who is a leading authority on chemical patent information.

REFERENCES

1. One pertinent publication is *General Information Concerning Patents—A brief introduction to patent matters,* which is for sale by the U.S. Superintendent of Documents, U.S. Government Printing Office, Washington, D.C. 20402. This is revised at frequent intervals.

2. IFI/Plenum Data Co., *Assignee Index,* Alexandria, VA, annual. IFI also produces online patent databases and publishes *Patent Intelligence and Technology Report,* which, for companies with 10 or more patents, categorizes these by Patent Office subclass and analyzes data for the last 5 years.

3. U.S. Patent Office, *Manual of Classification of Patents,* U.S. Government Printing Office, Washington, D.C. Revised periodically. Includes *Alphabetical Index of Subject Matter.*

4. G. O. Platau, "Documentation of the Chemical Patent Literature," *J. Chem. Doc.,* **7**(4), 250–255 (1967).

5. P. J. Pollick, "Patents and Chemical Abstracts Service," *Sci. Technol. Libr.,* **2,** 3–22 (1981).

6. P. J. Pollick, "Processing of Patent Bibliographic Data at CAS," *World Patent Inf.,* **3**(3), 128–131 (1981).

7. J. T. Maynard, *Understanding Chemical Patents,* American Chemical Society, Washington, D.C., 1978.

8. R. L. Costas, *Introduction to Patents,* American Chemical Society, Washington, D.C., 1974. (Audio tape course and manual.)

LIST OF ADDRESSES

Computer Horizons, Inc.
1050 Kings Highway North
Cherry Hill, NJ 08034

Derwent, Inc.
Suite 500, 6845 Elm Street
McLean, VA 22101

Derwent Publications, Ltd.
Rochdale House, 128 Theobald Rd.
London WCIX 88

IFI/Plenum Data Company
302 Swann Avenue
Alexandria, VA 22301
   (800-368-3093)

InfoLine®
12 Vandy Street
London EC2A1 2 DE, England

Kenneth Mason Publications, Ltd.
The Old Harbormaster's, 8 North
   Street,
Ensworth, Hampshire, P.O. 10
   7dd, England

Mead Corp.
P.O. Box 933
Dayton, OH 45401 (513-865-6800)

Noyes Data Corporation
Park Ridge, NJ 07656

Pergamon International Information Corp.
1340 Old Chain Bridge Road
McLean, VA 22101 (800-336-7575)

Public Search Room of the Patent Office, Crystal Plaza
2021 Jefferson Davis Hwy.
Arlington, VA

Search Check, Inc.
2001 Jefferson Davis Highway
Arlington, VA 22202

U.S. National Technical Information Service
Springfield, VA 22161

14 SAFETY AND RELATED TOPICS

14.1. INTRODUCTION

The amount of information on safety, occupational hygiene, toxicity, environmental impact and control, and related aspects of chemistry and chemical engineering has continued to grow rapidly during the last few years. This kind of information is especially important before beginning work on any new project or on new or less well known chemicals, but it also has longer-range significance. Pertinent information benefits not only persons directly connected with the investigation. It is equally important for colleagues in the same laboratory, pilot plant, or plant area and for all who may later use or otherwise come into contact with the chemicals.

Although the field is large and of worldwide interest and action, this chapter emphasizes primarily U.S. information sources. Moreover, safety rather than environmental information sources are emphasized, albeit the two fields are closely related and may overlap.

The kinds of safety information with which the chemist or chemical engineer must be familiar includes one or more of the following examples:

1. Safe laboratory, pilot plant, and plant practices in general (applies to all chemists and chemical engineers).
2. Hazardous chemical reactions.
3. Flash point and other flammability or explosive characteristics. Also, smoke generation, if any.
4. Effects, if any, of specific chemicals on human beings.
5. Effects on animals, especially if data on human beings is lacking.
6. Proper methods for pollution control, transportation, storage, and handling.
7. Effects on vegetation or aquatic life if discharged into the air or as liquid effluents.
8. Biodegradability.
9. Pertinent local, state, and especially federal standards and regulations issued by such agencies as the U.S. Environmental Protection Agency, Food and Drug Administration, Consumer Product Safety

TABLE 14.1. Some Key Environmental Events and Dates

Date	Event
1962	Rachel Carson's *Silent Spring*
1965	Solid Waste Disposal Act
1969	National Environmental Policy Act (NEPA)
1970	Clean Air Act Amendments
1970	Resource Recovery Act
1970	EPA formed
1972	Federal Water Pollution Control Act
1972	Safe Drinking Water Act
1975	Issue of chlorinated organics becomes priority topic to EPA
1976	Resource Conservation and Recovery Act (RCRA)
1976	Toxic Substances Control Act (TSCA)
1977	Clean Water Act (CWA)
1977	Love Canal news stories
1977–79	National Interim Primary Drinking Water Regulations
1980 to date	Dioxin spills news stories
1980	Comprehensive Environmental Response, Liability and Compensation Act (CERCLA)
1980	Solid Waste Disposal Act Amendments (to RCRA)
1982	Regulations on volatile organics in drinking water
1984	RCRA Reauthorization
1984	Bhopal disaster

Commission, Occupational Safety and Health Administration, and Federal Trade Commission.

The field is volatile. Standards, regulations, technology, and toxicity/safety/ environmental ramifications change rapidly. Because of this, chemists can expect presently available sources of safety and environmental information, as described in this chapter, to be supplemented or replaced continuously by new and improved sources. Organizations such as Chemical Abstracts Service (CAS) and BioSciences Information Service, federal agencies such as the National Library of Medicine, and for-profit information industry organizations can be expected to take the lead in developing new information tools in this field. Chemists should do their best to maintain appropriate contact with these organizations on a regular basis and to keep up with the newer sources. Table 14.1 lists some key environmental dates that may be of interest.

14.2. LOCATING PERTINENT SAFETY DATA—
INITIAL APPROACHES

In industry, and to an increasing degree in universities, much of the needed data may already be available from departments in one's own organization.

These departments are typically designated by such names as *Loss Prevention* or *Environmental Hygiene and Toxicology*. Assistance of experts in these departments is essential in evaluating any potential hazards. Toxicological data, because of its biomedical nature, needs to be evaluated by a professional toxicologist or a physician.

In smaller, or less well organized laboratories, the bench chemist may need to make a personal investigation of pertinent safety data and should consider the following suggestions:

1. The manufacturer of a chemical, if it is commercially available, is a prime source of safety data and should be contacted as a first step. If trade literature is already on hand, the chemist must check with the manufacturer to ensure that this literature contains the latest information. Material Safety Data Sheets (MSDS) are an important source of safety information. Under U.S. government regulations effective as of November 25, 1985, these are required to be available (and updated regularly) for almost all commercial chemicals. Some states also have regulations regarding the MSDS. MSDS forms are prepared by manufacturers for their products and are made readily available to customers and frequently to potential customers. These forms can also be a helpful source of physical property information, as mentioned in Chapter 15. A collection of nearly 30,000 MSDS forms from several hundred manufacturers is offered for sale (on microform) by Information Handling Services.* Indexes are provided by CAS Registry Number, brand and trade name, chemical name, and supplier/source. Updates are made available at frequent intervals. In addition, an MSDS collection is offered by the Genium Publishing Co.* This includes several hundred safety data sheets, all of which are said to be based on evaluated information. The collection was originally available directly from General Electric Co., but it is now available from Genium. Emphasis is on materials made or used by General Electric, but other products are also included. There are indexes by chemical, trade, and common names, by supplier/manufacturer, and an updating service is available. The sheets are distinguished by depth and breadth of coverage, and, because of the review process to which they are said to be subjected, probably have improved objectivity. As of the end of 1985, over 800 sheets were available, and more than 100 new or revised sheets were being added annually.

Another MSDS collection is offered by VCH Publishers.* This work (1), *Compendium of Safety Data Sheets For Research and Industrial Chemicals* covers 867 chemicals.

2. Since published literature in the field is voluminous, consultation of basic reference books is a good next step. An exception to this sequence applies if an online system, such as described later, is available to the chemist. If an online system is available, *this* is the preferred next step.

3. Federal, state, and local environmental officials may be able to provide advice and offer suggestions on safe handling.

4. The chemist can find the most recently published safety information, particularly on hazardous chemical reactions or laboratory procedures, in the *letters to the editor* and other sections of chemical news as well as in other chemical magazines and journals. The chemist should scan these sections regularly and carefully, on a timely basis, if possible in cooperation with other chemists in the same laboratory, plant, or R & D team. Examples of good sources for this information include *Chemical & Engineering News, Chemistry in Britain,* and the *Journal of Chemical Education.* In addition, the *Journal of Chemical Education* features a series of columns on safety entitled *Safety in the Chemical Laboratory,* currently edited by Malcolm M. Renfrew. The American Chemical Society's Division of Chemical Education has reprinted these articles in a series of four paperback books (2). Major chemical accidents, such as may occur on a plant scale or during transportation, are reported quickly by mass news media and then in more detail in the technical press. Some chemical organizations keep running tabs of these incidents, wherever they ocur, to improve overall safety of their *own* operations.

5. If safety data on the specific compound being worked with is not found, data on analogous or homologous compounds and other related structures can sometimes, but not always, provide useful guides to what can be expected. Such extrapolations are best made by a toxicologist or other trained safety expert as appropriate.

6. In watching trends and developments in areas of safety, toxicity, and environmental concerns, the chemist needs to pay special attention to information and news from countries such as Japan, Sweden, Canada, and West Germany. These countries have shown special concern for the environment and are regarded as bellwether nations in this field. For example, potential hazards of inappropriate use of mercury were first widely publicized in Sweden and Japan.

7. Close to the leading edge of environmental developments are well-organized groups of informed citizens. These groups have been among the first in the United States to publicize potential hazards. *Although not always correct in their assessment,* they can provide clues that may prove to be valid. Examples include:

Sierra Club
530 Bush Street
San Francisco, CA 94108

Scientists Institute for Scientific Information
560 Trinity Ave.
St. Louis, MO 63130

which publishes the bi-monthly magazine *Environment.*

Center for Science in the Public Interest
1501 16th Street, N.W.,
Washington, D.C. 20009

which issues periodic newsletters.

8. Trade associations such as the Chemical Manufacturers Association, Soap and Detergent Association, Chemical Specialties Manufacturers Association, and Synthetic Organic Chemical Manufacturers Association are important sources of safety and environmental information. See also Section 15.6.

9. If no published information on hazardous properties of a chemical is found, this does not mean that no hazard exists. The chemist must assume that the chemical may be hazardous unless positive and reliable information is found showing that the chemical has been proven to be safe under specific conditions.

Questions that may help the chemist, assisted and advised by toxicologists, in evaluating the health risk (or lack of risk) that a chemical poses include these examples:

1. Will there be substantial or significant human exposure in (a) production, raw materials, intermediates, product, and waste products; (b) distribution; (c) end use; and (d) disposal?

2. How does the chemical compare to existing products already in use that have similar properties and exposure characteristics?

3. How does it compare with such other existing products in physical properties that affect human exposure (vapor pressure, corrosiveness, etc.) and in toxicological properties?

4. Does it contain functional groups that are known carcinogens?

5. Does a toxicologist have any reason to suspect that the product may pose an unusual risk in manufacture, use, or disposal that could not be minimized by adequate warnings or reasonable standard procedure? Experience with products in similar distribution patterns, end use, and disposal characteristics will also be valuable.

6. Will there be substantial exposure to the environment?

a. Will it be produced in large quantities? What quantities (if any) could be released to the environment? In what forms?

b. How does it compare to existing products being distributed that have similar structure or toxicological effects and similar environmental exposure characteristics?

c. What are the comparative physical properties, molecular structure, persistence, solubility, toxicity, and degradation characteristics?

Because no single published source of safety data can be all inclusive, and for other reasons, it is a mistake to use only one such source. The source used could be unevaluated or biased, or data could be taken out of context to mean something entirely different. Multiple *original* sources should be obtained and evaluated by an appropriately qualified expert such as a toxicologist.

14.3.   BOOKS AND RELATED INFORMATION SOURCES

Basic reference books on safety, toxic effects, and related aspects of chemicals are often the quickest and most efficient way to gather needed information. Several reference tools attempt to bring together in one place data on the safety aspects of a number of chemicals. Examples are cited in References (3–11) and a few are described in the following sections.

There are caveats: Since any book can quickly become obsolete, particularly in safety-related matters, it is imperative that the chemist look at the most recent knowledge available and obtain as much pertinent detail as possible. In addition, no published compilation can cover the gamut of safety and related matters; none is complete.

1. The most widely known handbook on potentially hazardous chemicals is Sax's *Dangerous Properties of Industrial Materials* (10), which now contains over 3000 pages. Entries for about 20,000 materials are in the sixth edition.

Sax's book is excellent, but like all other books of its type, it is only a starting point. Not all materials or hazards are covered in the Sax book, and alternative information sources are sometimes more current or more complete. Sax's book and other similar published works do, however, perform the important function of helping alert the chemist to many hazards—alerts that are then best followed up in detail in more specialized and current sources, preferably in cooperation with safety or toxicity experts.

2. Another classic reference book is *Clinical Toxicology of Commercial Products* by Gosselin et al., now in its fifth edition (5). This includes toxicity data for 1642 substances, with extensive references to the literature. In addition, composition data for about 16,000 consumer products are included. Data from this book, frequently known simply as either *Gosselin* or *CTCP*, are available online through *CIS* (see Section 10.11.A.3).

3. Bretherick's *Handbook of Reactive Chemical Hazards* (11) contains stability data on specific compounds, data on possible violent interaction between two or more compounds, general data on a class of compounds or information on individual chemicals in a known hazardous group, and structures associated with explosive instability. References to sources of data are given.

4. The most complete list of toxic effects of chemicals is the *Registry of Toxic Effects of Chemical Substances* (*RTECS*) (12) published by the U.S. National Institute for Occupational Safety and Health (NIOSH).* This contains information for some 67,000 chemicals.

The *Registry* is available in three formats: a microfiche version, which contains the complete file updated quarterly; a printed book updated annually; and computer tape.

The file is searchable online through the National Library of Medicine. In addition, online access through the *Chemical Information System* is possible (see Section 10.11.A.3).

Online versions of *RTECS* contain information not previously available in the printed volume because they are updated quarterly. More importantly, online versions provide retrieval capabilities that permit the user to zero in on data that may be difficult to locate in other ways. For most substances listed in *RTECS*, the data in Figure 14.1 are provided.

Before using the *Registry*, chemists should read the introductory material carefully. Absence of a substance from *RTECS* does not imply that it is nontoxic and nonhazardous; the compilation is not complete. Presence of a substance does not necessarily mean that a significant toxic effect has been found, although all chemicals can be toxic, depending on dose. (Each page in the *Registry* carries a statement that "omission of a substance or notation does not imply any relief from regulatory responsibility.") However, the toxicity data in *RTECS* have *not* been critically evaluated, except for a small number of chemicals in the genetic toxicity program of EPA that are listed in the microform edition. Accordingly, involvement of a toxicologist in use of the *Registry* is highly recommended.

5. An important reference work in this field is the *Encyclopedia of Occupational Health and Safety*. The third edition of this two-volume work was published by the International Labor Organization of Geneva, Switzerland, in 1983. The editor is Dr. Luigi Parmeggiani, secretary to the Permanent Commission and International Association of Occupational Health.

6. The classic U.S. reference source in industrial hygiene and toxicology is *Patty's Industrial Hygiene and Toxicology* (3), now in its third edition and edited by G. D. Clayton and F. E. Clayton.

7. Another important book is *Hazardous and Toxic Materials* by H. W. Fawcett (13). It is mentioned here because of Fawcett's record of leadership in safety.

8. Two works of key importance to the environmental professional are *Standard Methods for the Examination of Water and Wastewater* (14) and *Methods of Air Sampling and Analysis* (15). Testing procedures and standardized methods in these manuals help provide the consistent and reproducible results needed to comply with current requirements. Continued revision of both books can be expected as standards and technology change. Both volumes are available from the American Public Health Association.*

(2) ► UPDT: 8209　　　　CAS: 9999–99–9　MW: 357.88　MOLFM: C·Re·S·Te

(6) ► SYN: NIOSHRTECSIUM • RTECS (DOT) • RTECSIAM (French) • TSL (OBS.)

(7) ► *IRRITATION DATA AND REFERENCES:*
　　　　skn-hmn 5%/1M nse MOD　　　　　　　　　　(12)►JAMAAP 135,1920,68
　　　　skn-rbt 10 mg/24H open MLD　　　　　　　　　　AMIHBC 10,61,54
　　　　eye-rbt 20 mg　　　　　　　　　　　　　　　　AMIHBC 10,61,54

(8) ► *MUTATION DATA AND REFERENCES:*
　　　　mma-sat 50 ug/plate　　　　　　　　　　　　PNASA6 72,5135,75
　　　　cyt-hmn: leu 100 umol/L/8H　　　　　　　　MUREAV 4,53,67
　　　　dlt-mus-orl 35 mg/kg　　　　　　　　　　　TXAPA9 23,288,72
　　　　hma-mus/sat 400 ppm　　　　　　　　　　　GISAAA 42(1),32,77

(9) ► *REPRODUCTIVE EFFECTS DATA AND REFERENCES:*
1T03　　　　orl-rat TDLo:270 mg/kg (5D male)　　　　　　TJADAB 19,41A,79
3T25T46T52　ihl-rat TCLo:100 ppm (3D pre/4-7D preg)　　　NTIS** AD-900-000
2T35T75　　mul-mus TDLo:900 mg/kg (1D male/6-15D preg/5D post)　TOXID9 1,125,81

(10) ► *TUMORIGENIC DATA AND REFERENCES:*
2V01M61　　orl-rat TDLo:90 gm/kg/78W-I　　　　　　29ZUA8 –,183,80
1V03　　　ihl-mus TCLo:400 ppm/8W-I　　　　　　JTEHD6 – (Suppl.2),69,77
2V02K60　　orl-rat TD :100 gm/kg/104W-I　　　　　　BJCAAI 16,275,62

(11) ► *TOXICITY DATA AND REFERENCES:*
　　　　ihl-hmn TCLo:500 mg/m3/7M　　　　　　　　AIHAAP 23,95,62
2J13K18　　orl-rat LD50:1500 mg/kg　　　　　　　　MarJV# 26 Apr 76
*　　　　scu-mus LDLo:1200 mg/kg　　　　　　　　FCTXAV 17(3),357,79

(13) ►　　AQUATIC TOXICITY RATING: TLm96:100-10 ppm　　WQCHM* 2,–,74

(14) ►　　REVIEW: CARCINOGENIC DETERMINATION: ANIMAL POSITIVE　IARC** 20,151,80
　　　　REVIEW: TOXICOLOGY REVIEW　　　　　　　　　PLMJAP 6(1),160,75
　　　　REVIEW: TLV-TWA 10 ppm; STEL 25 ppm　　　　　DTLVS* 4,358,80
　　　　REVIEW: TLV-SUSPECTED CARCINOGEN　　　　　DTLVS* 4,358,80

(15) ►　　STANDARDS & REGULATIONS:
　　　　OSHA STANDARD-air:TWA 50 ppm　　　　　　FEREAC 39,23540,74
　　　　DOT-ORM-A, LABEL: NONE　　　　　　　　FEREAC 41,57018,76

(16) ►　　CRITERIA DOCUMENT: OCCUPATIONAL EXPOSURE TO RTECSIUM recm　NTIS**
　　　　std-air:CL 10 ppm/60M

(17) ►　　STATUS: SELECTED BY NTP FOR CARCINOGENESIS BIOASSAY AS OF SEPT 1982
　　　　STATUS: NTP SECOND ANNUAL REPORT ON CARCINOGENS, 1981
　　　　STATUS: NIOSH MANUAL OF ANALYTICAL METHODS, VOL 3 S255
　　　　STATUS: NIOSH CURRENT INTELLIGENCE BULLETIN 41, 1980
　　　　STATUS: REPORTED IN EPA TSCA INVENTORY, 1982
　　　　STATUS: EPA TSCA 8(a) PRELIMINARY ASSESSMENT INFORMATION　FEREAC 47,26992,82
　　　　　　FINAL RULE

A.　RTECS Accession Number, a sequence number assigned to each substance in the Registry.
1.　Substance Prime Name.
2.　Date when substance entry was last revised.
3.　Chemical Abstracts Service Registry Number.
4.　Molecular Weight of this substance.
5.　Molecular or Elemental Formula of this substance.
6.　Synonyms, common names, trade names, and other chemical names for this substance.
7.　Skin and Eye Irritation Data.
8.　Mutation Data.
9.　Reproductive Effects Data.
10.　Tumorigenic Data.
11.　Toxicity Data.
12.　CODEN, an acronym for the references from which the data and other citations were extracted.
13.　Aquatic Toxicity Rating.
14.　Reviews of this substance.
15.　Standards and Regulations for this substance promulgated by a Federal agency.
16.　A Criteria Document supporting a recommended standard has been published by NIOSH.
17.　NTP, NIOSH, and EPA Status information about this substance.

Figure 14.1. An example of a typical substance entry in the 1981–1982 *Registry of Toxic Effects of Chemical Substances.*

9. The proceedings of the Purdue University Industrial Waste Conferences (16) are an invaluable collection of case histories on handling a wide variety of pollution abatement problems. These proceedings have been published annually for over 30 years.

10. Kristina Voigt and Hans Rohleder edited *A Compilation of Data Sources for Existing Chemicals* (17). This comprehensive 247 page volume is intended primarily as a guide to those concerned with setting of priorities for chemicals of concern for environmental and/or human health reasons. Content includes: handbooks and tables; monographs and reports; and computerized databases.

14.4. SOME ONLINE SOURCES

The computerized literature retrieval services of the National Library of Medicine (NLM) are available online through a system known as *MEDLARS*. This includes such components as *Toxline* and *HSDB*. See also Section 9.6.

Other online sources of safety data include these examples:

1. *Hazardline,* a product of Occupational Health Services, Inc.* This file is searchable online through *BRS®*. Content includes regulatory and safe handling data for many chemicals as well as other helpful information.

2. The databank of the Health and Safety Executive (HSE), which is the organization responsible for safety and health at work in Great Britain. This is searchable online through Pergamon *InfoLine®* and includes references to recent literature.

3. The Oil and Hazardous Materials/Technical Assistance Data System (OHM/TADS) of the U.S. Environmental Protection Agency. Searchable through the *Chemical Information System* (see Section 10.11.A.3), this has data on properties and disposal methods for some 1300 substances designated as oils or hazardous materials by EPA.

4. The *Chemical Regulations and Guidelines System (CRGS)* searchable through *DIALOG®* Information Services. This is an index to U.S. Federal regulatory material.

5. *Occupational Safety and Health,* the database of the National Institute of Occupational Safety and Health (NIOSH), also searchable through *DIALOG®* Information Services. This covers about 150 journals, as well as proceedings, from 1973 to date.

6. Data obtained from NIOSH on exposure of approximately 200,000 workers to workplace situations (chemicals) that could increase disease risk. This is offered by Information Consultants,* as part of their online *Integrated Chemical Information System* (see Section 10.11.A.3).

Examples of other tools containing safety information that can be searched online are mentioned elsewhere in this chapter. A recent article summarizes and reviews online sources of safety data (18). The *CIS* online system (Section 10.11.A.3) contains a large amount of both environmental and safety data in addition to that mentioned in item 3 above.

14.5.   GOVERNMENT LAWS, REGULATIONS, AND STANDARDS

Most chemists and engineers are aware that there are many government regulations and standards relating to chemicals and the chemical industry. These government actions have, in many cases, significant impact on the public and on the way chemists and engineers work. Federal, state and local agencies, legislation, and regulations are all important. The U.S. EPA (Environmental Protection Agency) and U.S. OSHA (Occupational Safety and Health Administration) are probably the best known agencies to most chemists.

The interpretation and implementation of regulations have both legal and technical ramifications. Attorneys, toxicologists, engineers, and chemists who specialize in this field are needed to assure appropriate compliance. As in the case of patents, this is an area that is often best handled by specialists, although all chemists should have some basic familiarity with important developments.

General sources of information on government regulations and actions include the following:

1. Chemical news magazines. Also, major national newspapers, especially the *Wall Street Journal.* For the chemist or engineer to whom regulatory affairs are not a full- or part-time job, these sources should suffice.

2. The daily *Federal Register,* the official U.S. Government announcement source for regulations of all types, including those pertaining to toxicity, other safety matters, and environmental affairs. This is voluminous, but several online and printed commercial services are available to facilitate use. For example, see BNA in the following paragraph, *CIS* (Section 10.11.A.3), and *CRGS* (Section 14.4).

There are numerous specialized information tools and sources on the many and rapidly changing regulations and standards. One of the most outstanding of these is *Environment Reporter,* published weekly by the Bureau of National Affairs (BNA). The intent of this publication is to help subscribers:

1. Keep abreast of the latest developments on the environmental scene.
2. Be alert to what must be known about federal laws and regulations.

3. Be provided with major state-by-state air, water, solid waste, and land-use laws and regulations.

4. Be given full text of court decisions in important environmental cases.

5. Be supplied with analysis of specific environmental protection topics.

Even more specific to the chemical industry is BNA's *Chemical Regulation Reporter,* which includes weekly reference information on important developments with special emphasis on the *Toxic Substances Control Act* and *FIFRA (Federal Insecticide, Fungicide, and Rodenticide Act).* Much of this BNA publication is online through Mead's *NEXIS/LEXIS®* system (see also Section 13.14). In addition, part of BNA's *Environment Reporter* is online through *LEXIS®,* and a BNA file on regulations, *Chemreg,* is searchable on *ICIS* (see Section 10.11.A.3).

Another good source is EIC®/Intelligence. Products include the *Environment Regulation Handbook* (covers major federal environmental laws and regulations) and state environmental laws and regulations in microform.

The tools mentioned here, especially the compilations of BNA, can also help the chemist make direct contact with appropriate high level government officials and technical experts by identifying names of these individuals and their functions. Recent government agency organization charts and directories are also important in keeping track of key personnel.

14.6. PROFESSIONAL SOCIETIES AND OTHER ASSOCIATIONS

Professional societies and trade associations are active in promoting safety and providing safety-related information:

1. The American Chemical Society's Committee on Chemical Safety provides advice on safety matters and has published the booklet *Safety in Academic Chemistry Laboratories* (19). A Division of Chemical Health and Safety has been formed and is operational. The Society operates a Health and Safety Referral Service, (202-872-4511), not intended to be a resource for direct answers, but rather to refer inquirers to resources that hopefully will provide the answers. In addition, the Society offers a pamphlet, *Health and Safety Guidelines for Chemistry Teachers.*

2. The American Institute of Chemical Engineers has publications on safety, including *Loss Prevention* (20), *Ammonia Plant Safety* (21), and *Dow's Fire and Explosion Index Hazard Classification Guide* (22). More recently, a Center for Chemical Process Safety has been established. The first publication is a manual on hazard evaluation. In addition, short courses and other services are to be offered. T.W. Carmody, formerly with Union Carbide, is the first director.

3. The Chemical Manufacturers Association (CMA), located in Washington, D.C., provides assistance in safe handling of chemicals. A toll free telephone number (800-424-9300) provides immediate access to CMA's Chemical Transportation Emergency Center (Chemtrec). This center was established in 1971 to handle accidents with chemicals in transportation or warehousing. The service (a) determines what chemicals are involved in an accident, (b) supplies preliminary advisory information retrieved from Chemtrec files, and (c) secures involvement of appropriate technical specialists. Another CMA office (Chemical Referral Center) refers inquirers to the names and phone numbers of manufacturers of chemicals as based on the chemical or trade names of these products. The telephone for this CMA office is 800-262-8200. Safety Data Sheets issued by CMA (formerly known as MCA or Manufacturing Chemists Association) covered about 100 heavily used chemicals of commerce. However, these were discontinued in 1979 and should not be used even if located because they are obsolete.

4. The National Safety Council* has a broad range of safety-related activities and publications. Many of these deal with safety in general or with other aspects of safety, but some, such as publications of their Chemical and R & D Sections, have material of specific interest to chemists.

5. The chemist will find the National Fire Protection Association (NFPA)* a valuable source of safety information (23). Of special interest is NFPA's *Manual of Hazardous Chemical Reactions* (23a). This is a compilation of 3550 mixtures of two or more chemicals reported to be potentially dangerous in that they may cause fires, explosions, or detonations as based on a survey of incidents cited in the literature. Before the chemist or engineer embarks on an R & D or other project, this NFPA publication is a key source to be examined. The NFPA publication *Hazardous Chemical Data* (23b) contains basic safety data on approximately 416 chemicals. These data include: description (including odor), fire and explosion hazards (including flash point, flammable limits, and ignition temperature), life hazards, personal protection, fire fighting, usual shipping containers, storage, and remarks (including representative standards and safety data sheets that may be applicable). These and other NFPA reports are brought together in a single volume, *Fire Protection Guide on Hazardous Materials,* now in its eighth edition published in 1984 (23c).

In addition NFPA has adopted a detailed fire protection standard for laboratories that use chemicals. This 54 page code (24) is one of the most comprehensive of its type. It deals with chemical handling and storage, compressed gases, fire prevention, ventilation, building construction, exit doors, and automatic fire-extinguishing systems.

6. The American Conference of Government Industrial Hygienists (ACGIH)* publishes threshold limit values (TLV®s) for chemical substances and physical agents in workroom air. ACGIH is not an official government agency, but its recommendations are highly regarded. This is a quick reference source for exposure limits of common substances. Many of these TLV® values serve as the basis for OSHA (Occupational Safety and Health Administration) permissible limits and are constantly under review. Another excellent source of industrial hygiene data is the American Industrial Hygiene Association, Akron, OH.

7. Published information on hazardous reactions is useful, but no matter how thorough, such materials cannot possibly include hazardous reactions that have never been tried. In response to this need, the American Society for Testing and Materials (ASTM),* developed and distributes a computer program (designated as CHETAH). The program provides estimates of the maximum possible energy release for covalent compounds of carbon, hydrogen, oxygen, nitrogen, and 18 other elements (see also Section 15.15).

8. The Royal Society of Chemistry, like its U.S. counterpart, is active in the field of safety and related topics. One example is publication of the third edition of the widely acclaimed *Hazards in the Chemical Laboratory* (25). This is regarded by many as a handbook essential to laboratory safety. Other Royal Society safety activity includes its printed monthly abstracting publications *Laboratory Hazards Bulletin* and *Chemical Hazards in Industry,* both of which, in addition, are searchable online via Pergamon *InfoLine*®.

14.7.   *CHEMICAL ABSTRACTS* AND OTHER ABSTRACTING AND INDEXING SERVICES

A vital source in this field of chemistry, as in all others, is *CA*. Some sections that can be scanned selectively on a current awareness basis include the following:

Section 1 Pharmacodynamics
Section 4 Toxicology
Section 50 Propellants and Explosives
Section 59 Air Pollution and Industrial Hygiene
Section 60 Waste Treatment and Disposal
Section 71 Nuclear Technology

All 80 sections, however, may contain some material of interest.

Pertinent safety-related headings in the *CA* General Subject Index include:

Accidents	Flammability
Alarm devices	Health hazard
Combustibles	Health physics
Disease, occupational	Implosion
Electric shock	Injury
Explosibility	Nuclear reactors
Explosion	Poisoning
Fire, fireproofing, fireproofing agents, fire-resistant materials	Respirators
	Safety
	Safety devices

These entries are helpful when looking for more generic information (such as safety in handling flammable organic liquids or the question of potential toxicity of surfactants to aquatic life). The same entries can also be helpful when looking for specifics (such as explosibility of *o*-nitrophenylsulfonyl-diazomethane), although for specific substances the chemist should also look under the name of that substance in the *CA* chemical substance index.

Beginning in 1976, keyword indexes to weekly issues of *CA* were extended to cover safety more broadly. However, this entry (the keyword safety) by design does not include toxicology. Therefore, to get a complete picture of safe-handling aspects, keyword index entries pertaining to toxicity must also be scanned. Because keyword indexes are not complete for toxicity data, the best places to look in *CA* are online or the six-month volume and five-year collective indexes. This is true particularly for toxicity studies reported incidentally to synthesis of a compound.

Also available is a subset grouping of *CA* abstracts, *Chemical Hazards, Health & Safety,* produced by computer search of the full *CA*, as part of the *CA SELECTS* series (see Section 6.11). Chemists and engineers find this a convenient way of keeping informed about safety on a reasonably current basis.

Every two weeks *Chemical Hazards, Health & Safety* presents abstracts on such subjects as:

- Health and safety of personnel working with or in the area of radiologically or chemically hazardous substances.
- Hazardous properties of chemical substances and chemical reactions.
- Effects of human exposure to and protection of humans from hazardous substances.
- Toxicity of hazardous substances to humans.
- Safety concerns in chemical laboratories and chemical industry.

Excluded from coverage are toxicity studies done on laboratory animals. Other *CA SELECTS* publications of related interest are *Flammability; Drug & Cosmetic Toxicity;* and *Food Toxicity*.

CA has recently made available about 50,000 abstracts in the areas of environment, safety, and toxicology via a CD—ROM (compact disk—read only memory) optical disk.

In addition to *CA,* or the subset groupings noted, the related CAS publication *Chemical Industry Notes* (*CIN*) (see Section 7.14) also needs to be consulted for extracts of reports on specific incidents such as chemical plant explosions, fires, spills, leakages, and the like. If this material is "newsy," and reported in trade and industry publications or a few major national newspapers, it is more likely to be reported in *CIN* than in the full *CA*.

As mentioned in Section 14.6, two important abstracting and indexing services on safety and hazards are published by The Royal Society of

Chemistry. As described in Section 9.4.B, the Hazardous Materials Technical Center issues an abstract bulletin and offers other services relating to hazardous materials and wastes. *Current Abstracts of Chemistry and Index Chemicus®* (see Section 8.3) can prove helpful in alerting chemists to potential biological hazards of newly reported compounds. Other products of the same publisher that can be helpful in following safety-related developments are their *ASCATOPICS®* services (see Section 4.3.C).

HEEP (Abstracts on Health Effects of Environmental Pollutants), published monthly since 1972 by the *BioSciences Information Service (BIOSIS)*,* is an up-to-date and comprehensive source with an emphasis on toxicity. Included are environmental chemicals and substances other than medicinals that affect human health, as well as general reviews and original papers reporting on potentially harmful effects on humans, lower vertebrates (used as indicators of the substances toxic to man), vertebrates and invertebrates in the food chain, and analytical methods for examining biological tissues or fluids. Monthly and annual indices facilitate its use.

This is one of the sources included in the online *TOXLINE* service. For chemists who do not have ready access to *TOXLINE*, *HEEP* is a prime alternative for toxicity data.

A unique advantage of *HEEP* is that it covers not only possible adverse effects of chemicals on humans, but also, to some extent, possible effects on wildlife, including fish. The importance of this point is readily apparent to any chemist attuned to contemporary environmental issues.

The full *Biological Abstracts,* also issued by *BIOSIS,* can contain some abstracts on biological activity that may not be reported elsewhere; coverage is from 1926 to the present. Manual, computer-based batch, and online searches can be conducted, and a current awareness service is available. One of the principal merits of this file is its broad scope.

Environmental Health and Pollution Control, published by Excerpta Medica, subsidiary of Elsevier Science Publishing, Amsterdam, The Netherlands, and Princeton, NJ, is a monthly abstracting service that includes coverage selected from more than 15,000 scientific journals and periodicals. Chemists and engineers will find this particularly useful in covering developments in Europe, although the coverage is reportedly worldwide. Other related abstract bulletins include *Microbiology, Public Health,* and *Occupational Health.* These publications are subsets of the comprehensive *Excerpta Medica* medical abstracting service that is also searchable for the more recent years online through both *DIALOG®* Information Services and *BRS®* in a file known as *Embase.* The online files contain about 25% more abstracts than the printed publications.

Environment Abstracts is another excellent source. It, too, is available in both printed form and online. The online version is *Enviroline®* and can be searched by users of both *DIALOG®* Information Services and SDC® Information Services.

The scope is broad. Categories covered include: air pollution; chemical and biological contamination; energy; environmental education; environmental design and urban ecology; food and drugs; general; international; land use and misuse; noise pollution; nonrenewable resources; oceans and estuaries; population planning and control; radiological contamination; renewable resources—terrestrial; renewable resources—water; solid waste; transportation; water pollution; weather modification and geophysical change; and wildlife. Copies of most of the documents included are available from the publisher, EIC/Intelligence Center.*

Pollution Abstracts published by Cambridge Scientific Abstracts,* contains references to worldwide technical literature covering some 3000 journals, conference proceedings and papers, and monographs in the areas of air and water pollution, solid wastes, noise, pesticides, radiation, and general environmental quality. It is available both in conventional printed form and also online via Lockheed *DIALOG*®.

There are at least two indexing and abstracting services that cover the science of safety defined broadly: *CIS Abstracts* and *Safety Science Abstracts*.

CIS Abstracts is produced by the International Occupational Safety and Health Information Center (*CIS*) of the International Labor Organization (ILO), which has headquarters in Geneva, Switzerland, and an office in Washington, D.C. Many find this publication especially useful because of its international scope. Coverage includes job safety, worker protection, and disease prevention. Worldwide laws, regulatory documents, and publications of intergovernmental organizations are a particular strength. It is also available online through *Questel* and ESA–IRS (see Section 10.4). A subset of this database is also available through *TOXLINE* (see Section 9.6.A) The card files of *CIS* began in 1960, but the online version begins its coverage in 1974. Despite its name, ILO is not a labor union organization; rather, it is an agency within the United Nations system.

Safety Science Abstracts is a compilation of abstracts that have been selected, classified, and indexed from the world's technical and scientific literature on the science of safety. Each issue contains extensive cross-references and indexes by subject, author, and acronym. It is published by Cambridge Scientific Abstracts, in association with the Institute of Safety and Systems Management, University of Southern California, Los Angeles.

14.8.   STRUCTURE–ACTIVITY RELATIONSHIPS

Structure–activity relationship (SAR) techniques propose to predict toxicity on comparisons of structures and properties with known substances. This approach is still controversial and is the subject of considerable discussion as to its ultimate validity and reliability for such crucial uses as issuing of government regulations. Studies now under way should help to resolve the conflicting viewpoints.

One attempt to apply SAR concepts utilizes the *Log P and Parameter Database* assembled by Corwin Hansch and colleagues at Pomona College, Claremont, CA. More than half of this database is devoted to partition coefficients—the equilibrium ratio of the concentrations of a solute distributed between an organic phase and water. The octanol/water partition coefficient, expressed in log terms, serves as the "standard" parameter for measuring hydrophobic character. The Pomona College *Medchem Database* currently lists over 27,000 log P values in 300 solvent/water systems, as well as 20,000 electronic and steric parameters for a variety of substituents. It is updated semiannually and is available in printed, microfiche, magnetic tape, and online formats from Technical Database Services (TDS).*

Using a Hammett type extra-thermodynamic approach, Hansch has developed over 2000 regression equations that relate these parameters to toxicity and other types of biological activity. The various fields where this concept has been successfully employed have been summarized in a paper by Albert J. Leo (26).

14.9. TOXICITY TO AQUATIC ORGANISMS; WATER RESOURCES

Potential toxicity of chemicals to fish and other aquatic organisms is of key environmental interest, especially if there is a chance that such organisms might come into contact with a chemical, or its by-products or waste streams, during manufacture or use. Government agencies and other sources have compiled much of earlier and more recent published literature. For example, see Reference 27. Recent data can also be obtained from some of the online sources and printed abstracting and indexing services mentioned earlier. In addition, there are several more specialized sources:

- Aquatic pollution is one of the fields covered in *Aquatic Sciences and Fisheries Abstracts* published monthly by Cambridge Scientific Abstracts.*

- *Water Resources Abstracts* is a monthly abstracting and indexing service published by the U.S. Department of Interior's Water Resources Scientific Information Center. This covers the subject of water quite broadly, including toxicity to aquatic life. It can be used in conventional (printed) form, and online access is available through Lockheed *DIALOG®*.

- An important tool is *Aqualine,* which is a product of the Water Research Center (WRC).* WRC is the focal point for water research in the United Kingdom; it was formed by a merger of the Water Pollution Research Laboratory, the Water Research Association, and the Technology Division of the Water Resources Board. The online database contains some 33,000 abstracts, which had been published by the former Water Pollution Research Laboratory's *Water Research Abstracts*. The latter began publication in 1927 and over the years gained a worldwide reputation. From 1974 to date, *Aqualine* contains material from *WRC Information,* now known as *Aqualine Abstracts,* an important biweekly abstracting service published since April, 1974 by WRC. The scope now includes not only the pollution aspects, but also the drinking water and supply aspects, which had been excluded by *Water Research Abstracts;* this is said to be the "world's first abstracting journal devoted exclusively to the whole sphere of hydrological activity." Online availability of *Aqualine* is via Pergamon *InfoLine®*. WRC's INSTAB service is described in Section 14.10. Another online file devoted to water is the *Aquanet*™ database of the American Waterworks Association, Denver, CO. *Aquanet*™ coverage begins with 1971.

Despite availability of these and other sources, there is a relative shortage of published data on effects of chemicals on aquatic life. Mammalian toxicity data are more readily available.

Direct mail or phone contact with government experts on the forefront of this work can provide additional leads beyond what is in published sources. Examples of leaders in the field are scientists at the U.S. Environmental Research Laboratory.*

14.10.   BIODEGRADABILITY

Biodegradability of chemical substances is of considerable environmental interest. One of the most extensive known files on this topic is maintained by the Water Research Center.* This special service is known as INSTAB (Information Service on Toxicity and Biodegradability). The collection, which is critically appraised during its compilation, covers the effects of chemicals on biological sewage treatment processes, both aerobic and anerobic. Also included are biological effects on freshwater invertebrates and fish. The service is mainly for members of the center, although nonmembers may pay to use the facilities.

Other information on biodegradability will be found in more general tools such as *CA* and *Biological Abstracts* and in specialized environmental tools such as *Pollution Abstracts*.

14.11.   WATER-QUALITY DATA

Water-quality data are important in appraisal and management of water sources, pollution surveillance and studies, monitoring of water quality criteria and standards, and development of energy resources.

One good source of water data is the National Water Data Exchange (NAWDEX). This is part of the U.S. Geological Survey.* NAWDEX was established in 1976 to assist users in identifying, locating, and acquiring needed data. It has been described as a confederation of all types of water-oriented organizations. To cite just one example, coordination is maintained with the Environmental Protection Agency's STORET file, which has water-ways quality data from more than 200,000 collection points. The general public can access NAWDEX through its more than 70 local assistance centers located throughout the United States.

14.12. EXAMPLES OF KEY JOURNALS

There are many journals in this field, but it may help the reader to mention a few examples:

- *Environmental Science and Technology* (American Chemical Society, Washington, D.C.)
- *Journal of the Water Pollution Control Federation* (Washington, D.C.); annual literature review issue is excellent.
- *Journal of Hazardous Materials* (Elsevier Publishing Co., Amsterdam, The Netherlands); this journal, of international scope, could prove especially important.
- *Chemical Engineering* (McGraw-Hill, New York); frequently contains articles and new notes of environmental interest.
- *Hazards Review* is a monthly international bulletin reporting on health and safety hazards associated with production, processing, and use of rubber and plastic materials. This is a cooperative venture of the Rubber and Plastics Research Association of Great Britain (*RAPRA*) and Elsevier Science Publishing Co. The publication includes news on government regulations and legislation, industrial codes of practice, safer products and processes, and conferences.

14.13. NEWSLETTERS

This field is characterized by a number of specialized newsletters. These may contain information that supplements and augments chemical news magazines and sources such as mentioned in Section 4.3.A. Emphasis in the newsletters is on rapid concise reporting. One good example is *Air/Water Pollution Report,* published weekly by Business Publishers, Inc.* Other related newsletters by the same publisher are *Clean Water Report, Toxic Material News, Hazardous Waste News,* and *State Regulation Report.*

14.14. OTHER TOPICS

Much of the next chapter has safety implications.

Despite all the toxicological data and sources described in this book and elsewhere, a report by a committee of the National Research Council concluded that, because little data from testing and experience is available on many thousands of chemicals, assessing the potential hazard of all of these to humans is impossible. Animal test data may be available, but extrapolation of these results to humans is highly uncertain. Fewer than 100,000 chemicals are regulated or inventoried by the U.S. Environmental Protection Agency and Food and Drug Administration, but there are more than 7.5 million known chemicals. This means that chances of finding specific toxicological data about a compound of interest, particularly if it is not commercial, are poor. For chemicals that have been studied, personal contacts with toxicologists (and other researchers in the field) and current primary literature are the best sources. [An example of a good primary source is the National Institute For Occupational Safety and Health (NIOSH).*] In some cases, cautious estimation or prediction may be possible. For chemicals not yet studied, similar structures may sometimes give clues as to potential toxicity, but in this connection, advice of a toxicologist is essential in attempting to extrapolate from what has already been proven to the unknown.

REFERENCES

1. L. H. Keith and D. B. Walters, Eds., *Compendium of Safety Data Sheets For Research and Industrial Chemicals*. VCH Publishers, Deerfield Beach, FL 1985.
2. N. V. Steere, Ed., "Safety in the Chemical Laboratory," Vols. 2–3, 1971, 1974; Vol. 4, ed. by M. M. Renfrew 1981; J. Chem. Ed., Easton, PA 18042.
3. G. F. Clayton and F. E. Clayton, *Patty's Industrial Hygiene and Toxicology*, 3rd ed., Wiley, New York, 1983–1985.
4. W. B. Deichmann and H. W. Gerarde, *Toxicology of Drugs and Chemicals*, Academic, New York, 1969.
5. R. E. Gosselin, H. C. Hodge, and R. P. Smith, Eds., *Clinical Toxicology of Commercial Products*, 5th ed., Williams and Wilkins, Baltimore, 1984.
6. A. Hamilton, H. L. Hardy and A. J. Finkel, *Industrial Toxicology*, 4th ed., John Wright–PSG, Inc. Littleton, MA, 1982.
7. N. V. Steere, *CRC Handbook of Laboratory Safety*, 2nd ed., CRC Press, Boca Raton, FL, 1971.
8. M. Sittig, *Hazardous and Toxic Effects of Chemicals*, Noyes Data Corp., Park Ridge, NJ, 1979.
9. M. Sittig, *Handbook of Toxic and Hazardous Chemicals*, Noyes Data Corp., Park Ridge, NJ, 1981.

10. N. Irving Sax, *Dangerous Properties of Industrial Materials,* 6th ed., Van Nostrand Reinhold, New York, 1984.

11. L. Bretherick, *Handbook of Reactive Chemical Hazards,* 3rd ed., Butterworths, London, 1985.

12. U.S. National Institute for Occupational Safety and Health, *Registry of Toxic Effects of Chemical Substances,* Superintendent of Documents, Washington, D.C. Published every few years with microfiche supplement; also available on-line.

13. H. H. Fawcett, *Hazardous and Toxic Materials,* Wiley, New York, 1984.

14. American Public Health Association, *Standard Methods for the Examination of Water and Wastewater,* 15th ed., Washington, D.C., 1985.

15. Morris Katz, Ed., *Methods of Air Sampling and Analysis,* 2nd ed., American Public Health Association, Washington, D.C., 1977.

16. *Proceedings of the Industrial Waste Conference,* Purdue University, Ann Arbor Science Publishers, Ann Arbor, MI, annual.

17. K. Voigt and H. Rohleder, Eds., *A Compilation of Data Sources for Existing Chemicals,* Gesellschaft für Strahlen-und Umweltforschung of Munich, West Germany, 1984.

18. D. M. Cipra and C. F. Damron, "Safety—A Guide to Nearly 50 Databases," *Database,* 8(2), 23–30 (1985).

19. *Safety in Academic Chemistry Laboratories,* American Chemical Society (Committee on Chemical Safety), Washington, D.C., 1985.

20. *Loss Prevention Series,* American Institute of Chemical Engineers, New York. Published every few years. Part of a series of *Technical Manuals.*

21. *Ammonia Plant Safety Series,* American Institute of Chemical Engineers, New York. Published every few years. Part of a series of *Technical Manuals.*

22. *Dow's Fire and Explosion Index Hazard Classification Guide,* 5th ed., Publication number T-71, American Institute of Chemical Engineers, New York, 1981.

23. National Fire Protection Association: (a) *Manual of Hazardous Chemical Reactions* (Publication 491M); (b) *Hazardous Chemical Data* (Publication 49); (c) *Fire Protection Guide on Hazardous Materials,* 8th ed., Quincy, MA, 1984.

24. National Fire Protection Association, *Fire Protection for Laboratories Using Chemicals* (Publication 45), Quincy, MA, 1975.

25. L. Bretherick, Ed., *Hazards in the Chemical Laboratory,* 3rd ed., Royal Society of Chemistry, 1981.

26. A. J. Leo, "The Octanol–Water Partition Coefficient . . . ," *J. Chem. Soc., Perkin Trans. II,* 825–838 (1983).

27. (a) J. E. McKee, Ed., *Water Quality Criteria,* 2nd ed., California State Water Quality Control Board, 1963; (b) U.S. Environmental Protection Agency Water Pollution Control Series, *Water Quality Criteria Data Book,* Vols. 3 and 5, EPA 18050 GWV05/71 and HLA09/73, U.S. Government Printing Office, Washington, D.C.; (c) and annual literature reviews of the *Journal of the Water Pollution Control Federation* (June issue) which include, among many other topics, reviews of aquatic toxicity such as to freshwater fish.

LIST OF ADDRESSES

American Conference of Governmental Industrial Hygienists
6500 Glenway Ave.
Cincinnati, OH 45211

American Public Health Association
1015 18th Street, N.W.
Washington, D.C. 20036

American Society for Testing and Materials
1916 Race Street
Philadelphia, PA 19103

BioSciences Information Service
2100 Arch Street
Philadelphia, PA 19103-1399

Bureau of National Affairs
1231 25th Street
Washington, D.C. 20037

Business Publishers, Inc.
951 Pershing Dr.
Silver Spring, MD 20910

Cambridge Scientific Abstracts
5161 River Road, Bethesda, MD 20816

EIC/Intelligence Center
48 W. 38th Street
New York, N.Y. 10018

Genium Publishing Co.
1145 Catalyn Street
Schenectady, N.Y. 12303-1836
(518-377-8855)

Information Consultants, Inc.
1133 15th Street N.W.
Washington, D.C. (202-822-5200)

Information Handling Services
Englewood, CO 80150

National Fire Protection Association
Batterymarch Park
Quincy, MA 02269

National Institute for Occupational Safety and Health
4676 Columbia Parkway
Cincinnati, OH 45226

National Safety Council
444 North Michigan Avenue
Chicago, IL 60611 (312-527-4800)

Occupational Health Services, Inc.
400 Plaza Drive
Secaucus, NJ 07094

Technical Database Services
10 Columbus Circle
New York, N.Y. 10019
(212-245-0384)

U.S. Environmental Research Laboratory
Duluth, MN 55804

U.S. Geological Survey
421 National Center
Reston, VA 22902

U.S. National Institute for Occupational Safety and Health (NIOSH)
1600 Clifton Rd.
Atlanta, GA 30333

VCH Publishers
303 N.W. 12th Avenue
Deerfield Beach, FL 33442-1705
(305-428-5566)

Water Research Center
P.O. Box 16, Henley Road
Medmenham, Marlow
Buckinghamshire SL72HD,
England

15 LOCATING AND USING PHYSICAL PROPERTY AND RELATED DATA

15.1. INTRODUCTION

Most chemists and engineers find it difficult to locate physical property and related data for new or little-studied chemicals or chemical systems. The task of locating such data can be tedious, expensive, and often fruitless. The same difficulty applies to location of critically evaluated data for almost any chemical—many chemists and engineers find these among the most difficult-to-locate kinds of chemical information. Location of evaluated, accurate thermodynamic data is especially challenging. If data are inaccurate or not fully understood, this can lead to erroneous conclusions or to syntheses or processes that do not work as expected. Proper use of accurate data is important in the design and engineering of pilot plants and plants that will safely produce chemicals of high quality at the desired yield. Laboratory chemists doing synthesis work find good physical property data equally important in their work. Further, accurate data and correct interpretation of data are vital to major policy decisions at the national level, such as on environment or energy matters.

The problem is complicated by the current policy of some journals not to publish extensive tables of data or full experimental details in an attempt to conserve printing and other costs. The trend is to make these data and details available in microfilm or other microforms that need to be purchased separately by readers if they are interested. Many scientists consider this a false economy. They believe that information can be *lost* or misinterpreted by this practice, especially if the data microfilmed are not subject to normal editorial review.

Another factor is that some scientists are reluctant to evaluate the results of their work critically, or even to provide sufficient information to permit *others* to recognize or evaluate the accuracy of their work. Some estimates are that for about 50% of data reported in scientific literature too little information has been reported for independent evaluation. Even with all the sources described in this chapter, as well as many other sources, it is probably correct to speculate that most properties that scientists need are not available in the literature. This is because there are over 7.5 million known chemical substances and an uncounted number of mixtures. Nevertheless, informed use of the literature can be extremely helpful in any investigation of properties.

15.2.   CATEGORIES OF DATA

The source and methods reviewed in this chapter can be grouped into several categories:

1. *Original Sources.*   These are organizations or systems that generate or critically evaluate data. The National Standard Reference Data System is an example, specifically the system's data centers.

2. *Secondary and Tertiary Sources.*   These are compilations of data from original sources and news about projects and activities. Examples include handbooks, journals, and books, including compilations and reports. Abstracting and indexing services, especially *CA*, are valuable for accessing pertinent data, and there are several special bulletins and newsletters in the field.

3. *Evaluation and Estimating.*   In this field particularly, qualified chemists and engineers may need to do some of their own evaluation and perform some calculations using appropriate computer programs. Some organizations have designated specialists who concentrate on this activity.

15.3.   RECORDING PROPERTY DATA

When any one property of a chemical is being searched for and is located, the chemist should make note of other properties (or at least the source of such properties). Experience has shown that this can save considerable future work, because several properties of a single chemical are often grouped together in one paper or other source.

Many chemists, in addition to recording physical property data in their laboratory notebooks, also make use of a specially designed property data form. Such a form should be kept on file for each chemical and should have separate columns for recording kinds of properties, values obtained, source, and date of source.

If a chemist is part of a team or group working together on the same project, all completed data sheets should be kept in a central place. This helps avoid repetitive efforts in which different individuals may look for the same values. A central repository for data is especially important in industry. For example, completed data sheets often become part of a process manual used by engineers to design and build pilot plants and plants. For almost all commercial chemicals, U.S. government regulations require use of a standard form (Material Safety Data Sheet) with emphasis on properties related to safe handling. This form can be adapted (expanded) to include a complete range of properties about which an organization wishes to maintain a verified, centralized record for all chemicals of interest.

15.4. NATIONAL STANDARD REFERENCE DATA SYSTEM

The National Standard Reference Data System (NSRDS) is one of the key sources available to the chemist and engineer for obtaining the latest and best physical property and related data. NSRDS was established in 1963 as a means of coordinating, on a national scale, production and dissemination of critically evaluated reference data in the physical sciences.

Under the Standard Reference Data Act, the National Bureau of Standards (NBS) of the U.S. Department of Commerce has the primary responsibility in the federal government for providing reliable scientific and technical reference data. The key to the program is decentralization. NBS headquarters are located in large, campuslike facilities in Gaithersburg, MD, near Washington, D.C.

The Office of Standard Reference Data of NBS coordinates data evaluation centers located in university and other laboratories as well as within NBS. The data centers compile and critically evaluate numerical data on physical and chemical properties and also provide services such as answering inquiries for specific data and production of custom bibliographies.

The NSRDS comprises the set of data centers and other data evaluation projects administered or coordinated by the NBS. The primary aim of this program is to provide critically evaluated numerical data, in convenient and accessible form, to the scientific and technical community of the United States. A second aim is to advance the level of experimental measurements by providing feedback on sources of error in various measurement techniques. Through both these means, the program strives to increase the effectiveness and productivity of research, development, and engineering design.

The technical scope of the program is restricted to well-defined physical and chemical properties of substances and systems that are well characterized. Although this definition leaves some borderline cases, the intent is to concentrate the effort on intrinsic properties that are clearly defined in terms of accepted physical theory. Properties that depend on arbitrarily defined characteristics of the measurement technique are generally excluded. Likewise, materials of uncertain or variable composition are not included. Biological properties and data relating to large natural systems (e.g., the atmosphere, the oceans) also fall outside the program.

A. Program Structure

Current activities in the Standard Reference Data program are carried out in 25 data centers and approximately 30 short–term projects located in technical divisions of NBS and in academic and industrial laboratories. Each activity collects and evaluates available data on a specified set of properties and substances. Activities are grouped into three discipline–oriented program areas:

1. *Physical Data*—includes data on atomic, molecular, and nuclear properties, and spectral data utilized for chemical identification. Program manager is Sharon Lias.

2. *Chemical Data*—covers primarily kinetic, thermodynamic, and transport properties of substances important to the chemical and related industries. Program manager is Howard J. White, Jr.

3. *Materials Properties Data*—includes structural, electrical, optical, and mechanical properties of solid materials of broad interest. Program manager is John R. Rumble, Jr.

Principal output of the program consists of computer readable and hard copy compilations of evaluated data and critical reviews of the status of data in particular technical areas. Evaluation of data implies a careful examination, by an experienced specialist, of all published measurements of the quantity in question, leading to the selection of a recommended value and a statement concerning its accuracy or reliability. The techniques of evaluation depend on the data in question but generally include an examination of the method of measurement and the characterization of the materials, a comparison with relevant data on other properties and materials, and a check for consistency with theoretical relationships. Adequate documentation is provided for the selections of recommended values and accuracy estimates.

As noted earlier, the NSRDS is managed by the Office of Standard Reference Data (OSRD) of the NBS. This office has the responsibility of allocating that part of the NBS budget that is spent on critical data evaluation, both within the NBS technical divisions and through contracts with outside organizations. The staff of the office act as monitors for all supported projects. Management of the dissemination program of NSRDS is also in the hands of OSRD, and an information service is operated on a limited scale. In addition, OSRD maintains close contact with other data compilation activities, both in the United States and abroad. It attempts, both domestically and internationally, to avoid needless duplication and to encourage coverage of important technical areas. Evaluated data are disseminated through the following mechanisms:

- *Journal of Physical and Chemical Reference Data*—a quarterly journal containing data compilations and critical data reviews, published for the NBS by the American Institute of Physics and the ACS.
- NSRDS–NBS Series—a publication series distributed by the Superintendent of Documents, U.S. Government Printing Office.

- Appropriate publications of technical societies and commercial publishers.
- Magnetic tapes, online networks, and other computer–based formats (see Section 15.13 and Table 15.1).
- Response by OSRD and individual data centers to inquiries for specific data.

The most recent publications list of NSRDS can be scanned to find out what is available in printed form. Also, NSRDS publishes an informal newsletter, *Reference Data Report*, which has news and ideas about data centers, publications, meetings, and other activities related to data evaluation and dissemination. Further information on NSRDS publications, sources of data, and support of data compilation activities can be obtained from the Director, David R. Lide, Jr.*

B. Data Centers and Projects

Listed in the following sections are titles, locations, and project leaders for all continuing data centers and short-term projects that receive at least a part of their support from the OSRD. The source of this tabulation is NBSIR–84 2986, December, 1984, which provides a status report on NSRDS and a brief description of each center or project. Data centers are listed in alphabetical order by name. Short-term projects are categorized by the OSRD program area under which they are managed.

1. Data Centers

Alloy Phase Diagram Data Center, Center for Materials Science
NBS, Gaithersburg, MD 20899—Kirit Bhansali.

Aqueous Electrolyte Data Center, Center for Chemical Physics
NBS, Gaithersburg, MD 20899—David Smith–Magowan.

Atomic Collision Cross Section Data Center, Center for Basic Standards
NBS, Gaithersburg, MD 20899—Jean W. Gallagher.

Atomic Energy Levels Data Center, Center for Radiation Research
NBS, Gaithersburg, MD 20899—W. C. Martin.

Atomic Transition Probabilities Data Center, Center for Radiation Research
NBS, Gaithersburg, MD 20899—Wolfgang L. Wiese.

CINDAS (Center for Information and Numerical Data Analysis and Synthesis)
Purdue University, West Lafayette, Indiana 47907—C. Y. Ho.

Chemical Kinetics Data Center, Center for Chemical Physics
NBS, Gaithersburg, MD 20899—J. T. Herron.

Chemical Thermodynamics Data Center, Center for Chemical Physics
NBS, Gaithersburg, MD 20899—David Garvin.

Corrosion Data Center, Center for Materials Science
NBS, Gaithersburg, Md 20899—G. M. Ugiansky.

Crystal Data Center, Center for Materials Science
NBS, Gaithersburg, MD 20899—A. D. Mighell.

Diffusion in Metals Data Center, Center for Materials Science
NBS, Gaithersburg, MD 20899—John Manning.

Fluid Mixtures Data Center, Center for Chemical Engineering
NBS, Gaithersburg, MD 20899—Neil A. Olien.

Fundamental Constants Data Center, Center for Basic Standards
NBS, Gaithersburg, MD 20899—Barry N. Taylor.

Ion Kinetics and Energetics Data Center, Center for Chemical Physics
NBS, Gaithersburg, MD 20899—Sharon G. Lias.

JANAF Thermochemical Tables, Dow Chemical Company
Midland, MI 48640—Malcolm W. Chase, Jr. See Section 15.12 for detail.

Molecular Spectra Data Center, Center for Chemical Physics
NBS, Gaithersburg, MD 20899—Frank J. Lovas.

Molten Salts Data Center, Rensselaer Polytechnic Institute
Troy, N.Y. 12181—George J. Janz.

National Center for the Thermodynamic Data of Minerals, U.S. Geological Survey
Reston, VA 22070—John L. Haas, Jr.

Phase Diagrams for Ceramists Data Center, Center for Materials Science
NBS, Gaithersburg, MD 20899—L. Cook.

Photon and Charged Particle Data Center, Center for Radiation Research
NBS, Gaithersburg, MD 20899—M. J. Berger.

Radiation Chemistry Data Center, Radiation Laboratory
University of Notre Dame, Notre Dame, IN 46556—Alberta B. Ross.

Thermodynamics Research Center, Texas A & M University
College Station, TX 77843—K. N. Marsh.

Thermodynamics Research Laboratory, Washington University
St. Louis, MO 63130—Buford D. Smith.

2. Projects (Physical Data Program)

Critical Compilation of Mass Spectral Data, Mass Spectral Data Centre
Nottingham, England. Source of support: NBS.

Compilation of Atomic Wavelengths Below 2000 Angstroms, Spectroscopic Data Center
Naval Postgraduate School, Monterey, CA 93940—Raymond L. Kelly.

K Shell Ionization by Hydrogen and Helium Ions, East Carolina University
Greenville, NC 27834—Gregory Lapicki.

Digitization of the Coblentz Society Infrared Database, Johns Hopkins Applied Physics Laboratory
Columbia, MD—William Strauss. Source of support: EPA, NIH.

Soft X–Ray Interactions With Matter, Center for Radiation Research
NBS, Gaithersburg, MD 20899—Martin Berger. Source of support: DOE, NBS.

Medical Physics Data Book (2nd ed.), Medical Physics Data Group
AAPM (American Association of Physicists in Medicine) New York, NY 10017. Source of support: AAPM, NBS.

3. Projects (Chemical Data Program)

Data on Aqueous Electrolytes, University of Delaware
Newark, DE 19711—R. Wood. Source of support: NSF.

DIPPR Data Projects, American Institute of Chemical Engineers
New York, NY 10017. Source of support: NBS, Industry. (see Section 15.15 for detail).

Equilibrium and Transport Properties of Polyatomic Gases and their Mixtures at Low Density, Brown University
Providence, RI 02912—J. Kestin, E. A. Mason. Source of support: NSF.

Preparation of Pourbaix Diagrams for Selected Systems, Belgian Center for Corrosion Study
Brussels, Belgium—Marcel Pourbaix. Source of support: DOE.

Properties of Polar Fluids, Center for Chemical Engineering
NBS, Gaithersburg, MD 20899—L. Haar, J. M. H. Levelt-Sengers.
  Source of support: NBS.

Thermophysical Properties Data for Fluids, University of Maryland
College Park, MD 20742—J. V. Sengers. Source of support: NSF.

**Critical Evaluation of High Temperature Kinetic Data, Aerospace
  Corporation**
Los Angeles, CA 90009—N. Cohen. Source of support: DOD and NBS.

**Solubility of the Sparingly Soluble Salts of Zinc, Cadmium, and Mercury in
  Water and Aqueous Salt Systems, Emory University**
Atlanta, GA 30322—H. Lawrence Clever. Source of support: NSF.

**Critical Evaluation of Chemical Kinetic Data for Metastable Electronically
  Excited Species, ChemData Corporation**
Santa Barbara, CA 93103—K. Schofield. Source of support: DOE.

**A Critical Review of Polychlorinated Biphenyls: Physical—Chemical Data of
  Environmental Relevance, University of Toronto**
Ontario, Canada M5S1A4—D. Mackay. Source of support: DOE.

4. Projects (Materials Properties)

**Cation–Nitrogen Distance in Nitrides of Crystalline Compounds, University
  of Illinois—Chicago Circle**
Chicago, IL 60680—W. H. Baur. Source of support: NSF.

**Crystallographic Data for Organic Materials, Cambridge Crystallographic
  Data Center**
Cambridge, England. Source of support: NBS.

5. OSRD Binary Phase Diagram Evaluation Projects

Rare Earth Alloys, Iowa State University
Ames, IO 50011—K. A. Gschneidner. Source of support: DOE

Copper Alloys, Carnegie–Mellon University
Pittsburgh, PA 15213—D. E. Laughlin. Source of support: DOE.

**Vanadium and Niobium Alloys, Ames Laboratory of DOE, Iowa State
  University**
Ames, IO 50011—J. F. Smith. Source of support: DOE.

Titanium Alloys, National Bureau of Standards
Gaithersburg, MD 20899—J. Murray. Source of support: ONR.

Alkali Metals, Thermfact
Montreal, Canada—A. Pelton. Source of support: DOE.

Chromium Alloys, University of Alabama
University, AL 35486—J. P. Neumann. Source of support: DOE.

Actinide Alloys, Los Alamos National Laboratory
Los Alamos, NM 87544—D. Peterson. Source of support: DOE.

6. Other Projects

Reference Data on the Refractive Index of Selected Optical Materials, CINDAS, Purdue University
West Lafayette, IN 47907—H. H. Li. Source of support: DOE.

ESCA Data Base Project, Surfax Company
Oakland, CA 94618—Charles Wagner. Source of support: NSF.

Corrosion Rate Data for Stainless Steel, Georgia Institute of Technology
Atlanta, GA 30332—M. Marek. Source of support: NBS.

CINDAS, Purdue University
West Lafayette, IN 47907—Pramod D. Desai. Source of support: NSF.

Standards For Computerization of Mechanical Property Data, Sci-Tech Knowledge Systems
Scotia, NY 12302—Jack Westbrook. Source of support: Army.

Surface Sputtering Yield Project, National Bureau of Standards
Gaithersburg, MD 20899—Cedric Powell. Source of support: DOE and NBS.

Polymer Solution Molecular Weight and Viscosity Project, National Bureau of Standards
Gaithersburg, MD 20899—H. Wagner. Source of support: NBS.

Some details on the centers at Texas A & M University and at Purdue University, presented in the next sections, may help illustrate typical data center activity.

See also Section 15.13 on computer–readable tapes available through NBS.

C.   Thermodynamics Research Center

One of the best known of the data centers affiliated with NSRDS is the Thermodynamics Research Center (TRC), Texas A & M University, College Station, originally directed by Bruno J. Zwolinski and now by Kenneth N. Marsh. This center is notably strong in data for organics; inorganics are also covered but not extensively. Some 25 technical and support staff conduct the work.

Two major efforts underway at TRC are described here:

1. Hydrocarbon Project (formerly API Research Project 44) *Selected Values of Properties of Hydrocarbons and Related Compounds,* is one of the most complete compilations of physical, thermodynamic, and spectral data available on these compounds. The project began in 1942 at the NBS with financial support from the American Petroleum Institute (API), and under the direction of F. D. Rossini. In 1950 it was moved to the Carnegie Institute of Technology, Pittsburgh. In 1961, Bruno J. Zwolinski assumed the directorship, and the project was moved to Texas A & M University. The material is published in looseleaf form and there are two supplements per year. There are a total of approximately 3000 valid data sheets.

2. The Thermodynamics Research Center Data Project (TRCDP) (*Selected Values of Properties of Chemical Compounds*) began in 1955 at the Carnegie Institute of Technology, under the auspices of the Manufacturing Chemists Association. Sponsorship changed July 1, 1966 to the Texas A & M TRC. The principal objective is the preparation, publication, and distribution of spectral data and data on physical and thermodynamic properties of all classes of organic compounds (except hydrocarbons and related compounds covered in the Hydrocarbon Project), as well as industrially important nonmetallic inorganic compounds. This, too, is one of the most complete compilations of spectral and physical thermodynamic data to be found on these compounds.

A merger of the two projects just described is planned by TRC.

The *Comprehensive Index of API44—TRC Selected Data on Thermodynamics and Spectroscopy* (1) is a useful aid in locating data for about 10,000 compounds in the 12 publications of TRC. The categories of data found are IR, UV, Raman, Mass Spectroscopy, NMR, and physical and thermodynamic properties from the beginnings of the project, more than 30 years ago through 1973. This index is being updated.

The Hydrocarbon Project and the TRCDP tables include data for many chemical compounds of interest and importance to the chemical and petroleum industries, and to science in general. The tables are based on the best experimental data available. This information is critically evaluated and presented in a convenient form for ready use. Furthermore, data are estimated in temperature and pressure ranges that are not easily accessible in the laboratory. Data are often estimated for new compounds not yet synthesized but that may possess desirable properties.

Since the early 1970s, TRC has worked on a project designated as International Data Series. Part of this series relates to original data on binary organic nonaqueous and nonelectrolyte mixtures. Data presented include physical and thermodynamic properties. The editor is Henry Kehiaian of Paris, France. Furthermore, there is a B (aqueous mixtures) part of this series, but only a few reports have been issued so far pending identification of a suitable publisher.

TRC also has under way data generation efforts sponsored by NBS. Results of these appear in the *Journal of Physical and Chemical Reference Data*. Another effort is production of the organic section of the *Bulletin of Chemical Thermodynamics* (see Section 15.8) under NBS funding.

A new series, *Key Chemical Data Books,* has been initiated by TRC. These are to be similar to the American Petroleum Institute (API) Monograph series booklets that have dealt mostly with aromatic hydrocarbons and heterocyclics and which TRC has been preparing for publication and distribution by API.

TRC is involved in several computerized databases. One is the thermodynamics module of *CIS* (see Section 10.11.A.3) for which TRC provides the organic component while NBS provides the inorganic component. Online calculations of data are possible in this file. Negotiations have been completed with Technical Database Services (see also Section 14.8) to provide an online file covering vapor pressure and other data from the International Data Series. Finally, it is understood that a contract has been signed with Fein–Marquart Associates, Inc., Baltimore, MD, to produce a computer readable database and associated software on the thermodynamic properties of fluids. This is intended to permit calculation of properties using a microcomputer.

D. Center for Information and Numerical Data Analysis and Synthesis

Another major data analysis center associated with NSRDS is the Center for Information and Numerical Data Analysis and Synthesis (CINDAS) located at Purdue University.* The center was founded in 1957 by Dr. Y. S. Touloukian, and the current director is Dr. C. Y. Ho. In addition to its affiliation with NSRDS, it is a Department of Defense Information Analysis Center (see Section 9.4).

CINDAS serves as a focal point for authoritative expertise on thermophysical, electronic, electrical, magnetic, and optical properties of materials. One of the centers located within CINDAS is the Thermophysical Properties Research Center (TPRC). Information and data on more than 60,000 different substances and materials are currently coded for 14 properties:

Thermal conductivity

Accommodation coefficient

Thermal contact resistance

Thermal diffusivity

Specific heat at constant pressure

Viscosity

Emittance

Reflectance

Absorptance

Transmittance

Solar absorptance to emittance ratio

Prandtl number

Thermal linear expansion coefficient

Thermal volumetric expansion coefficient

The Properties Research Laboratory, an affiliate of CINDAS, measures the thermal, electrical, and optical properties of materials over wide temperature ranges.

An important publication from CINDAS is their *Thermophysical Properties Research Literature Retrieval Guide: 1900–1980* (2). This covers the world literature on 14 properties of 44,300 substances in seven volumes. A related publication is *Thermophysical Properties of Matter: The TPRC Data Series (3)*. More recently, CINDAS has launched a new series, *Data Series on Material Properties* (4); the first volumes appeared in 1981, and the series is still in progress.

15.5. COMMITTEE ON DATA FOR SCIENCE AND TECHNOLOGY

At the international level, the Committee on Data for Science and Technology (CODATA) was created in 1966 as a special committee of the International Council of Scientific Unions to aid in international coordination of data compilation, retrieval, and evaluation. Material available from CODATA includes its *Bulletin*, which contains key sets of recommended data guidelines for presentation of data, and articles on methodology of data evaluation. Other products of chemical interest from CODATA are *Recommended Key Values for Thermodynamics* (see Section 15.12) and *Evaluated Kinetic and Photochemical Data for Atmospheric Chemistry* (Bulletins 33 and 45).

CODATA also compiled a *Compendium* (5), published in 1969, which lists, among other things, continuing numerical data projects and their publications. An attempt is being made to supplant the *Compendium* through the *CODATA Directory of Data Sources for Science and Technology,* 10 sections of which have already appeared. Those of interest to chemists include Chemical Kinetics, Atomic and Molecular Spectroscopy, and Chemical Thermodynamics. Individual chapters appear as issues of the *CODATA*

Bulletin. Major conferences are held every two years and proceedings are published. The address for CODATA is in Paris, France.*

15.6. *JOURNAL OF CHEMICAL AND ENGINEERING DATA* AND OTHER JOURNALS

The *Journal of Chemical and Engineering Data* (*JCED*) published since 1956 by the ACS, although not related to NSRDS, is a good example of a current and original source of reliable physical property data for both pure compounds and mixtures. A quick scan of indexes to this publication and of the table of contents of its more recent issues can sometimes produce more useful information than an equivalent amount of time spent perusing the printed subject indexes to *CA*. No search for the most recent physical property data for specific compounds can be considered complete without consulting *JCED* either directly or through the use of indexing and abstracting services.

In addition to containing experimental or derived data relating to pure compounds or mixtures covering a range of states, *JCED* includes:

1. Papers based on published experimental information that have made tangible contributions.
2. Experimental data that aid in identifying or utilizing new compounds.
3. Papers relating to newly developed or novel syntheses of organic compounds and their properties.

JCED sets a high standard. Its guide for authors specifies that experimental methods should be referenced or described in enough detail to permit duplication of the data by others familiar with the field. Published or standardized procedures and their simple modifications need not be described, but a readily available reference should be cited. The data should be presented with such precision that information may be easily obtained from within the paper, within the stated limits of uncertainty of the experimental background. This journal is cited as just one example of several in the field.

Another leading example is the *Journal of Chemical Thermodynamics*. Others are published by the American Institute of Chemical Engineers, the American Institute of Physics, the American Chemical Society, and other sources. Note that while both scientific journals and trade magazines are rich sources of data, they often offer different kinds of data. Trade magazine data tend to deal more with applications.

Articles in scientific journals concerned primarily with preparative chemistry are good sources of property data, primarily *basic* properties such as melting point or refractive index. If access to such articles is made through abstracting and indexing services, the chemist must note that these properties are not necessarily indexed or abstracted, since their presence is often implicitly assumed. On the other hand, trade-oriented magazines frequently have articles with information on properties affecting end use or applications, such as data relating to handling or processing. Location of these data using magazine indexes is unreliable because properties are frequently not indexed.

15.7.   MANUFACTURERS' LITERATURE

The trade and the other product literature of chemical manufacturers often gives full and accurate treatment of physical and chemical properties, especially those values that relate to use and to safe handling such as data contained in Material Safety Data Sheets (see Section 14.2). Original data, which may not be available elsewhere, are frequently presented here. In using this literature, the chemist needs to make careful note of the grade or purity for which the properties are specified. For the most recent data, and that not found in trade literature, the chemist should write or telephone the manufacturer's technical representative who specializes in the product or product line.

15.8.   *BULLETIN OF CHEMICAL THERMODYNAMICS*

A key to recent thermodynamic data and literature is the annual *Bulletin of Chemical Thermodynamics*, edited by Robert D. Freeman.* It is issued under the auspices of the International Union of Pure and Applied Chemistry and is published by Thermochemistry, Inc. This current awareness *Bulletin* has three main sections: bibliography, reports on current research worldwide, and index.

The bibliography is a listing of papers with chemical thermodynamic content published during the preceding year. The reports section provides terse summaries of studies completed but not yet published. Some 35 countries are covered. Actual data are not given (except for such information as range of variables), but there is sufficient detail to permit users to decide whether to make contact with the cited investigators. The index provides access to the other two sections by substance or property. Other information included is a calendar of meetings, a list of pertinent books, and review articles.

15.9.   *LANDOLT–BÖRNSTEIN*

One of the most extensive printed sources of physical properties and related data is *Landolt–Börnstein* (6). This work was originally compiled by Hans

Landolt and Richard Börnstein in 1883. The latest (sixth) edition began in 1950 and was completed in 1979, and additional volumes (*New Series*) are now in process. The sixth edition contains 28 volumes and almost 100 volumes of the *New Series* are available.

The main divisions of the sixth edition are

1. *Atomic and Molecular Physics*—basic properties and interactions of nuclei, atoms, ions, molecules, and crystals (5 vol.).

2. *Properties of Matter in its Aggregated States*—mechanical and thermal properties, equilibria, interfaces, transport phenomena, general physicochemical data (8 vol.). Electrical, optical, magnetic properties (5 vol.).

3. *Astronomy and Geophysics*—(1 vol.).

4. *Technology*—nonmetallic materials (1 vol.), metallic materials (3 vol.), heat technology (4 vol.), and electrical engineering, light and X-ray technology (1 vol.).

The main groups of the *New Series* are

1. *Nuclear and Particle Physics*—energy levels, radii, decay, reactions of nuclei. Angular correlations. Properties, production, scattering of elementary particles etc. . . . . (12 vol. to date).

2. *Atomic and Molecular Physics*—structure and molecular constants of free molecules, magnetic properties of free radicals and of coordination and organometallic transition metal compounds, luminescence, molecular acoustics, etc. . . . . (16 vol. to date).

3. *Crystallography and Solid State Physics*—crystal structure data of elements, intermetallic, inorganic and organic compounds. Linear and nonlinear elastic, dielectric, optic and coupled properties. Ferro- and antiferroelectric substances. Magnetic oxides and related compounds. Electron and phonon states of metals, etc. . . . . (26 vol. to date).

4. *Macroscopic and Technical Properties*—thermodynamic properties of liquid systems. High pressure properties of matter (5 vol. to date).

5. *Geophysics and Space Research*—(2 vol. to date).

6. *Astronomy and Astrophysics*—(4 vol. to date).

All new volumes are published as part of the *New Series*.

Many American, British, and Canadian chemists have found older volumes of the sixth edition difficult to use because of the predominantly German text, lack of an overall index to date, and complex organization of the material. Consistent with the trend for all major German reference works, such as *Beilstein* and *Gmelin*, the newer volumes of *Landolt–Börnstein* are to be in English or, at a minimum the tables of contents and introductions are to be in both English and German. In addition, a general subject index to the sixth edition and the *New Series* volumes is in preparation for 1986 publication. The indexes previously available for individual volumes have helped somewhat in location of data, but are no substitute for an overall index.

15.10.   HANDBOOKS

The most heavily used sources of physical property data are the *CRC Handbook of Chemistry and Physics* (7), originally published in 1913 and updated approximately every year or two since then, *Lange's Handbook of Chemistry* (8), and *Perry's Chemical Engineer's Handbook* (9a). These provide rapid, convenient, desk-top access to many different kinds of data and can be purchased for prices within the budgets of most practicing chemists and engineers.

These and other desk handbooks usually contain more than just physical properties—a feature that further enhances the value of handbooks. For example, the *Handbook of Chemistry and Physics* includes such other material as an extensive section on mathematical tables, information on sources of critical data, and rules for nomenclature of organic chemistry. The *Chemical Engineers Handbook* includes a section on mathematical tables, data on materials of construction and corrosion, and information on cost and profitability. Another handy desk reference volume is *The Chemist's Companion* (9b).

Chemists will find the 10th (1983) edition of *The Merck Index* (10) exceptionally useful. This source was first published in 1889. The tenth edition has descriptive information on more than 10,000 chemicals, drugs, and biologicals arranged alphabetically by generic or nonproprietary name. Although the publisher is a major pharmaceutical company, coverage is across the board for the most important chemicals of many types. The volume includes organic name reactions, a comprehensive cross index of approximately 56,000 synonyms (chemical, trivial, and generic names as well as trademarks and code numbers), a formula index, and a variety of tabular information. CAS Registry Numbers are given when available, and there are carefully selected references to key journal and patent literature for many of the chemicals listed. More than 50% of the chemicals are illustrated with stereochemical structural formulas. For many chemicals included, there is information on general, medical, or veterinary uses as well as toxicity. The extensive scope and low price of this book make it a purchase of unusual value for any chemist. It is published on a nonprofit basis as a service to the scientific community. Martha Windholz is the editor. This publication is now searchable online (see Section 10.4) in up-dated form through *Questel,* Chemical Information System, Inc., and *BRS®.* The *Merck Index Online* contains revisions of more than 300 of the 10,000 plus monographs found in the printed volume. The online file is to be updated every six months.

The volumes cited here are examples of the many kinds of handbooks of value to chemists. The *CRC* Press, probably the best known publisher of handbooks of interest to chemists and engineers, now offers an online file *Superindex*, which is available through *BRS®* (see Section 10.4). This provides a merged and cross-referenced file of back of the book indexes to several hundred English language reference books. It includes information on the exact location of a specific reference including title of the book, page number, and index entry. Actual data are not provided. Many chemists find this file useful in helping to decide quickly which handbook or other volume has the required data.

Chemists and engineers should use the data in desk handbooks—although valuable, convenient, and indispensable from a practical day-to-day working standpoint—only with some appropriate cautions:

1. New data are constantly being generated and published in journals and other sources. Handbooks, especially those published in conventional bound form, cannot (and do not) claim to provide the most recent data. Thus desk handbooks can be described as incomplete and out of date but not necessarily incorrect. The *Merck Index* is an exception since it is updated online.

2. Tables printed in many handbooks have not been subjected to direct and recent critical review and evaluations. Hence the data are not necessarily the *best* data—an important factor in many chemical investigations.

3. Handbook data are usually based on, or *copied* from, other sources. Typographical and other errors are more likely to occur than in original sources.

4. Details of conditions under which data were originally generated are not always given in handbooks. Limitations are not always evident. If complete references to the original sources of the data are also missing, this further complicates attempts at data evaluation by handbook users.

5. The indexes may not be sufficiently detailed to locate the desired information without considerable digging. This can be a special problem because of the widespread use of synonyms and other alternate names. Occasional outright loss of data can occur because of this difficulty.

6. Some chemical engineers believe that single values are not too useful for engineering calculations and that curves of values, not just single points, are desirable. Tabular values are, however, frequently not data points, but rather *smoothed* and interpolated values.

7. The latest editions of some handbooks omit useful data found in earlier editions. It is worthwhile to keep on hand or consult at least one prior edition.

Even with these and other limitations, the chemist or engineer will probably continue to make heavy use of handbooks as convenient sources of information for investigations in which the highest degree of up to dateness, accuracy, or precision is not essential. Handbooks, such as those mentioned, serve as good starting points when the latest, best, and most complete data are not needed.

15.11.   *INTERNATIONAL CRITICAL TABLES*

One of the most extensive compilations of physical property data is the *International Critical Tables* (11), often referred to as the *ICT*. This eight volume set, published in the late 1920s and early 1930s, is now out of print, although still on the shelves of many chemistry libraries.

The chemist will find some data here still valid, but the compendium has been largely superseded by newer works. Today, *ICT* can be regarded as a *court of last resort*, with much of its utility limited to unusual or *exotic* property data not likely to be found elsewhere. It can sometimes be a good starting point.

The history of *ICT* and related thermodynamics efforts at NBS and elsewhere is described by Garvin et al. (12).

15.12.   EXAMPLES OF OTHER REFERENCE SOURCES

Some key reference sources are listed here. Most descriptions are as given in *NBS Special Publication 454* (13a).

S. W. Benson, *Thermochemical Kinetics—Methods for the Estimation of Thermochemical Data and Rate Parameters*, 2nd ed., Wiley, New York, 1976. Molecular approach to techniques for rapid and relatively quantitative estimation of thermochemical data and reaction-rate parameters for chemical reactions in the gas phase. An appendix contains a collection of tabulated group values for different classes of organic compounds and thermochemical data for selected elements, radicals, and molecules. Also included are tables for estimates of $S°$ and $C_P^°$ contributions from vibration and torsional degrees of freedom as a function of temperature, frequency, and barrier.

CODATA Recommended Key Values for Thermodynamics 1977 [*CODATA Bulletin* No. 28 (1977)]. Recommended values for the quantities $\Delta H_f^°$ (298.15 K), $S°$(298.15 K), $H°$ (298.15 K) $-$ $H°$ (0 K) for 102 of the thermochemically more important elements and compounds, including some aqueous species. These bulletins supersede earlier CODATA reports of this group. The recommended values are not fully consistent with any previously published thermodynamic tables, but the values are intended to form the basis of future generations of compilations (see also annotation on compilation by Parker, Wagman, and Garvin later in this section).

J. D. Cox and G. Pilcher, *Thermochemistry of Organic and Organometallic Compounds*, Academic, London, New York, 1970. A critical compilation of thermochemical data for the title field published since 1930. The enthalpies of formation of some 3000 substances are listed, with estimates of error. Where enthalpies of vaporization are known or can be reliably estimated, these are listed, and in these cases, the enthalpies of formation of both gaseous and condensed phases are given. Extensive introductory material presents experimental procedures for reduction of experimental data of the type found in the book. Applications of thermochemical data are given, and there is a section on methods of estimating enthalpies of formation of organic compounds.

E. S. Domalski, W. H. Evans, and E. D. Hearing, *Heat Capacities and Entropies of Organic Compounds, J. Phys. Chem. Ref. Data*, Suppl. 1, **13** (1984). Data on approximately 1400 organic compounds.

L. Haar, J. S. Gallagher, and G. S. Kell, *NBS/NRC Steam Tables*, Hemisphere Publishing Co., New York, NY, 1984. Accurate representation of data for thermodynamic properties of water and steam from triple point to 1000°C and for pressures up to at least 10 bars.

C. H. Horsley, Ed., *Azeotropic Data—III*, American Chemical Society, Washington, D.C., 1973. Data on azeotropes, nonazeotropes, and vapor–liquid equilibria for more than 17,000 systems. No attempt has been made to evaluate accuracy of the data, most of which are from the original literature. This volume is a revision of *Azeotropic Data I* and *II* and includes new data collected since 1962.

G. J. Janz, *Thermodynamic Properties of Organic Compounds Estimation Methods, Principles, and Practice* (revised ed.), Academic, New York, 1967. Computation of thermodynamic properties such as heat capacities, entropies, enthalpies, and Gibbs energies by statistical mechanical methods and by methods of structural similarity, group contributions, group equations, and generalized vibrational assignments. The chemical properties enthalpy of formation and enthalpy of combustion are treated in terms of bond energies and group increments. Some 78 tables are given of increments, group contributions, and bond contributions as specifically needed for estimation of particular properties.

T. E. Jordan, *Vapor Pressure of Organic Compounds*, Interscience, New York, 1954. A comprehensive compilation of vapor pressure data for organic compounds. Included are tables on the hydrocarbons, alcohols, aldehydes, esters, ketones, acids, phenols, and metal organic compounds. Data for each compound are shown in graphical form, that is, vapor pressure as a function of temperature. References to the data sources in the literature are given.

A. S. Kertes, Ed., *Solubility Data Series*, Pergamon, Oxford, various years. Series of critically evaluated solubility data volumes planned to cover all physical and biological systems. Sponsored by IUPAC (International Union of Pure and Applied Chemistry).

J. A. Larkin, Ed., *International Data Series, Series B, Data on Aqueous Organic Systems,* National Physical Laboratory, Teddington, Middlesex, United Kingdom. A relatively new series for dissemination of selected data on mixtures. Similar in format, scope, and policies to Series A (*Thermodynamic Properties of Binary Systems of Organic Substances*).

W. F. Linke and A. Seidell, *Solubilities: Inorganic and Metal-organic Compounds— A Compilation of Solubility Data From the Periodical Literature*, Vol. I: A–Ir, Van Nostrand, Princeton, N.J., 1958; Vol. II: K–Z, American Chemical Society, Washington, D.C. 1965. Comprehensive compilation of mostly unevaluated solubility data for inorganic and metal-organic compounds. Both aqueous and nonaqueous solvent systems are included. References are given to the data sources.

W. J. Lyman, W. F. Reehl, and D. H. Rosenblatt, *Handbook of Chemical Property Estimation Methods*, McGraw-Hill, New York, 1982. The 26 properties for which estimation methods are given are octanol/water partition coefficient; solubility in water; solubility in various solvents; adsorption coefficient for soils and sediments; bioconcentration factor in aquatic organisms; acid dissociation constant; rate of hydrolysis; rate of aqueous photolysis; rate of biodegradation; atmospheric residence time; activity coefficient; boiling point; heat of vaporization; vapor pressure; volatilization from water; volatilization from soil; diffusion coefficients in air and water; flash points of pure substances; densities of vapors, liquids, and solids; surface tension; interfacial tension with water; liquid viscosity; heat capacity; thermal conductivity; dipole moment; index of refraction.

V. B. Parker, D. D. Wagman, and D. Garvin, *Selected Thermochemical Data Compatible With the CODATA Recommendations*. National Bureau of Standards Report No. 75–968, Washington, D.C. 1976. Selected thermochemical properties, ΔG_f°, ΔH_f°, S°, C_p° (all at 298.15 K), ΔH_f° (0 K), and H (298–0 K), are given for 384 substances (almost entirely inorganic) including many of the more commonly encountered aqueous species. The selected values are intended to be compatible with the current CODATA recommendations on key values for thermodynamics. (see *CODATA Recommended Key Values for Thermodynamics*).

H. Stephen and T. Stephen, *Solubilities of Inorganic and Organic Compounds* (5 vols.), Pergamon, London, 1963. Selection of data on the solubilities of elements, inorganic compounds, and organic compounds in binary, ternary, and multicomponent systems. References are given to sources of data in the literature. The data are unevaluated.

T. S. Storvick and S. I. Sandler, Eds., *Phase Equilibria and Fluid Properties in the Chemical Industry—Estimation and Correlation,* American Chemical Society, Washington, D.C., 1977. A symposium volume containing state of the art reviews.

D. R. Stull, "Vapor Pressure of Pure Substances. Organic Compounds," *Ind. Eng. Chem.,* **39,** 517–540 (1947); "Vapor Pressure of Pure Substances. Inorganic Compounds," *Ind. Eng. Chem.,* **39,** 540–550 (1947). Evaluated vapor pressure data on over 1200 organic and 300 inorganic compounds.

D. R. Stull and H. Prophet, *JANAF Thermochemical Tables*, 2nd ed. National Standard Reference Data Series, National Bureau of Standards, Report No. 37, U.S.. Government Printing Office, Washington, D.C. 1971. The JANAF (Joint Army–Navy–Air Force) tables contain data on thermodynamic properties for over 1000 chemical species. This is a particularly good source for workers in inorganic chemistry; some simple organics are also included. Supplements are published in the *Journal of Physical and Chemical Reference Data* beginning in 1974. The third edition is being completed under editorship of W. Malcolm Chase (Dow Chemical Co.) and others. It is to appear as a supplement to the *Journal of Physical and Chemical Reference Data* in 1986, and it is also to be available in magnetic tape form as an NBS Standard Reference Database.

D. R. Stull, E. F. Westrum, and G. C. Sinke, *The Chemical Thermodynamics of Organic Compounds*, Wiley, New York, 1969. Monograph divided into three parts. The first gives the theoretical basis and principles of thermodynamics and thermochemistry, some experimental and computational methods used, and some applications to industrial problems. The second part gives thermal and thermochemical properties in the ideal gas state from 298 to 1000 K. In this section the sources of data are listed and discussed, and standardized tables are presented for 918 organic compounds. Values of C_p°, S°, $-(G - H_{298}^\circ)/T$, $H^\circ - H_{298}^\circ$, ΔH_f°, ΔG_f°, and log K_p are given at 100 K intervals. In the third section are listed selected values of enthalpy of formation, entropy, and consistent values of ΔG_f° and log K_p

of organic compounds at 298 K. More than 4000 compounds are listed, including a few inorganic compounds. A chapter briefly discusses methods of estimating thermodynamic quantities.

D. D. Wagman, W. H. Evans, V. B. Parker, R. H. Schumm, I. Halow, S. M. Bailey, K. L. Churney, and R. L. Nutall, *The NBS Tables of Chemical Thermodynamic Properties—Selected Values for Inorganic and C_1 and C_2 Organic Substances in SI Units,* American Chemical Society, Washington, D.C., 1983. Published as Supplement 2 to Volume 11 of the *Journal of Physical and Chemical Reference Data,* this volume lists 26,000 values for chemical thermodynamic properties of over 14,000 substances, counting each substance in each phase and each concentration listed for a solution. This is a comprehensive, updated edition of the *NBS Technical Note 270* series that appeared in eight parts between 1965 and 1981 and is a modern version of the classic *NBS Circular 500* that was published in 1952. An estimated 60,000 references were used in development of the tables, which include standard state data on enthalpy, entropy, Gibbs energy, and heat capacity. Properties of aqueous solutions and pure compounds are also included. All data have been critically evaluated and checked for consistency with thermodynamic constraints by specially developed computer programs. The purpose of presenting these tables of evaluated data is to provide scientists and engineers with reliable values of chemical thermodynamic properties of the elements and their compounds, from which calculations can be made of equilibrium constants and changes in enthalpies, entropies, and heat capacities for chemical processes. These data can be used in such areas as chemical engineering design, environmental modeling, and chemical research. The tables contain values, where known, of the enthalpy and Gibbs energy of formation, entropy and heat capacity at 298.15 K (25°C), the enthalpy difference between 298.15 and 0 K, and the enthalpy of formation at 0 K, for inorganic substances and for organic substances containing one or two carbon atoms. In some instances, such as complexes with organic ligands and metal–organic compounds, data are given for substances in which each organic ligand contains one or two carbon atoms.

15.13. COMPUTER–READABLE NUMERIC DATABASES AVAILABLE FROM THE NATIONAL BUREAU OF STANDARDS

In 1983 the NBS Office of Standard Reference Data expanded its program for numeric database dissemination by making available to the general public four databases in magnetic tape form as shown in Table 15.1. The tapes, which can be mounted on an interactive system, are the NBS/NIH/EPA/MSDC Mass Spectral Database, NBS Thermophysical Properties of Hydrocarbon Mixtures Database, NBS Chemical Thermodynamics Database, and the NBS Crystal Data Identification File. The Mass Spectral, Chemical Thermodynamic, and Crystal Identification databases are also available on the Chemical Information System (see Section 10.11.A.3). The databases permit identification of chemical unknowns encountered in different environments, prediction of chemical reaction equilibria, and analysis of heat transfer processes.

TABLE 15.1 NBS Databases in Magnetic Tape Format

NBS Standard Reference Database	Online Availability
NBS/NIH/EPA/MSDC Mass Spectral Database	CIS and Questel (see Section 10.11.A)
NBS Chemical Thermodynamics Database	CIS
NBS Crystal Data Identification File	CIS (see Section 10.11.A.3), CISTI (Canadian Institute For Scientific and Technical Information, Ottawa)
NBS Thermophysical Properties of Hydrocarbon Mixtures (TRAPP)	
Thermophysical Properties of Helium[b]	
Interactive Fortran Program to Calculate Thermophysical Properties of Six Fluids[b]	
Electron and Positron Stopping Powers of Materials	
X-Ray and Gamma-Ray Attenuation Coefficients and Cross Sections	
Thermophysical Properties of Water[a]	
Activity and Osmotic Coefficients of Aqueous Electrolyte Systems	
DIPPR (see Section 15.15)	
JANAF Thermochemical Properties	
Interactive Fortran Program to Calculate Thermophysical Properties of Fluids[a]	

[a] Available on diskette for personal computer use.
[b] Subsumed within "Interactive Fortran Program . . ." below.

The Mass Spectral Database contains electron ionization mass spectra of almost 40,000 different compounds. Each spectrum is described by compound name, molecular formula, molecular weight, and CAS Registry Number. A Quality Index value is assigned to each spectrum.

The mass spectral data have been collected from a variety of sources through the cooperation of NBS, National Institutes of Health, Environmental Protection Agency (EPA), and United Kingdom Mass Spectrometry Data Center. Updates are to be made available.

Recent (1986) additions to the NBS database program include the JANAF Tables (see Section 15.12) and data from the DIPPR program (described later in Section 15.15).*

The Thermophysical Properties of Hydrocarbon Mixtures Database (TRAPP) predicts density, viscosity, and thermal conductivity of hydrocarbon mixtures of arbitrary composition using a tested model and its associated computer software. The model is valid over a wide range of temperatures and pressures and can be applied to hydrocarbons ranging up to C_{20}. The model is also applicable to other chemical compound types and to thermodynamic states ranging from dilute gas to compressed liquid. Multicomponent systems are covered. Data for 60 individual chemical components are contained in the database. New chemical components will be added.

The NBS Chemical Thermodynamics Database, described earlier in Section 15.12, contains recommended values of selected thermodynamic properties for over 15,000 inorganic substances and C_1 and C_2 hydrocarbons. Properties covered include enthalpy of formation from elements in their standard state, Gibbs (free) energy of formation from elements in their standard state, entropy, enthalpy content ($H_{298} - H_0$), and heat capacity at constant pressure (C_p), all at 298.15 K and 1 bar. Enthalpy of formation at 0 K is also given. Each data entry gives the chemical name and formula as well as the physical state.

The NBS Crystal Data Identification File contains data characterizing over 60,000 crystalline substances. Data include reduced cell parameters, reduced cell volume, space group number and symbol, experimental and calculated density, chemical type classification, and chemical name and formula. The database will be updated.

15.14. DATA QUALITY AND EVALUATION

Considerably more attention than ever before is now being paid to data quality, in large measure because of recent intense concern about the environment and continued growth of online databases. It is all too easy to assume that just because data may come from a computerized database it is accurate and reliable. This may or may not be the case. The speed and ease of online access can lead to considerable misinterpretation.

In February 1982, the Chemical Manufacturers Association, jointly with U.S. government officials from such agencies as the NBS and the EPA, sponsored a definitive workshop on data quality. One conclusion was that *data indicators* are needed to describe such factors as methods used to obtain the data; how the data has been evaluated; description of accuracy; and source. These data indicators, it is recommended, should accompany all data sources, especially online databases, so that users can make appropriate decisions. If there is an estimate of accuracy or certainty, this can be especially helpful.

On May 8–9, 1984, the Chemical Manufacturers Association, again jointly with U.S. government officials, held another workshop on the issue of quality under the chairmanship of Curtis Elmer, Monsanto Company. This time the focus was to develop indicators of data *documentation* quality, rather than data quality itself, because it was felt that the latter might be too ambitious at this time.

The consensus was that the following need to be documented:

1. Scope, purpose, and rationale.
2. Experimental design and strategy.
3. Conditions of test, site, and test system.
4. Substance, agent, test chemical, and test organism (by accurate name, composition, analysis, and source).
5. Sampling procedures, methods, equipment and conditions for those areas where materials are sampled from the environment or the workplace.
6. Method descriptions, including protocol, dosage in the case of animal tests, route of administration and duration, and controls used.
7. Analytical methods, including standards, equipment type, validation, and calibration.
8. Results, tables/graphs, statistical analysis methods, and other data anlyses.
9. Quality assurance (QA) statement, that is, was there a quality assurance program? Was the study signed off by QA?
10. Was there a peer review of the studies and/or publication?

Several types of indexing or rating scales were proposed, depending on how well a database, report, or paper met the criteria. Data that are important and critical to one chemist may not be to another. Thus, the significance of data quality depends on the chemist and the use to which the data will be put.

One good example of an evaluated online database is the *Hazardous Substances Data Bank* of the National Library of Medicine. Before the complete records are made available to the public, they are reviewed by a *Scientific Review Panel* to accurately convey what is known about the toxicity of the chemicals being considered.

Maizell (13b) and Luckenbach et al., (13c) are among those who have published important papers on methods of data assessment. Some of the highlights of Maizell's paper are summarized later in this Section. The paper by Luckenbach et al. is a detailed report on screening procedures followed by editors at *Beilstein's Handbook of Organic Chemistry*. Among other things, they point out the importance, occasionally overlooked by some, of consistency with known chemical knowledge and principles. In some cases, 50% of the literature being considered can be excluded when the procedures described are employed.

What if rigorously evaluated data such as that provided by NSRDS or other sources is not available? What if adequate verification of the data in the chemist's laboratory is not possible?

In such cases chemists may need to make their own relatively subjective assessment of published data on physical and chemical properties. Data from several sources may coincide or may vary significantly. Data from a single source may be suspect. In determining reliability of data found in the literature, chemists need to consider factors such as these examples:

1. Is the chemical identity of the substance unambiguously specified? This is a basic starting point, because it helps avoid pitfalls due to variations in nomenclature and in the many commercial trade names and grades.

2. Is there agreement between several independent sources? When there is lack of agreement, one choice is to take the safest, most conservative value. For example, in considering different flash-point values, the chemist could select the lowest value to provide the greatest possible safety margin (*note the importance of checking in several different sources for values of the same constant*). Even if two measurements agree, however, both may have systematic errors.

3. Is the source of the data a paper written to determine specific physical constants rather than a paper in which the constant is determined only incidentally as in the course of a laboratory preparation? The former is a more reliable source.

4. How recent is the source? Recently determined values (and the more recent the better) are usually preferred over older values because of improvements in techniques and apparatus and advances in knowledge—if everything else is equal. But this does not mean that older sources of information should be totally discounted or neglected.

5. Is the source a specialized book such as a monograph on specific compounds or classes of compounds? Such books are frequently preferred as sources of physical property data over more general treatises, which are less specialized.

6. Do the sources being compared refer to the same original source for their data? In comparing secondary (i.e., unoriginal or second-hand) sources, the chemist should beware of the possibility that the secondary sources are all based on the same original source, which may be erroneous.

7. Is the author (investigator) a specialist in the chemical or property being studied, and what is that person's reputation for good work?

8. What is the reputation of the laboratory or research center where the work was done? For example, high confidence is placed in work done at the U.S. National Bureau of Standards.

9. What is the reputation of the publisher or publication? For example, values appearing in a source such as the *Journal of the American Chemical Society* would ordinarily be given more credence than a value appearing in a journal of unknown or questionable reputation.

10. Do the data appear in tables or graphs in which even a single error is detected? If the answer is yes, such data need to be looked on with extraordinary scrutiny.

11. How much experimental detail is given? Inclusion of full detail enhances confidence in the results. Lack of detail diminishes confidence. Some specifics for which detail is helpful and which should be looked for include:

a. Purity of material on which the determination is made.
b. Source of the material (supplier).
c. Type of apparatus used in making the determination and precision of that apparatus.
d. Reliability and reputation of methods used.
e. Limits of error or confidence limits.

The lack of such information could reduce or even entirely negate the placing of confidence in data reliability. Inclusion of such information increases confidence and facilitates evaluation.

12. Are all pertinent facts reported, including those unfavorable to the author's position or theory? To what extent are opposing interpretations and views included?

13. Do the results appear to be overprojected and overextended?

14. Are the data internally consistent?

15. Is the work written so that it can be correlated, repeated, or verified by others?

16. Are interpretations clearly labeled as such?

17. Is there an attempt to evaluate and assess the reliability of the results critically?

15.15. DETERMINING OR ESTIMATING PROPERTIES; SOURCES OF PREDICTION

If a careful search of literature and other related information sources does not reveal needed data of high reliability, one or more of the following steps can be taken:

1. The chemist may find it quicker, easier, and more accurate to determine the data in the laboratory—if the required instrumentation and expertise are available—than to spend additional days searching the literature. This decision is a trade-off, based on estimated time and costs involved, as well as importance of the data. Reliable experimental determinations are ordinarily the choice over estimating or predicting, if time and funds permit.

2. Data may be available through use of the resources of DIPPR (Design Institute for Physical Property Data). Some 32 industrial firms and NBS have joined together in DIPPR. The purpose is to assemble, determine, and evaluate data on physical properties and predictive methods for selected chemicals and mixtures employed in the chemical industry. Operation is under the American Institute of Chemical Engineers (AIChE). The investigators are Ronald P. Danner and Thomas E. Daubert, both professors at Pennsylvania State University. Data are released to member companies first, and one year later, these are to be available for purchase by others through AIChE. For example, a *Data Prediction Manual* became publicly available in 1983. These looseleaf volumes are intended to provide prediction methods, correlations, and graphical data needed to predict physical, thermodynamic, and transport properties of chemicals. Chapters in this manual are: general data; critical properties, vapor pressure; density; thermal properties; phase equilibria; surface tension; viscosity; thermal conductivity; diffusivity; combustion; and measures of environmental impact.

The DIPPR database now consists of several hundred compounds (of which almost half are available to the public) with an initial goal of 1000 industrially important compounds. For each compound, data are given for some 26 constants and 13 sets of temperature-dependent property correlation coefficients. Information is provided in printed form and on magnetic tapes. The project includes estimates of accuracy of each property value and references to sources of any measured or predicted data values employed in selecting recommended data. Interactive computer programs have been developed to facilitate access to the data, and nationwide online access is anticipated.

3. The data can be obtained using other computer programs, which may or may not have some predictive capabilities. Examples of programs (or sources) include:

  a. FLOWTRAN, available through Monsanto on a license basis or to U.S. and Canadian students through CACHE (Computer Aids for Chemical Engineering Education).
  b. The Chemical Thermodynamic and Energy Release Evaluation Program (CHETAH) developed by a task group of the American Society for Testing and Materials (ASTM), Philadelphia, PA. It helps predict key thermodynamic properties. As noted in Section 14.6, CHETAH can be helpful in identifying the relative hazard potential of chemicals.
  c. CHEMTRAN® is a product of ChemShare Corp.* Another product of this firm is MixMaster®, which is claimed to offer online access to the world's vapor–liquid and liquid–liquid equilibrium data, along with online correlation of these data.
  d. PPDS (Physical Property Data Service) developed by the British Institution of Chemical Engineers.* This system is now online through Technical Database Services, New York, N.Y.
  e. The thermodynamic components of the online *CIS* system (see Section 10.11.A.3) and the microcomputer package from Fein-Marquart.
  f. The DETHERM database of the Deutsche Gesellschaft für Chemisches Apparatewesen (DECHEMA), which has headquarters in Frankfurt, West Germany. DETHERM is a databank for physical property data of chemical compounds and mixtures. Some 3000 com-

pounds believed to be of most importance to chemical engineers are included. Unusual compounds of purely scientific interest, materials of all types, alloys, polymers, and so on are included. System capabilities include data retrieval, algebraic manipulation of data that has been retrieved, and data analysis. The databank includes both numerical and nonnumerical data. The online file can be accessed through the STN^{SM} or *INKA* systems (see Section 10.12.C).

Alternatively, inquiries may be addressed directly to DECHEMA* or to Fachinformationszentrum Chemie GMBH.* The latter was founded in 1981 by DECHEMA, Gesellschaft Deutscher Chemiker, and others to provide a broad range of information service in chemistry and chemical technology. The U.S. contact is Gerhard Stoeckl at Scientific Information Systems.*

A related effort is DECHEMA's *Chemistry Data Series*, which consists of compilations of thermophysical properties with evaluated data (14). Volumes include vapor–liquid equilibrium data, liquid–liquid equilibrium data on organic systems, a publication on low–temperature vapor–liquid equilibrium data, and compilations on heats of mixing and critical constants of pure substances.

g. *Aspen Plus*®, a process simulation program offered by Aspen Technology, Inc.* *Aspen Plus*® is also accessible on a pay as you use basis through the facilities of University Computing Co.* Physical property models and data in *Aspen Plus*® include: equations of state; fugacity coefficients; molar volume; enthalpy–entropy–free energy; vapor pressure; vapor–liquid equilibrium ratio; Henry's constant; complex solids density; complex solids enthalpy; thermal conductivity; surface tension; viscosity; and diffusion coefficient. Some of this work was originally sponsored by the U.S. Department of Energy and is still available from the National Energy Software Center, Argonne, IL in the original, but unenhanced version. Activities also include bioprocess simulation, simulation of hazardous waste disposal, and flowsheet optimization.

h. *PROCESS*^{SM}, a simulation program offered by Simulation Sciences Inc.* This includes methods for predicting physical and thermodynamic properties and also phase equilibrium prediction, as well as general purpose simulation. *PROCESS*^{SM} has an extensive (930) pure component library with all of the basic data required for simulation, including physical, thermodynamic, and transport properties. Users may override and supplement these data as desired. The program is available worldwide on a network basis and may also be installed on in-house computers.

i. CHEMEST, a computer software program developed by Arthur D. Little, Inc., Cambridge, MA. This presently contains 32 estimation methods for 10 properties (water solubility, soil adsorption, bioconcentration factor, acid dissociation constant, activity coefficient, boiling point, vapor pressure, water volatilization rate, Henry's Law constant, and melting point), and it may be purchased for use in house or accessed online through Technical Database Services.*

Each of these has unique capabilities and emphasis; the developers can be contacted directly for details.

V. E. Hampel and colleagues at Lawrence Livermore Laboratory* have compiled an extensive list of databases on thermodynamic and other properties (15). This will apparently be updated at intervals.

In addition to those listed in this Chapter, there are many other computer programs that could be used. Some are listed in the *Chemical Engineering* compilation *Microcomputer Programs for Chemical Engineers*, which is now available from the magazine. The AIChE has published a survey of programs for personal computers, and there are also other such compilations. The seven volumes of computer programs assembled by CACHE are another example of the many sources available.

4. It may be possible to infer roughly what the property might be by: (a) studying data for similar classes of related compounds; and/or (b) examining related data for the same compound; and/or (c) studying other compounds in the same homologous series.

5. The chemist can call or write the NSRDS either at their headquarters or at one of their specialized data centers.

6. The chemist can use one of the books written to aid in estimation of properties. The most recent of these are the previously mentioned DIPPR *Data Prediction Manual* and the highly regarded work by Reid, Prausnitz, and Sherwood (16). The latter is a critical review of the estimation programs and includes specific recommendations. The appendix is a data bank with property values for over 450 pure chemicals. Other books can be selected from the list in Section 15.12.

15.16. PROPERTIES OF METALS, PLASTICS, AND OTHER MATERIALS

In the field of metals information, specialized sources in addition to *CA* include, for example, *Metals Abstracts, Alloys Index,* and *World Aluminum Abstracts*. These are available online through such sources as *DIALOG®* Information Services. Printed versions of these and other related publications may be obtained through the American Society for Metals,* or through The Metals Society, London. Sources such as these are particularly useful in locating data about specific alloys or alloy components.

The principal reference book in the field of metals and metallurgy is the multivolume *Metals Handbook* (17), which is a publication of the American Society for Metals. This is an excellent starting point for any study of metals. There are extensive discussions, numerous data compilations, and frequent references to other pertinent literature.

Corrosion properties and data are of special importance because some estimates are that corrosion costs the United States over $100 billion per year. Corrosion data are extensive, but are often widely scattered over many publications and an extensive period of time.

Publications and databanks are available to help chemists and engineers cope with this scatter. Use of literature data, however, can be at best a guide or first approximation towards selecting appropriate materials of construction. Engineers usually need to conduct their own experimental studies (with candidates selected from the literature) under the precise conditions under which the proposed material of construction is to be used.

One of the most valuable literature sources in this field is *Corrosion Data Survey*, published by the National Association of Corrosion Engineers (18). NACE also publishes *Corrosion Abstracts* and many other publications*. Considerable additional information of value will be found in *Chemical Abstracts*, *Metals Index*, *Engineering Index*, and other sources.

Entries pertinent to corrosion can be found in different indexes under a number of scattered headings, as indicated in Section 10.8. It is especially important to be aware of this type of scatter in corrosion literature searching efforts and to utilize all pertinent alternative index entries as required.

In addition, in 1983, NBS and NACE signed a three-year agreement to provide evaluated corrosion data on alloys and other materials. The program, a series of pilot projects, emphasizes development of evaluation criteria for corrosion data and evaluates certain specific corrosion process data. The projects are coordinated by NBS' new Corrosion Data Center and NACE. Dr. Gilbert Ugiansky heads the Data Center at NBS, and Dr. Robert Baboian, Chairman of the Research Committee of NACE, is coordinator for that group.

The Office of Standard Reference Data of NBS and NACE are jointly funding the first pilot project on corrosion data for austenitic stainless steels in aqueous chloride solutions.

Marcel Dekker, a publisher located in New York* and in Basel, Switzerland, has initiated a computerized database of corrosion data, searchable online through SDC® Information Services. This is the online companion to the same publisher's book *Corrosion Resistance Tables* (19). The online file *Corrosion Database* is designed to assist in selection of proper construction materials. It provides a single source from which components of a system may be selected. The database contains data on effects of over 600 agents on 20 metals, carbon, glass, 26 plastics, and 13 rubbers, over a wide temperature range. The database is updated periodically.

DECHEMA has published a series of *Corrosion Data Sheets* (*Die Werkstoff Tabellen*) which now covers some 1000 corrosive materials and first published some years ago. This is available in looseleaf form and updated periodically, but it is not yet online. DECHEMA also offers a corrosion consulting service.

Finally, a report by Ronald D. Diegle and Walter K. Boyd of Battelle Memorial Institute,* attempts to identify the best sources of data on corrosion of metals. The report is entitled *Critical Survey of Data Sources: Corrosion of Metals* and can be obtained as NBS Publication 396–3.

A number of sources are available for other types of materials. For example, in 1984 the publishers of the magazine *Plastics Technology* (New York, N.Y.) introduced a computerized databank for plastics, *Plaspec*. This is based initially on the magazine's annual *Manufacturing Handbook and Buyer's Guide*, supplemented by extensive input from manufacturers of plastics and associated processing equipment. It is intended to help those working with engineering plastics to more effectively identify and select commercial materials with the range of properties and performance desired (e.g., heat distortion temperature or impact strength). Over 6800 grades of about 80 different families of plastics are included, and new materials are added on a regular basis. (Despite this impressive scope, not all plastics or commercial polymers are included.) Results may be ranked or sorted online, for example, from high to low for a specific property. The file also contains data on equipment and can facilitate locating prices and other marketing data. This databank is online through the facilities of Autonet® and can also be accessed through *Tymnet*® or *Telenet*®.

A competing online source of information about plastics is *Plastiserv*[SM], which also may be accessed through the CompuServe network, as well as *Tymnet*®, or *DataPac Canada*. Information covered includes products, properties, manufacturers, suppliers, trade shows, and industry news and economic trends. *Plastiserv*[SM]* offices are located in Columbus, OH and the project is a joint project of CompuServe, Inc. and Borg–Warner Chemicals, Inc.

Another useful source of physical properties of plastics and related materials is published by the D.A.T.A. subsidiary of Cordura Publications.* A series of handbooks covers plastics; foams and adhesives; and films, sheets, and laminates. Data given for each material include generic type, commercial name, manufacturer, key properties and characteristics, and features and uses. Properties, uses, and materials are all arranged in ranked tables to facilitate selection of materials with the required characteristics. At this time, most of these books are several years old. A revision of the volume on thermoplastics and thermosets is to appear in 1986. A new volume on adhesives is also to appear soon.

In the field of plastics and their properties, a knowledgeable source is the *Plastics Technical Evaluation Center* (Plastec). This is at the U.S. Army Armament R & D Command installation located in Dover, N.J. Extensive computerized and other files are maintained, and there are a number of publications. However, this operates as a Department of Defense Information Analysis Center, and accordingly, services are given primarily to employees of that department and to qualified defense contractors. See also Section 9.4.C.

General Electric Company has announced a databank relating to engineering plastics. The file is called *Eris* (Engineering Resins Information System) and is to include data on resins made by both General Electric and other companies. General Electric marketing representatives can access this databank for inquiries.

An online database produced by INSPEC (see Section 8.11) is Electronic Materials Information Service (EMIS). INSPEC is a division of the Institute of Electrical Engineers in England. The content consists of numeric physical property data. Materials covered are silicon, amorphous silicon, gallium arsenide, indium phosphide, lithium niobate, and quartz. For each of these, extensive and continually up-dated factual text and data are provided. Some other electronic materials are also included. Time period of coverage is from late 1980 to the present. In addition, some earlier review articles and books are included, and there is a directory of suppliers of materials important to solid-state technology. EMIS is searchable through the *Telenet®* and General Electric communications networks.

15.17. COORDINATION OF MATERIALS DATA FILES AND SOURCES

In 1983, the report of a workshop examining problems in materials data systems was published under the editorship of J. H. Westbrook (at that time with General Electric Co., Schenectady, NY), and J. R. Rumble, Jr. (U.S. National Bureau of Standards, Washington, D.C.) (20). A useful appendix to this report lists some 60 machine-readable files of engineering data on materials. This has since been updated and extended in the previously mentioned 1984 compilation by Viktor E. Hampel et al. (15), which is highly recommended. The workshop report describes the current state of the art in materials engineering data retrieval as a multiplicity of handbooks, card files, databases, and data centers. It points out that no single comprehensive file of properties required in engineering exists, and it makes a case for the initiation of such a file.

Thus, far, very little of interest to chemists has been accomplished in connection with these recommendations. The most advanced efforts are those of the Metal Properties Council, Inc., which proposes to link existing databanks in a National Cooperative Materials Property Data Network. For such a system to be useful, more databases need to be developed from existing compilations. Implementation depends on availability of funding.

Recent papers and reports helpful in understanding materials data have been prepared by Selover, AGARD, and Westbrook. These are listed as References 21–23.

15.18. TRENDS

In an important 1981 article *Critical Data for Critical Needs*, Lide of NBS identified the most likely future trends (24). These included:

1. Increasing need for reliable data.
2. Increased use of computer-based dissemination of data.
3. Increased need for coordinated development of computer systems.

Just a few years after these forecasts were made, their validity became even more apparent. No quick and easy solution to data problems was anticipated, nor is it apparent, even with enormous progress in computer capabilities. The road ahead will be difficult and complex.

REFERENCES

1. B. J. Zwolinski, R. C. Wilhoit, C. O. Reed, G. K. Estok, F. L. Rissinger, and E. V. Dose, Eds., *Comprehensive Index of API 44—TRC Selected Data on Thermodynamics and Spectroscopy*, Publication 100, 2nd ed., Thermodynamics Research Center, Department of Chemistry, Texas A & M University, College Station, TX, 1974.

2. "Thermophysical Properties Research Literature Retrieval Guide: 1900–1980," Purdue University, Center for Information and Numerical Data Analysis and Synthesis, 1982.

3. Purdue University Center for Numerical Data Analysis and Synthesis, "Thermophysical Properties of Matter: The TPRC Data Series," Plenum, New York, 1970–1979. The printed volumes are understood to be out of print, but microform versions are said to be available from University Microfilms, Ann Arbor, MI, or from the Defense Technical Information Center, Alexandria, VA.

4. Purdue University Center for Numerical Data Analysis and Synthesis, "Data Series on Material Properties", McGraw-Hill, New York, 1981 to date.

5. International Council of Scientific Unions Committee on Data for Science and Technology (CODATA), *International Compendium of Numerical Data Projects*, Springer-Verlag, Berlin, 1969. (Currently being supplanted by *CODATA Directory of Data Sources for Science and Technology*.)

6. *Landolt–Börnstein's Zahlenwerte und Funktionen aus Physik, Chemie, Astronomie, Geophysik und Technik*, 6th ed., Springer-Verlag, Berlin, Heidelberg, New York, 1950–1980; *New Series* in progress.

7. R. C. Weast, Ed., *CRC Handbook of Chemistry and Physics*, 67th ed., CRC Press, Boca Raton, FL, 1986. Revised at frequent intervals.

8. J. A. Dean, Ed., *Lange's Handbook of Chemistry*, 13th ed., McGraw-Hill, New York, 1985.

9. (a) R. Perry and D. W. Green, Ed., *Perry's Chemical Engineers Handbook*, 6th ed., McGraw-Hill, New York, 1984; (b) A. J. Gordon and R. A. Ford, *The Chemist's Companion*, Wiley, New York, 1973.

10. M. Windholz, Ed., *Merck Index*, 10th ed., Merck & Co., Rahway, NJ, 1983. Most recent data available in online version.

11. E. W. Washburn, Ed., *International Critical Tables of Numerical Data, Physics, Chemistry and Technology*, Vols. 1–7 and Index, McGraw-Hill, New York, 1926–1933.

12. D. Garvin, V. B. Parker, and D. D. Wagman, "Thermodynamic Data Banks" *Chemtech., 12*, 691–697 (1982).

13. (a) G. T. Armstrong and R. N. Goldberg, *An Annotated Bibliography of Compiled Thermodynamic Data Sources for Biochemical and Aqueous Systems* (1930–1975), NBS Special Publication 454, U.S. Government Printing Office, Washington, D.C., 1976; (b) R. E. Maizell, "Techniques of Data Research in Chemical Libraries," *J. Chem. Ed., 32*, 309–311 (1955); (c) R. Luckenbach, R. Ecker, and J. Sunkel, "The Critical Screening and Assessment of Scientific Results without Loss of Information—Possible or Not," *Angew. Chem. Intl. Ed. Engl., 20*, 841–849 (1981).

14. (a) J. Gmehling, U. Onken, W. Arlt, et al., Eds., *Vapor–Liquid Equilibrium Data Collection*, (Vol. I), 8 parts DECHEMA, Frankfurt, 1984; (b) K. H. Simrock, Ed., *Critical Data of Pure Chemical Compounds* (Vol. II) in preparation; (c) C. Christensen et al., Ed., *Heats of Mixing Data Collection*, (Vol. III), 2 parts, 1984; (d) M. Schonberg, Ed., *Estimated Vapor Pressures of Pure Organic Compounds*, Vol. IV, about 10 parts, in preparation; (e) J. M. Sorensen, et al., Ed., *Liquid–Liquid Equilibrium Data Collection*, Vol. V, 3 parts, 1979–1980; (f) R. Doring et al., Ed., *Vapor–Liquid Equilibrium for Mixtures of Low Boiling Fluids*, Vol. VI, 1981. These are all part of DECHEMA's *Chemistry Data Series* published in Frankfurt.

15. V. E. Hampel, W. A. Bollinger, C. A. Gaynor, and J. J. Oldani, Eds., *An Online Directory of Databases for Materials Properties*, UCRL-90276, Rev. 1, Lawrence Livermore Laboratory, Livermore, CA, 1984. Another related publication by the same editor is V. E. Hampel et al., *International Information Networks for Material Properties*, UCRL-90942, Rev. 1, Lawrence Livermore Laboratory, Livermore, CA, 1984.

16. R. C. Reid, J. M. Prausnitz, and T. K. Sherwood, *The Properties of Gases and Liquids*, 3rd ed., McGraw-Hill, New York, 1977.

17. K. Mills, Ed., *Metals Handbook*, 9th ed., American Society for Metals, Metals Park, OH, 1978 to date.

18. N. Hamner, Ed., *Corrosion Data Survey*, National Association of Corrosion Engineers, Houston, TX. Metals section, 5th ed. (1974), 6th ed., (1985), D. L. Graver, Ed.; Nonmetals section (published in 1975).

19. P. A. Schweitzer, Ed., *Corrosion Resistance Tables*, Dekker, New York, 1976; a related book by the same author and publisher is *Corrosion and Corrosion Protection Handbook*, 1983.

20. J. H. Westbrook and J. R. Rumble, Jr., *Computerized Materials Data Systems*, 1983. Publisher unknown.

21. T. B. Selover, Jr., and M. Klein, Eds., American Institute of Chemical Engineers Symposia Series, *Awareness of Information Sources*, **80** (237), (1984); another important related volume in this Symposium Series is (201) (1981).

22. AGARD-Lecture Series-130, *Development and Use of Numerical and Factual DataBases*, North Atlantic Treaty Organization, 1983.

23. J. H. Westbrook, *Information for Material Scientists and Engineers*, to be published in *Encyclopedia of Materials Science and Engineering*, M. Beever, Ed., Pergamon, London, 1986; distributed in the United States by MIT Press, Cambridge, MA.

24. D. R. Lide, Jr., Critical Data for Critical Needs, *Science*, **212**, 1343–1349 (1981).

LIST OF ADDRESSES

American Society for Metals
Metals Park, OH 44073

Aspen Technology, Inc.
251 Vassar St.
Cambridge, MA 02139

Battelle Memorial Institute
505 King Avenue
Columbus, OH 43201

Chem Share Corp.
2500 Transco Tower
Houston, TX 77056

CODATA
Secretariat, 51 Boulevard de
  Montmorency
75016, Paris, France

Cordura Publications, International
  Plastics Selector Inc.
Box 101, 8660 Mirimar Road,
San Diego, CA 92126-4330
  (1-800-854-7030 or
  1-800-421-0159)

CRC Press, Inc.
Boca Raton, FL 33431

DECHEMA
Postfach 970146, 0-6000 No. 1
Frankfurt (Main) 97, West
  Germany

Dr. David R. Lide, Jr.
Director, Office of Standard
  Reference Data
NBS, Gaithersburg, MD 20899
  (301)-921-2467

Dow Chemical Co.,
Midland, MI 48640

Fachinformationszentrum Chemie
  GMBH
Steinplatz 2, D–1000
Berlin 12, West Germany

Institution of Chemical Engineers
165–171 Railway Terrace
Rugby, CV21 3HQ England

Lawrence Livermore Laboratory
Livermore, CA 94550

Marcel Dekker, Inc.
270 Madison Avenue
New York, N.Y. 10016

National Association of Corrosion
  Engineers
P.O. Box 218340
Houston, TX 77218 (713-492-0535)

Oklahoma State University
Bulletin of Chemical
  Thermodynamics
Stillwater, OK 74074

Plastiserv
P.O. Box 20933
Columbus, OH (614-457-8126)

Purdue University
West Lafayette, IN 47907

Scientific Information Systems
7 Woodland Avenue
Larchmont, N.Y. 10538

Simulation Sciences Inc.
1051 West Bastanchury Road
Fullerton, CA 92633 (800-854-3198)

Technical Database Services
10 Columbus Circle
New York, N.Y. 10019
    (212-245-0044)

University Computing Co.
1930 Hilene Dr.
Dallas, TX 75207

16 CHEMICAL MARKETING AND BUSINESS INFORMATION SOURCES

16.1. INTRODUCTION

Chemical marketing and business information is important to almost all chemists. Particularly in industry, the chemical researcher has been brought out of the relative anonymity and alleged "ivory tower" of the laboratory and into the main stream of the business world. Industrial researchers are or should be accepted as full partners by their marketing counterparts. Technological and marketing decision making have become highly interdependent.

The boundary between marketing information sources and technical or scientific information sources is blurred, and the two types of sources are complementary. Important scientific information may appear for the first time in chemical marketing or business sources. Similarly, scientific and technical literature can provide important leads to chemical marketing specialists.

This chapter summarizes what is available and how to access it. For more detail, the reader should consult books in the field, such as Giragosian's classic *Chemical Marketing Research* (1) or *Industrial Marketing Research* (2) by Donald Lee.

As compared to most scientific or technical data, some marketing data, especially estimates or projections into the future, are relatively "soft" or less certain. Laboratory verification of marketing data is not possible. As chemists become more and more involved with the business aspects, they can expect to make increasing use of the tools mentioned in this chapter. In so doing, they should call on the expertise of chemical marketing specialists, especially when there are questions of interpretation or extrapolation to the future.

Many types of market data are highly proprietary to individual companies. For example, if three or fewer companies make a product, the government does not include in its published statistics the total production of this chemical.

16.2.   INFORMATION ABOUT MANUFACTURERS

In addition to determining who makes a product, chemists may be interested in chemical manufacturers for other reasons:

1. If a company is a possible employer.
2. If a company is being considered for personal or other financial investments.
3. If the company represents present or potential competition.
4. If a joint research or other business venture is being contemplated.
5. If the company is a potential source for a research grant or other funding.

Information about chemical manufacturers that is usually found most valuable includes:

1. Products made by the company.
2. Locations (principal offices and plants).
3. Names, titles and backgrounds of officers and other key personnel.
4. Historical background, present status, and future plans.
5. Financial data such as sales and profits, preferably broken out by major product line.
6. Patents and other publications.
7. Research and development staff and facilities.

A variety of information sources is available. For example:

1. Reports issued by companies to stockholders, especially annual reports. These have information about present and historical financial status, new products and other major research achievements, names of key officials, principal locations, and other information.

2. Reports filed by companies with the Securities and Exchange Commission (SEC) in Washington, D.C. These filings are required by law for all companies that are publicly owned. Most reports filed with the SEC are readily available at SEC offices, directly from the company, or from private organizations that provide this kind of material, for example, Disclosure, Inc.* These reports are important because they often contain more detailed financial and marketing information than is found in the annual report issued to stockholders.

3. Reports from stockbrokers, and handbooks and services issued by the financial and investment community (see references 3–5). This material is useful in obtaining a description and evaluation of the financial picture of the company—past, present, and future—but it also frequently contains other information.

4. Patents assigned to the organization and other publications by persons connected with the organization. These can be identified by methods described in Chapter 13. A study of patent publications helps indicate areas of interest as well as strength of technical effort.

5. News about the company as reported in chemical news magazines, general business and trade periodicals, and newspapers.

6. Buyer's guides and other tools discussed in the following sections.

7. Biographical data about officers and other key personnel as found in *Who's Who* sources such as those published by Marquis, Who's Who, Inc.*, e.g., *Who's Who in America* (6); related sources covering various regions of the country; and related or similar sources covering the business community. Data about executives can also sometimes be found in reports to stockholders or to the government, in directories such as Dun & Bradstreet's *Reference Book of Corporate Managements* (7), and in Dun & Bradstreet and other financial reports. Biographical data about scientists may be found in such sources as *American Men and Women of Science* (8a), which is now available in both printed and online forms. A newer source is *Who's Who in Frontier Science and Technology* (8b), first published in 1984. Absence of a person from biographical directories does not necessarily mean that a person is unimportant; some persons decline to provide the data requested or shun publicity of all types. Similarly, inclusion of a person does not necessarily connote exceptional expertise. Additional sources of data about people include their publications, patents, and news items about them. The Industrial Research Institute* publishes a directory of industrial research executives, but this is available only to employees of companies that belong to the Institute. Finally, a personal letter or phone call to the person of interest can frequently obtain the needed biographical information more accurately and quickly than any printed or other secondary source. At a minimum, a list of publications of a chemist can usually be easily obtained either directly or indirectly, and this alone can be very informative.

Many of the types of data listed here are readily available online.

16.3. BUYER'S GUIDES AND RELATED TOOLS

Finding out what company makes a chemical of interest is one of the most frequent informational needs of the chemist or engineer. Fortunately, a number of useful tools are available.

These include the SRI International *Directory of Chemical Producers— USA* (9). Published since 1961, this volume is far more than just a buyer's guide. In many cases (for over 200 major products), information is given on capacities at specific plant locations, and in some cases, process or route used is also given. Access is threefold: by product (who makes what, where, how much, and by what route, if known); by company (divisional structure, plant locations, and product produced at each site); and by region (what companies and plants are within specific zip code locations). The location feature is helpful to chemical salespersons and to others concerned with business aspects of the chemical industry (e.g., locating potential sales prospects within a geographical region or locating the nearest contact point or plant of the manufacturer to help expedite delivery). Telephone numbers are included.

The annual directory is kept up to date by two cumulative supplements. These note changes in company names and addresses, corporate structure, products produced at each plant location, and mergers and acquisitions. The supplements also contain information on new plants planned or under construction. Subscribers can contact SRII directly with any questions about listings.

The current edition of the directory is an invaluable source of information to all concerned with the U.S. chemical industry. Previous editions should be kept because they provide a good historical reference to the state of the chemical industry each year. As indicated in the title, coverage is limited to the United States.

Primary emphasis is on commercial chemicals and on actual producers (distributors and formulators are not included). Commercial quantity is defined as at least 5000 pounds or U.S. $5000 in value annually. The compilation includes about 10,000 chemicals and some 1500 companies. For products made in other countries, chemicals made in developmental or research quantities, and names of distributors, other directories need to be consulted. The chemist will find this (and all other buyer's guides) occasionally incomplete because of frequent additions to and deletions from product lines throughout the chemical industry. In addition, some companies are not willing to freely share data about their product lines.

Among the other buyer's guides available, the two best known are

1. *Chemical Week Buyer's Guide Issue* (10)—this annual publication includes "major producers and sources for more than 6,000 products." There are also addresses and telephone numbers for company offices, and there is a listing of trade names. Similar information is given for companies that provide packages and containers for the shipping of chemicals.

2. *OPD Chemical Buyer's Directory* (11)—like *Chemical Week Buyer's Guide*, this annual volume lists sources of supply for "chemicals and related process materials." *OPD* is now known as *Chemical Marketing Reporter* which is noted in Section 4.3.A. The historical meaning of *OPD* is *Oil, Paint, and Drug Reporter*.

Neither of these publications contains all the information found in the SRI *Directory*, nor is either as complete or up to date. But they are much more widely available and are heavily used.

Some chemicals are made in small or research quantities only and for this, or other reasons, may not be listed in the more widely used guides. The best list for such chemicals is *Chem Sources—USA* (12).

Another source of producers is the *Fine Chemicals Directory* (13) which is searchable online through Pergamon *InfoLine*® (see Section 10.12.B). This permits access to suppliers of commercially available research chemicals, including organics, inorganics, biochemicals, and dyes and stains. Data in the file also include molecular formulas and Wiswesser Line Notations. The file is accessible through the system developed by Molecular Design, Limited, which permits graphic searching of structures (see Section 10.11.A.4). The publisher is Fraser Williams Ltd.*

A number of specialized guides are available to assist in purchase or identification of chemicals and formulated products intended for specific end uses. There are also special guides for laboratory and other kinds of equipment. Some examples include:

1. *Soap Cosmetics Chemical Specialties Blue Book Issue* (14).
2. *Lockwood's Directory of the Paper and Allied Trades* (15).
3. *Farm Chemicals Handbook* (16). Includes plant foods (fertilizers) and pesticides. Considerable other information given beyond that needed for purchasing.
4. *Analytical Chemistry LabGuide* (17).
5. *Chemical Engineering Equipment Buyer's Guide* (18).

A. International Buyer's Guides

Buyer's guides such as those previously mentioned cover primarily U.S. manufacturers. Because chemistry is international, the chemist should be aware of guides and directories covering other countries. For example:

1. SRII *Directory of Chemical Producers—Western Europe* (19). This comprehensive book includes over 13,000 chemicals and more than 3000 manufacturers.
2. *Japan Chemical Directory* (20).
3. *Chem Sources—Europe* (21). Discontinued in 1980.
4. *Chemistry and Industry Buyer's Guide* (22) A publication of the Society of Chemical Industry; includes information on chemicals and manufacturing plant and laboratory equipment.
5. *Worldwide Chemical Directory* (23) A worldwide "address book" of chemical industry; includes addresses, telephone numbers, and brief

descriptions of activities; contains four sections: list of companies, trade and professional associations, chemical plant contractors, and chemical tanker and storage companies.

6. Another publication of worldwide scope is *Directory of World Chemical Producers (24)*.

See also the discussion on Chemical Data Services in Section 16.10.

B.   More General Buyer's Guides

On occasion, the chemist or engineer needs to identify sources of equipment or materials not necessarily related directly to chemistry. For these occasions, a comprehensive overall guide for U.S. manufacturers is *Thomas Register of Manufacturers & Thomas Register Catalog File* (25). This is also available online through *DIALOG*® Information Services. The *VSMF*® files noted in Section 16.9 may also be useful.

16.4.   OTHER SOURCES FOR LOCATING CHEMICALS

Chemists who cannot locate sources for chemicals of interest in buyer's guides need not despair. Several options remain open.

Some companies either specialize in stocking hard-to-find chemicals or will consider making these on request perhaps in small (research) quantity as well as in somewhat larger quantities. Examples include:

Aldrich Chemical Co.
940 W. Saint Paul Ave.
Milwaukee, WI 53233
Primarily organics and biochemicals: off the shelf, custom synthesis
    service, or in larger-than-laboratory quantities.

Eastman Organic Chemicals
343 State St.
Rochester, NY 14650

Alfa Products Division of Ventron Corp.
Beverly, MA 01915
Specializes in inorganics.

Sigma Chemical Company
P.O. Box 14508
St. Louis, MO 63178
Specializes in biochemical and organic compounds for research and
    diagnostic and clinical reagents.

Carbolabs, Inc.
Fairwood Rd.
Bethany, CT 06525
Specializes in "difficult" reactions.

In addition, there is a directory (26) that lists companies that have capability in custom processing. Finally, the chemist can contact manufacturers of related compounds or of precursors to see if they can provide the chemicals needed.

16.5. PRICES

The best regularly published source of list prices for chemicals, specifically those used in commerce on a large scale, is the weekly *Chemical Marketing Reporter*. This is an extensive listing, but it is not all inclusive. Even if a chemical is not listed here, it may still be available for purchase. Buyer's guides need to be consulted, as previously recommended. Because prices depend on quantity, location of buyer and manufacturer, and other variables, an individualized quotation from the manufacturer is highly desirable for the most accurate pricing data.

16.6. GENERAL CHEMICAL BUSINESS INFORMATION

To help the chemist or engineer locate the latest chemical business information, several good abstracting and indexing services and other tools are available.

One of the most complete services is *PROMPT* (*Predicasts Overview of Markets and Technology*) and its associated subsections known as *Predi–Briefs*. This tool is distinguished by in-depth abstracting and versatile indexing access. Online retrieval is possible through facilities of *DIALOG*® Information Services, *BRS*®, and Data Star (see Section 10.4). *PROMPT* and other business information tools of values, are published by Predicasts, Inc.* Mentioned earlier in this book is *Chemical Industry Notes*, a significant tool in this field published by CAS, but not as complete as *PROMPT*.

An information service aimed primarily at covering the commercial and marketing aspects of the chemical industry in Europe was initiated by the Royal Society of Chemistry in January 1985. The service, *Chemical Business NewsBase*, is accessible via three online host systems and is updated weekly.

The *Business Periodicals Index* is simply an index (no abstracts), and it is not specifically oriented to chemistry. It does refer to many articles of chemical business interest, is easy to use, and can be found in many libraries. The publisher is the H. W. Wilson Co. and it is now available online.

Chemical Products Synopsis is a reporting service on 200 individual major chemical commodities. The succinct reports provide general marketing information, including present situation, short- and long-term outlook, pricing, producers, uses, and brief information on the marketing and environmental aspects. The publisher is Mannsville Chemical Products.* A similar publication is *Chemical Marketing Reporter's* service *Chemical Profiles*. This is published weekly as part of the parent magazine and can also be purchased separately.

Trends in End Use Markets is an especially valuable indexing and abstracting source on plastics markets with indexes by markets and materials. The publisher is Springborn Group.*

In addition to what is mentioned elsewhere in this chapter, there are several good weekly news sources that cover the chemical industry as a whole, for example *Chemical Marketing Reporter* and *Chemical Week*. For the Canadian chemical industry, one of the best news sources is *Corpus Chemical Report*. For European chemical industry news, an excellent source is *European Chemical News*. For those who require more frequent updates than the weekly magazines, an especially good source is the daily *Chemical Week Newswire*, which is transmitted electronically to subscribers to this service by the publishers.

16.7. *CHEMICAL ECONOMICS HANDBOOK*

For business, marketing, and some technical information on almost all major commercial chemical products and product groups, the 32-volume *Chemical Economics Handbook* (*CEH*), published by SRI International®, is a key source of enormous value. A companion bimonthly volume, *Manual of Economic Indicators*, updates many of the tables.

The *CEH* program has been a continuing research activity since 1950, devoted to development of detailed information about the economic progress of the U.S. chemical industry. The objective of the program is to serve participants' needs to be informed accurately and currently about the present and future status of raw materials, primary and intermediate chemicals, and product groups (such as plastics and fertilizers). Program emphasis is on providing insight into future markets—both in terms of technological requirements of future demand and quantities of chemicals that will be consumed. Background information is also supplied on economic aspects of the chemical industry as a whole, on chemical consuming industries, and on national economic trends. Data given include: producers, plant capacities when available, prices, markets by category, and technology overview. Arrangement is by product. This is based primarily on government data, although some other sources are also used.

The *CEH* program is sponsored by subscription from over 300 corporations and other organizations. Research results are available to program participants in the following ways:

- A looseleaf set of the *CEH*, featuring comprehensive product reports and statistical data sheets.
- Monthly installments of new and revised material to be added to *CEH* volumes.
- Bimonthly issues of the *Manual of Current Indicators*, which update the statistical data series in the body of *CEH*.
- An inquiry and consulting service.

CEH is oriented primarily to U.S. developments, although significant worldwide events are considered.

In 1984, SRII began to offer subscribers an online index to *CEH*. This included: a master list of complete contents by title, in page number order; a list of the reports, product reviews, and data summaries in the current hard copy installment; a list of what is to be included in next month's installment; and all report and product review tables of contents including their hard copy and online page references. The online file is updated on a monthly basis. In 1986, 85 percent of the entire *Handbook* was available online to subscribers to the printed version. The online version may be made available to nonsubscribers in 1987 via *SDC*®; this would include all text and tables. *CEH* reports may be purchased separately by firms not participating in the total program.

A separate but related SRII program of importance is their *Specialty Chemicals Update Program* (*SCUP*), which includes detailed analysis of a number of important specialty chemical markets. Another equally important program on specialty chemicals is available from Strategic Analysis, Inc.*

In 1978, SRII initiated *World Petrochemicals* (formerly *World Hydrocarbons*). This is international in scope and is limited to petrochemicals, as indicated by the title.

Another SRII product is *Chemical Conversion Factors and Yields— Commercial and Theoretical* (7). For each product listed, data given includes: process and/or coreactants; mol ratio, raw material to product; units raw material consumed per unit of product (theoretical); yield, percentage of theoretical; units raw material consumed per unit of product (commercial process factors).

16.8. MULTICLIENT STUDIES

Multiclient studies are prepared by consulting organizations. The studies cover the marketing aspects in detail, and often include valuable information on technology of products or product groups. The basis of most multiclient studies includes numerous field interviews with producers, users, and others; study of patents and other literature; and sometimes computer modeling. Since many studies cover topics in depth, the price of a single study can easily range up to several thousand dollars.

Frequent producers of multiclient studies include:

1. **Battelle Memorial Institute**
   505 King Avenue
   Columbus, OH 43201

2. **British Sulphur Corp., Ltd.**
   Parnell House, 25 Wilton Road
   London, SW1V 1NH, England
   (Concentrates on world
   fertilizer industry.)

3. **Business Communications Co.**
   471 Glenbrook Road
   Stamford, CT 06906

4. **Frost and Sullivan, Inc.**
   106 Fulton Street
   New York, NY 10038

5. **Charles H. Kline and Co., Inc.**
   330 Passaic Avenue
   Fairfield, NJ 07006

6. **Arthur D. Little, Inc.**
   35 Acorn Park
   Cambridge, MA 02139

7. **Predicasts, Inc.**
   200 University Circle Research
   Center
   11001 Cedar Avenue
   Cleveland, OH 44106

8. **Peter Sherwood Associates**
   20 Haarlem Avenue
   White Plains, NY 10603

9. **Skeist Laboratories, Inc.**
   112 Naylon Avenue
   Livingston, NJ 07039

10. **Springborn Group**
    Springborn Center
    Enfield, CT 06082

11. **SRI International**
    333 Ravenswood Avenue
    Menlo Park, CA 94025–3477

12. **Strategic Analysis, Inc.**
    Box 3485, R.D. 3
    Fairlane Road
    Reading, PA 19606

These are examples of the many active organizations. They vary widely in size, quality, and emphasis. In this list 1, 5–7, 11, and 12 are probably the largest.

Directories of multiclient studies are available. These include:

Findex–The Directory of Market Research Reports, Studies and Surveys
Find/SVP, 500 Fifth Avenue
New York, NY 10110.

Directory of U.S. and Canadian Marketing Surveys and Services
Charles H. Kline and Co., Inc.
330 Passaic Avenue
Fairfield, NJ 07006

The 5th ed. (1985) was published by Rauch Associates, Bridgewater, NJ, 08807 who have now acquired rights to this publication. Rauch has also taken over other Kline studies such as their guides to the United States ink, paint, and packaging industries.

In addition, the *Find/SVP* index to multiclient studies is available online through *DIALOG®* Information Services.

16.9. PRODUCT DATA

When the manufacturer of a product has been located, the chemist can obtain extensive information about the product by requesting product bulletins and other trade literature from the manufacturer. Some of this material contains information not readily available from any other source, as mentioned earlier. Examples of information found in trade literature include: (a) specifications; (b) physical and chemical properties in detail; (c) any handling precautions (safety data); (d) analytical methods; (e) methods of appropriate use in specific applications; and (f) list of pertinent articles and other references. Some manufacturers (e.g., DuPont) maintain centralized product information centers that can provide considerable data about their products over the telephone.

A convenient source of product data is *Visual Search Microfilm File* (*VSMF®*), a product of Information Handling Services.* A number of chemicals and other products are included in this extensive collection of product catalogs.

Several tools are available to help chemists and engineers match materials with desired properties. Some of these tools are described in Chapter 15 on properties.

16.10. OTHER SOURCES AND TOOLS

A source of chemical marketing and business information on virtually a worldwide basis is Chemical Data Services.* This organization publishes numerous reports and directories about specific products, companies, countries, and other topics. The following are some examples.

1. *Chemical Plant and Product Data* Provides, on a replacement basis, published statistics of production, trade, consumption, and capacity for over 100 basic chemicals covering leading industrial countries, together with lists of producers and other information. The data are reportedly taken from a variety of original sources in many languages, including official statistics, national trade returns, and data from published journals as well as correspondence. Data include names of producers, plant locations, existing and planned plant capacities, processes, feedstocks, licensors and contractors, production figures, annual trade statistics, and other. Online access is available.

2. *Chemfacts—Country Series*   A series of books published country by country with information on products, plants, and companies. Countries covered include: Portugal, United Kingdom, Spain, Canada, Japan, The Netherlands, Scandinavia, Belgium, Italy, and Federal Republic of Germany.

3. *Chemfacts—Product Series*   A new series of reports, each of which deals with world manufacturing activities, production, and trade in individual products.

4. *Chemical Company Profiles Western Europe*   Provides facts and figures for over 1200 companies in 19 Western European nations. Similar publications are available for other continents and for plant contractors.

5. *Worldwide Chemical Directory*   See previous mention in Section 16.3.A.

16.11.   OTHER REMARKS

Chemical marketing data are well suited to computerization and specifically to online access. Much material is already available in this form, and chemists can expect further, more sophisticated developments.

The original source of much U.S. marketing data is the federal government. There is a wide variety of publications, and in addition, some government offices will conduct special investigations for individual companies and others on a fee (cost recovery) basis. It is important that users of the data make known to appropriate officials their needs for this information. Expressions of interest can sometimes help ensure continued availability of what is needed, although the recent trend is for the government to provide less marketing data than in the past.

REFERENCES

1. N. H. Giragosian, Ed., *Chemical Marketing Research*, Reinhold, New York, 1967.
2. D. Lee, *Industrial Marketing Research*, 2nd ed., Van Nostrand Reinhold, New York, 1984.
3. *Moody's Industrial Manual*, Moody's Investors Service, New York. Published annually with supplements.
4. *Dun & Bradstreet® Million Dollar Directory®*, Dun & Bradstreet, New York. Published annually.
5. *Standard and Poor's Register of Corporations*, Standard and Poor's, New York. Published annually.

6. *Who's Who in America*, Marquis, Chicago. Also available online. Published every two years.

7. *Reference Book of Corporate Management*, Dun & Bradstreet, New York. Published annually.

8. (a) *American Men and Women of Science*, R. R. Bowker Co., New York. Updated every three years. Also available online; (b) *Who's Who in Frontier Science and Technology*, Marquis, Chicago, 1985. Also available online.

9. *Directory of Chemical Producers—USA*, SRI International, Menlo Park, CA. Published annually with supplements.

10. *Chemical Week Buyer's Guide Issue*, McGraw-Hill, New York. Published annually.

11. *OPD Chemical Buyer's Directory*, Schnell Publishing Co., New York. Published annually.

12. *Chem Sources—U.S.A.*, Directories Publishing Co., P.O. Box 1824, Clemson, SC. Published annually. An international volume is planned for 1986 publication.

13. *Fine Chemicals Directory*, Fraser Williams (Scientific Systems Ltd.), Cheshire, U.K. Available online.

14. *Soap Cosmetics Chemical Specialties Blue Book Issue,* McNair-Dorland Co., New York. Published annually.

15. *Lockwood's Directory of the Paper and Allied Trades*, Vance Publishing Corp., New York. Published annually.

16. *Farm Chemicals Handbook*, Meister Publishing Co., Willoughby, OH. Published annually.

17. *Analytical Chemistry LabGuide*, American Chemical Society, Washington, D.C. Published annually.

18. C. S. Cronin, Ed., *Chemical Engineering Equipment Buyer's Guide*, McGraw-Hill, New York. Published annually.

19. *Directory of Chemical Producers—Western Europe*, SRI International, Menlo Park, CA. Published annually.

20. Japan Chemical Week, *Japan Chemical Directory*, The Chemical Daily Co., Ltd., Tokyo, Japan. Published annually.

21. *Chem Sources—Europe*, Directories Publishing Co., Clemson, SC. Published annually. Discontinued in 1980; to be replaced by international volume mentioned in Reference 12.

22. *Chemistry and Industry Buyers' Guide*, Society of Chemical Industry, London, England. Published annually.

23. *Worldwide Chemical Directory*, 4th ed., Chemical Data Services, London, England, 1982.

24. *Directory of World Chemical Producers*, Chemical Information Services, Oceanside, NY 1985/86.

25. *Thomas Register of Manufacturers & Thomas Register Catalog File*, Thomas Publishing Co., New York. Published annually. Also available online.

26. *Custom Processing Services Guide*, 6th ed., Custom Guide Co., Closter, NJ, 1986.

27. *Chemical Conversion Factors and Yields— Commercial and Theoretical*, 3rd ed., SRI International, Menlo Park, CA, 1984.

LIST OF ADDRESSES

Chemical Data Services
Quadrant House, The Quadrant
Sutton, Surrey SM2 5AS, England

Disclosure, Inc.
5161 River Road
Bethesda, MD 20816

Fraser Williams Ltd
London House, London Road
   South
Poynton, Cheshire, SK12 1YP,
   England

H. W. Wilson Co.
950 University Avenue
Bronx, NY 10452

Industrial Research Institute
100 Park Avenue
New York, NY 10017,
   (212-683-7626)

Information Handling Services
P.O. Box 1154
Englewood, CO 80110

Mannsville Chemical Products
Cortland, NY 13045

Marquis Who's Who, Inc.
200 East Ohio Street
Chicago, IL 60611

Predicasts, Inc.
200 University Circle Research
   Center
11001 Cedar Avenue
Cleveland, OH 44106
   (800-321-6388)

Springborn Group
1 Springborn Center
Enfield, CT 06082

SRI International
Menlo Park
CA 94025

Strategic Analysis, Inc.
Box 3485, R.D. 3, Fairlane Road
Reading, PA 19606

17 PROCESS INFORMATION

17.1. INTRODUCTION

How individual chemical products are best made in full-scale commercial practice is of utmost interest and importance. Understandably, this information can be difficult or impossible to obtain from the literature. The reason is that efficient manufacture is essential to profits and therefore regarded as proprietary.

Although the chemist or engineer may find the broad features of a process and even considerable detail described in nonproprietary literature, the fine detail and ongoing improvements that contribute to optimum production efficiency are rarely published in journals and books. Exceptions to the proprietary approach are organizations such as government agencies (e.g., Tennessee Valley Authority in its work on fertilizer technology.) These agencies often make fullest possible disclosure and provide virtually any reasonable assistance (information) necessary. Furthermore, more details are likely to be available for industrial products and processes that are *mature* (older) such as basic heavy inorganics like sulfuric acid. Other areas in which process detail is likely to be more readily available are those that affect the public good, such as pollution abatement and energy conservation. Despite the limitations noted, the astute chemist or engineer who knows the technology can sometimes glean valuable insight by careful study of pertinent literature.

A first step is often to study what is written in the major chemical encyclopedias such as *Kirk–Othmer* (see Section 12.2.A). These often provide any basic background required. Briefer treatments of process details for the best known products can be found in such excellent single volume works as *Faith, Keyes, and Clark's Industrial Chemicals* (1), *Shreve's Chemical Process Industries* (2), and *Riegel's Handbook of Industrial Chemistry* (3). These are handy for desk use.

Patents are often an invaluable source of information on processes, as explained in Chapter 13, and often contain details not found anywhere else. However, patents usually give *several* examples or specify a *range* of pressures, temperatures, or other conditions. The *optimum* conditions for the process can, therefore, be difficult to determine from patents, except by educated guessing.

Accounts of processes, or related pertinent information, may appear in such journals as *AIChE Journal*, *Chemical Engineering*, *Chemical Engineering Progress*, *Hydrocarbon Processing* (especially the November issue), *Industrial and Engineering Chemistry* (process design and development edition), and *International Chemical Engineering*.

These sources and others can be accessed directly or via such tools as *Chemical Abstracts* and *Derwent*, both of which are described earlier in this book. But the chemical engineer or chemist will frequently find the most comprehensive detail on process technology and economics for key chemical products in the services described in Section 17.2.

17.2. SPECIALIZED SERVICES

Several highly specialized services provide recent process technology and economics information, primarily for important chemicals of commerce, in considerable detail. The evaluations of processes and comparisons of competing processes are significant features. These sources are relatively expensive (typical costs are more than $20,000 per year), but they can be found in most major chemical companies and in some major universities. The price is well worth it. Because the information is private to clients of the services, reference to reports from these sources is not ordinarily found, unless the information is made publicly available by the source. The leading services currently available are

SRI International
333 Ravenswood Avenue
Menlo Park, CA 94025—*Process Economics Program (PEP)*.

Chem Systems, Inc.
303 S. Broadway
Tarrytown, NY 10591—*Process Evaluation/Research Planning (PERP)*.

The two services cover some of the same products. When this is the case, the chemical engineer or chemist should consult the corresponding reports of both services, because differences in information content, evaluations, and economic estimates may be found.

The SRI *PEP* service, currently directed by Harold Kyle, started in 1963 and now totals over 250 reports. Reports include an intensive technical review and analysis of each basic process. The aim is to establish commercially feasible operating conditions and to assess technical factors underlying present limitations as well as prospects for improvement. Sufficient details are presented to permit verification of design calculations and cost estimates.

Most *PEP* studies relate to industrial chemicals and polymers that are produced in substantial and growing volumes and are experiencing rapid changes in technology. Both well-established and newer products are included. About one half of the newly issued reports relate to specialty chemicals and the balance to commodities.

The program includes the following components:

- A series of reports on important chemical and refinery products, based on studies of process technology and cost. This is the core of the program.
- *Process Economics Reviews*, a highlighting of implications of certain new developments in the industry (about 12 issues per year).
- A bulletin describing the status of current process studies issued at four-month intervals.
- License monitoring.
- Consultation with members of program staff on reports and other matters of interest to individual clients.
- An annual *Yearbook* that updates cost figures presented earlier in the reports. The *Yearbook* includes economics of chemical production in Europe (West Germany) and Japan, as well as the United States (Gulf Coast). The most recent yearbook includes data for about 350 processes.

The Chem Systems *PERP* service, which is under the direction of Sol J. Barer, is intended to provide:

- A continuing evaluation of commercial significance of both existing and emerging technological developments—specific products, product groupings, product applications, and technologies.
- Realistic planning information on manufacturing economics of chemicals, polymers, and technologies.
- Information on status of chemical process, product, and technology development efforts of companies throughout the world.
- Identification of opportunities for purchasing or licensing partially developed technology.

Both the technology and commercial status are examined in arriving at forecasts. An analysis of company activities is important for certain reports since plant size will be a function of market concentration as well as engineering and economic factors. Geographic coverage usually includes the United States, Western Europe, and Japan. The potential commercial impact of technology developed in other geographic regions is considered in any examination of a product area.

The service is designed to be useful in new project planning, R&D, licensing evaluation, purchasing analysis, and technological forecasting.

It provides subscribers with:

- A series of six major reports per year, each covering current and future technology, economics, and commercial profiles for key chemicals, chemical processes, product applications, and technologies.
- Four *Quarterly Reports* discussing potential commercial impact of important recent technological developments. Each report develops and presents an economic and commercial analysis for five developments, which Chem Systems technical and consulting staffs believe to be significant.
- A consulting service (specified number of days per year), so that subscribers can discuss items of interest with staff.
- Program Review Meetings given by Chem Systems' staff for all subscribers. These are held annually in the United States, Western Europe, and/or Japan for Japanese subscribers, for the convenience of subscribers in each area.
- Members of *PERP* subscriber companies may attend a course on *Industrial Organic Chemistry and Its Economics*, which provides an overview of the industry.

17.3.   OTHER SOURCES OF PROCESS INFORMATION

Among the most overlooked sources of process information are environmental studies. In the course of describing pollution problems and solutions, detailed information on processes is often provided. Government publications are excellent examples of this type of treatment, for example, some of the studies sponsored by EPA. Other examples are found in the proceedings of the annual Purdue University meetings on industrial waste pollution abatement. Accordingly, although the main thrust of pollution abatement studies is environmental, chemists and engineers seeking process information should consult these studies.

REFERENCES

1. F. A. Lowenheim and M. K. Moran, *Faith, Keyes, and Clark's Industrial Chemicals*, 4th ed., Wiley, New York, 1975.
2. G. T. Austin, Ed., *Shreve's Chemical Process Industries*, 5th ed., McGraw-Hill, New York, 1984.
3. J. A. Kent, Ed., *Riegel's Handbook of Industrial Chemistry*, 8th ed., Van Nostrand Reinhold, New York, 1984.

18 ANALYTICAL CHEMISTRY—A BRIEF REVIEW OF SOME OF THE LITERATURE SOURCES

Tools of potential interest to analytical chemists are discussed in several preceding chapters. In this chapter, some of the more specialized sources that are of particular interest to analytical chemists are briefly discussed.

18.1. SPECTRAL DATA

Analytical chemists are fortunate to have available several large and important published collections of spectral data. The most important are those from Aldrich Chemical Company and Sadtler Research Laboratories. Hummel's collection is more modest but is very useful for polymers and plastics. Brief descriptions of each of these sets follows.

A. Aldrich Chemical Company

Aldrich Chemical Company* has published four popular and valuable spectral *libraries*, all edited by Charles J. Pouchert, their Vice President for Quality Control. The following are available:

   1. *Aldrich NMR Library*—includes over 8500 spectra in two volumes. The second edition appeared in 1983.
   2. *The Aldrich Library of Infrared Spectra*—over 12,000 spectra are included in one volume. The third edition appeared in 1981.
   3. *The Aldrich Library of FT–IR Spectra*—includes over 10,600 FT–IR spectra in two volumes. A digitized personal computer version is also available through efforts with Nicolet Instrument Corporation.* This searches peak tables from the digitized Aldrich–Nicolet FT–IR database and reports the Aldrich FT–IR volume page numbers for spectra that match the unknown's spectral features. Visual comparison with the Aldrich Library confirms identity of the unknown.
   In developing this file, emphasis is on a database that will be useful both today and in the future. For example, future mass storage devices and computer power should permit locating large quantities of data at lower cost and

at higher resolution than is possible today. All data are digitally recorded to help ensure accuracy and reproducibility. Data are also reviewed by both a general spectroscopist and by a specialist in the field.

4. *Aldrich Microfiche Library of Chemical Indexes*—includes seven physical property indexes in which regular Aldrich Chemicals are listed alphabetically and in order of increasing formula weight, melting point, refractive index, density, boiling point, and wavelength of maximum IR absorption. There are molecular formula and alphabetized listings of the Aldrich inventory.

In addition, a new carbon-13 library volume is being edited at Aldrich for publication in late 1987.

The Aldrich spectral libraries are distinguished by their user friendly and careful organization into chemical categories based on increasing complexity of chemical structures. For example, the IR volume is divided into 51 functional group categories such as nonaromatic, aromatic, aromatic heterocyclic, and so on. Each of these main categories is further divided by functionality. Indexes vary depending on volume, but typically include indexes by chemical name, molecular formula, and CAS Registry Number or Aldrich catalog number.

A feature of the bound volumes is their relative affordability; each library of spectra may be purchased for only a few hundred dollars each. Because several spectra appear on each page, valuable laboratory space is saved by these compact volumes. Other advantages include the fact that all compounds listed are commercially available from Aldrich. Every effort is made to list only those chemicals most likely to be used by other chemists, and Aldrich believes that 95% of all such chemicals are now included in their files. In addition, unlike some other sources, the full chemical name is given for all structures; this eliminates the problem of working with sometimes obscure trade names. Aldrich was founded in 1955, and the first collections of its spectra were published in 1970.

B. Sadtler Research Laboratories

Another very well-known source of spectral data is Sadtler Research Laboratories.* The Technical Director is J. Sprouse and the Managing Editor is M. Scandone. Sadtler was founded in 1874, and in 1947 they decided to publish their first collection of spectra. Sadtler is now a division of Bio-Rad Laboratories, Inc., Richmond, CA.

Sadtler says that is has available over 342,000 reference spectra. These can be obtained in printed (book) form and, in addition, much of the data is available for use in other ways, for example, with specific instruments or with personal computers.

C. Hummel

Hummel has written a three-volume treatise on polymer and plastics analysis (1). Volume 1 is a collection of about 1900 infrared spectra of polymers. Volume 2 gives a general introduction to spectra and methods of identification of plastics, fibers, rubbers and resins, and it provides spectra and other data on the most important commercial products. Volume 3 deals with additives and processing aids.

D. Cornell University and Related Spectral Files

Cornell University Computer Services* has available online a Mass Spectral Identification System based on the efforts of Fred W. McLafferty and his research group. This is one of the most extensive mass spectra libraries available anywhere (nearly 80,000 spectra for over 67,000 compounds as of early 1985 and being added to on a regular basis). The Wiswesser Line Notation (WLN) is provided for all compounds included.

Two programs are available, one for matching unknown spectra against the reference system, Probability Based Matching System or PBM, and the other for interpretation of unknowns if a sufficient match is not found (Self-Training Interpretive and Retrieval System or STIRS).

PBM has two unique features: *reverse searching*, which improves identification capability for mixture components, and *weighting* of the mass and abundance data, which improves retrieval performance and provides a quantitative confidence in the degree of match.

The STIRS program is said to be the only generally applicable computer program that identifies the substructures of an unknown compound directly from its mass spectrum. STIRS utilizes 15 classes of mass spectral data that are indicative of different compound types or substructures, and it selects those reference compounds whose spectral behavior most closely resemble those of the submitted unknown. Automatic analysis of functionalities to the retrieved compounds is available for 177 substructures. Examination of STIRS can even make possible reconstruction of part or all of the unknown structure even when the parent compound is not in the database.

A version of this file, consisting of a merged version of NBS's *NIH–EPA Mass Spectral Database* and *Wiley's Registry of Mass Spectral Data*, may be purchased for inhouse use through John Wiley & Sons.* This is known as the *Wiley/NBS Mass Spectral Database*. Included are 79,560 spectra from 68,128 different compounds. Under an agreement with NBS, Wiley is responsible for worldwide marketing of the database, which is being offered in a variety of industry-standard tape formats. For a further discussion of this and other NBS data files, see Chapter 15.

A valuable source of mass spectral, carbon-13 NMR, infrared, and other analytical data is the online *Chemical Information System* (see Section

10.11.A.3). One producer of *CIS* (Fein–Marquart) also has available free-standing, microcomputer search systems, for example, *Powder Diffraction Search/Match System*; this uses the entire database of the International Center for Diffraction Data. Another component of *CIS* is the *Wiley Mass Spectral Search System* (added in 1986).

E.   Other Spectral Files

The *C-13 Nuclear Magnetic Resonance* online database contains numerical values for over 50,000 NMR spectra of organic compounds, that is, chemical shifts, and as far as can be determined, multiplicities, relaxation times, and coupling constants. Molecular and structural formulas are given. Options available include: search of reference spectra by entering lines of a measured spectrum; chemical name or fragment of chemical name; molecular formula; search of similar spectra (Saho Spectra); search of spectra with defined structures or substructures; and estimation of chemical shifts expected on the basis of postulated structures. The producer of this database is the well-known German chemical manufacturer Badische Anilin und Soda Fabrik (now better known simply as BASF). It is available exclusively through the online file known as *INKA* (see Section 10.12.C).

Mass spectra are searchable in the *DARC–Questel* online system (see Section 10.11.A.2) by structure, substructure, and spectral data (peak and intensities, and fragments).

18.2.   *INORGANIC CRYSTAL STRUCTURE DATA*

Inorganic Crystal Structure Data is an online database that contains data on crystal structures of inorganic compounds (and is scheduled to contain, in addition, data on metals and their alloys). Some 23,000 structures are included, and 70% of these are described completely. Data include name, formula, bibliographic data, lattice parameters, space group, atomic position coordinates, temperature factors, *R* factor, measuring conditions, and other pertinent data. Interatomic distances and angles for compounds found in a previous search may be calculated, and pictures of crystal structures may be displayed on terminals with graphic capabilities. This database can be searched through *INKA* (see Section 10.12.C).

18.3.   *ANALYTICAL CHEMISTRY*

As part of its second April issue (issue No. 5 of the year), *Analytical Chemistry* (ACS journal) publishes an extensive series of annual reviews on analytical chemistry. This series started shortly after World War II to help chemists catch up with developments during the war years.

There are two approaches: during even-numbered years, *Fundamentals* reviews are published, with emphasis on techniques; during odd years, *Applied* reviews appear, with emphasis on applications to specific products and industries. These reviews are very comprehensive, but they are not all inclusive. Rather, they are intended to critically review advances and new developments. Authors of the reviews are drawn from universities, industry, and government.

18.4. THE ROYAL SOCIETY OF CHEMISTRY

The Royal Society of Chemistry (RSC) is a major source of analytical information. RSC's Mass Spectrometry Data Center offers a *Mass Spectrometry Bulletin*, which is a monthly current awareness bulletin containing references to the latest published literature on mass spectroscopy and related subjects such as instrument design and techniques, isotopic analysis, chemical analysis, organic chemistry, atomic and molecular processes, surface phenomena, solid state studies, and thermodynamics and reaction kinetics. Some 800 articles are listed and indexed in each issue, but there are no abstracts. The file began in November, 1966. A computer-readable version is also available, and a photocopy service is offered to subscribers by RSC. Online access is possible through Pergamon *InfoLine*® (Section 10.12.B) and the European Space Agency's *IRS* system (Section 10.4).

A related RSC product available in both computer-readable or printed form is *The Eight Peak Index of Mass Spectra* (3rd ed., 1983), which is intended as an aid for identifying unknown mass spectra. Data included are m/z of the eight most abundant ions; intensity of the eight most abundant ions; intensity of parent ion; compound name and formula; spectrum reference number and letter; and CAS Registry Number when available. Spectra from outside sources are included, for example:

1. Imperial Chemical Industries
2. API/TRC (American Petroleum Institute/Thermodynamics Research Center)
3. John Wiley & Sons
4. Dow Chemical Co.
5. Mass Spectral Data Center
6. NIH/EPA Chemical Information System
7. General Mass Spectrometry Literature
8. ASTM

The third edition includes 66,720 mass spectra covering 52,332 compounds. The printed form is in seven volumes, and the tape version is available in a single reel.

In addition to the *Mass Spectrometry Bulletin*, The Royal Society of Chemistry has other periodical publications in the field of analytical chemistry, for example: *Analytical Abstracts*, *The Analyst*, and *Analytical Proceedings*.

The end of 1983 marked 30 years of publication of *Analytical Abstracts*, the only major abstracting publication dedicated solely to analytical chemistry. This important monthly service offers worldwide coverage; over 12,000 abstracts appear each year. Keyword indexes appear in each issue and there are annual subject and author indexes. Cumulative indexes cover 1954–1963 and 1964–1968. Since this publication is now computer typeset, online availability is possible through *InfoLine*® and *SDC*®.

RSC has published the *Annual Reports on Analytical Atomic Spectroscopy* since 1972. These volumes provide critical reports on advances in analytical atomic emission, absorption, and fluorescence spectrometry. Articles, patents, and lectures are among the type of documents included.

Publications of RSC also include a series of monographs in the analytical sciences. Some of their *Specialist Periodical Reports* (annual or biennial review volumes) deal with other analytical topics, for example, mass spectrometry, molecular spectroscopy, NMR, and spectroscopic properties of inorganic and organometallic compounds. In 1982, RSC began publishing an annual index of review articles in analytical chemistry. The first volume contained approximately 580 references to review articles in English, German, and French. The series was discontinued after the 1983 volume.

18.5. AMERICAN SOCIETY FOR TESTING AND MATERIALS

Another major source of analytical methods and information is the American Society for Testing and Materials, now better known as ASTM.* Their *Annual Book of Standards* contains over 7500 standards or methods published in more than 60 volumes. Each year, about 1/3 of all ASTM standards are new, revised, or reapproved. Revisions are accomplished by expert technical committees that are composed primarily of representatives from industry. The standards are published in various forms, for example, specifications, test methods, classifications, definitions, and practices. Many of the standards relate to chemicals. There are 15 sections and a complete index to all volumes. Individual copies of standards may also be obtained, and ASTM also has a number of other pertinent publications.

18.6. OTHER SOURCES OF ANALYTICAL INFORMATION

The most definitive analytical standards for prescription pharmaceuticals, over-the-counter drugs, and other medicines are the *United States Phar-*

macopeia (USP) and the *National Formulary (NF)*. Both *USP* and *NF* are published in one volume by The U.S. Pharmacopeial Convention.*

One of the most important books on analytical chemistry is the extensive *Treatise on Analytical Chemistry* (2), edited by the late Philip J. Elving who taught at the University of Michigan for 32 years, along with I. M. Kolthoff.

Another important volume is the late Al Steyermark's reference book on microchemistry, *Quantitative Organic Microanalysis* (3). Steyermark's last assignment was at Rutgers University, and prior to this he was at Hoffman–LaRoche.

The Association of Official Analytical Chemists, or AOAC,* has as a goal the development, testing, and adopting of standard methods for foods, drugs, cosmetics, pesticides, feeds, fertilizers, hazardous substances, water, forensic materials, and any other product affecting public health and safety, economic protection of the consumer, or the quality of the environment.

One of AOAC's principal publications is *Official Methods of Analysis of the AOAC* (4). This compendium includes over 1700 analytical methods relating to products of interest to the association. It is revised every five years.

An important work is *Fieser and Fieser's Reagents for Organic Synthesis* (5). This well-known, multivolumed series covers the reagent literature. As of 1986, 12 volumes had appeared covering the reagent literature from 1966 through 1984.

An important reference source is the ACS's *Reagent Chemicals–ACS Specifications* (6). The sixth edition appeared in 1981.

Note: Stuart James and N. B. Chapman have written a 225-page book entitled *Analytical Literature* (Wiley, 1986).

REFERENCES

1. D. O. Hummel and F. Scholl, *Atlas of Polymer and Plastics Analysis*, 2nd ed., Verlag Chemie, Weinheim and Deerfield Beach, FL 3 vols. 1978–1984.

2. I. M. Kolthoff, P. J. Elving, and M. M. Bursey, Eds., *Treatise on Analytical Chemistry*, 2nd ed., Wiley, New York, 1978–1983; See also P. J. Elving and J. D. Wineforder, Eds., *Chemical Analysis: A Series of Monographs on Analytical Chemistry and its Applications*, Wiley, New York, 1972 to date.

3. A. Steyermark, *Quantitative Organic Microanalysis,* 2nd ed., Academic, New York, 1961.

4. S. Williams, Ed., *Official Methods of Analysis of the AOAC*, 14th ed., Association of Official Analytical Chemists, Arlington, VA, 1984.

5. *Fieser and Fieser's Reagents for Organic Synthesis*, Vols. 1–11, Wiley, New York, 1967–1983.

6. American Chemical Society, *Reagent Chemicals–ACS Specifications,* 6th ed., Washington, D.C., 1981.

LIST OF ADDRESSES

Aldrich Chemical Company
P.O. Box 355
Milwaukee, WI 53201
  (800-558-9160)

ASTM
1926 Race Street
Philadelphia, PA 19103
  (215-299-5585)

**Cornell University Computer
  Services**
439 Day Hall
Ithaca, NY 14853 (607-256-4981)

John Wiley & Sons
605 Third Avenue
New York, NY 10158

Nicolet Instrument Corporation
5225 Verona Road
Madison, WI 53711

**Association of Official Analytical
  Chemists**
1111 N. 19th Street
Suite 210, Arlington, VA 22209

Sadtler Research Laboratories
3316 Spring Garden Street
Philadelphia, PA 19104

**U.S. Pharmacopeial Convention,
  Inc.**
12601 Twinbrook Parkway
Rockville, MD 20852

19 SUMMARY OF REPRESENTATIVE MAJOR TRENDS AND DEVELOPMENTS IN CHEMICAL INFORMATION

This chapter discusses and briefly summarizes a few of the major representative trends and developments in chemical information. Most are described earlier in this book.

Any forecasts for the future in this field need to be made cautiously, only about five years ahead at the most, for at least three reasons:

1. Because of the rapid changes in the various modes of delivery of chemical information, including who does what and where, large sections of the present edition of this book will probably be outdated in a few years, just as its previous (1979) edition has become now.

2. It is easy to be carried away with speculation about laser, optical, and magnetic disks, and the paperless office or laboratory, without a full understanding of the potential impact, or timing, and likelihood of implementation on a very large scale. While there has been progress in such directions, widespread use of optical disks in chemical information, for instance (which might well change the trend from the centralized worldwide online databases discussed in this book to *local* stores of information), is not yet around the corner.*

3. Traditional forms of chemical information, such as the printed indexes to *Chemical Abstracts* and the printed volumes of *Beilstein* and *Gmelin*, have shown a resilience (popularity) that some find surprising. Whether this will continue unabated into the future is not known for certain, although it is believed to be likely in the absence of indicators to the contrary.

In any event, one of the premises of this book is that effective use of chemical literature and patents, and the associated secondary tools, will continue to be the key to developing and maintaining a competitive edge.

19.1. PROLIFERATION—ONLINE DATABASES

The proliferation of online databases has been remarkable. Online database searching has been greatly expanded and improved. More kinds of data and

*A recent review of CD-ROM databases and technology was published in High Technology, October 1986, pp. 44–51.

files can be searched, and search techniques have become more versatile and flexible. In addition, the number of important online host systems (vendors) has more than doubled (see Chapter 10).

The number of choices available is bewildering. Use of all the sophistication that is now available requires full-time specialization, training courses, and ready access to several different kinds of expensive equipment. This could eventually lead to inappropriate use of some files, other misuse, and unnecessary dilution of resources. Even some specialists are dismayed at this proliferation and the need to learn so much about systems that often overlap and compete. However, others claim that competition will result in lower costs and further improvement of services.

The end result, over the long term, is expected to be a shake out in which only better databases and vendors survive, and others are either merged in, or go out of business. On the other hand, the amount of chemical information that is available online is expected to continue to increase significantly.

19.2. PROLIFERATION—JOURNALS

Proliferation of journals, although far less spectacular than the growth of databases, has been going on for much longer. More journals means more opportunities to have papers published in the first place, and to have them published more quickly.

One potential disadvantage of journal proliferation is that the chemist who wants to be fully informed has more to read. Good papers may become widely scattered in a number of journals, and fewer individual chemists or libraries are likely to be able to afford to subscribe to everything that is potentially important to them. Journals that can compete financially will tend to survive. But some other good journals may tend to be watered down.

19.3. CHEMICAL STRUCTURE SEARCHING

Computer-based searching of chemical structures is now much more fully developed. Major computer-based systems for graphic searching of structures are offered by *CAS ONLINE*, *Questel–DARC*, and Molecular Design, Limited (see Section 10.11.A). These permit chemists to actually draw chemical structures, or to type representations of them, into computer systems. This makes for precise identification, independent of nomenclature, and facilitates searching by structural concepts.

In addition, there are other systems that permit substructure access, but in other ways, usually by means of a special coding system. Examples include:

1. *GREMAS* (see Section 10.11.B.1)
2. Wiswesser Line Notation (very important in its time, but now expected to decline significantly in popularity except for some input; (see Section 10.11.B).
3. Fragmentation and manual codes of the Derwent patent information files (see Section 13.11).
4. IFI/Plenum indexes to U.S. patents, specifically their *Comprehensive Data Base* (see Section 13.13.A).

Finally, generic searching is beginning to be made available by CAS, *DARC*, Molecular Design, Limited, and others. Beginnings of more vigorous, healthy competition among the various systems in this regard are already clearly evident. The next few years will probably see considerable refinement and perfection of generic searching (including full Markush searching capability), coordinated with improved and expanded reaction searching. The appropriate computer programs and equipment seem to be available for such advances, and both the interest and training of chemists are now strong enough to make it happen. Keen interest in potentially relating structures to environmental effects (if any) may be one driving force, as may the already strong and increasing interest in searching patents as generically as possible.

19.4. EFFECTS ON SEARCHING OF FULL-TEXT DATABASES AND OTHER NEW TYPES OF DISSEMINATION

Several years ago there was discussion of computer-based automatic indexing and abstracting. This has now been almost totally replaced by the concept and availability of full-text online databases. Examples of such databases now available include:

ACS primary journals (see Section 10.13).
Kirk–Othmer Encyclopedia of Chemical Technology (see Section 12.2.A)
Mead *LEXPAT* (U.S. patents) (see Section 13.14).

Full text means that every significant word of original text of the journals, patents, or other literature covered is available for online searching, display, and printing (although almost always excluding graphics or diagrams and nontabular figures).

One major advantage is convenient availability of full text in the first place. Another is that, in searching or retrieval efforts, the user is not totally dependent on potential variabilities of terms or concepts selected by professional indexers or abstractors. This can be a matter of some uncertainty no matter how expert the analysis may have been. In searching the full text of a chemical document, the user should be able to more readily find information on a subject that is not the major focus of the paper. Abstracting and indexing by secondary services tends to concentrate on the central thrust and/or novel aspects of the paper. Another advantage is the ability to retrieve such information as details of analytical and other specific techniques, specific numerical data (thermodynamic, biological, etc.), or sources and sponsors for financial support, none of which is normally indexed by most major chemical information services. Some services do provide pointers, tags, and so on, to the primary papers, but not the data themselves.

One disadvantage of the full-text approach is that authors of original documents use a variety of words to express the same concept or describe the same compound. Original document authors may not pay attention to concepts, and hence retrieval of these can be difficult. Another potential disadvantage is possible retrieval of nonpertinent information. Whatever is mentioned in the text, however casually, whether in reference to prior experiments, speculative conclusions, or future planning, is retrieved out of context without any benefit of prior screening such as that performed by a trained indexer.

Except for full-text online databases, advances in electronic dissemination of documents have been modest. One can speculate, for instance, whether Chemical Abstracts Service (CAS) will continue to produce its 12th (1987–1991) or 13th (1992–1996) *Collective Indexes* in their traditional printed forms, but these types of "crystal-ball" predictions can be very wrong. There were those who, just a short time ago, had predicted that CAS would not produce the 11th (1982–1986) *Collective Index*, yet advance, prepublication sales of that index testify to the contrary. However, a significant increase in the volume of chemical literature available in full-text form can be anticipated over the next few years.

19.5. MICROCOMPUTER USE

The full potential of micro- (or personal) computers is only beginning to be realized. It is already clear that the long term impact will be enormous. The exact nature of the impact is not certain, but it will surely include improved information storage and retrieval for both external and internal (personal) files.

With regard to external online databases, two major online vendors (*DIALOG®* Information Services and *BRS®*) now offer special off-hours (night) systems with lower rates especially to encourage use by those who have access to personal (and other) computers during these hours (see Section 10.15). This trend can be expected to continue and expand, if the economics are sufficiently attractive.

Some companies and other organizations have started to provide their chemists with personal computers to encourage home use of both external and internal (company files) databases, as well as to facilitate the concept of electronic mail. This may be a bonanza for compulsive hard workers (workaholics) who will not need to spend long hours in the office, if they are not involved full time with the experiments in the laboratory. On the other hand, home, with its potential distractions, may not provide a good environment for productive work.

Significant progress is beginning to be made in facilitating *downloading* (or electronic capture) of pertinent subsections from very large online databases, such as those of *Chemical Abstracts*, into microcomputer memories. Computer programs are now available to facilitate easy logging on (entry) to external online databases and to aid in their use once log on is completed (see Section 10.15). The *Institute for Scientific Information*® is an example of an organization that has developed this type of software.

Computer-readable subsets, or the complete contents, of selected reference works and databases are beginning to be made increasingly available for storage and searching on personal or minicomputers, completely independently of centralized online systems. Among the organizations now offering this approach are Molecular Design, Limited and Fein–Marquart Associates, Inc. (see Section 10.15).

As to internal files (a user's own records), computer programs are now available to help chemists index these files and thereby keep track of their research notes, reprints, and other papers. However, the real progress is the ability to input one's own data (text as well as numerical data) into computer storage and to manipulate, rearrange, extract, and so on, in addition to the ability to retrieve or simply keep track of that information. In connection with this, increasing availability of word-processing systems is important. Even today, many chemists keyboard, edit, and reedit not only their reports and papers, but also correspondence. Machine transfer (electronic mail) of papers and correspondence from one chemist to another, not necessarily in the same location or even in the same organization, is already an attractive option available to many. However, extensive personal (face-to-face) contacts among chemists will remain essential to scientific advances.

19.6. CHEMICAL ABSTRACTS SERVICE AND OTHER ABSTRACTING AND INDEXING SERVICES

Chemical Abstracts Service has moved in a number of important and well-coordinated directions at once, as described in more detail in Chapters 6 and 7. These moves put CAS in the position of the only true *full-service* organization in chemical information and make it the most dominant force in the chemical information industry.

On the one hand, CAS has expanded and improved its extremely popular current alerting series, *CA SELECTS*. It now offers quick copies or loans of virtually all recent documents it covers. In addition, it has initiated *CAS ONLINE*, which provides both graphic structure searching capability and full-fledged chemical substance, subject, and author index searching.

On the other hand, pricing for the full *CA*, by ACS policy, is now geared primarily to budgets of well-financed institutions, and optimum use of *CAS ONLINE* virtually requires that the user obtain special training. Thus, CAS is becoming at once both more personalized and more institutionalized. At the same time, a number of important enhancements to *CA* performance have been made, most notably improvement in speed of issue of printed indexes.

The most likely outlook for *CA* is further emphasis on use of computers in its production and especially in its use. Expansion and enhancement of *CAS ONLINE* and of STN^{SM} International, the Scientific and Technical Information Network, (see Section 10.12.C) is likely. Further expansion and improvement of *CA SELECTS*, and of customer-defined services for selective dissemination of information (see Section 4.3.C), can be anticipated.

Chemists can hopefully expect continued evolution and reasonable proliferation of additional smaller competing abstracting and indexing services (especially online), which are intended to meet the needs of formal or informal *special interest* groups, such as safety and health, chemical–biological activities, patents, and in branches of chemistry distinguished by rapid growth and high interest (see Chapter 8). If this forecast is valid, improved coverage of chemical information should result, but the increasing number of tools could potentially prove both too complex and expensive for the average chemist, except in the largest companies. Individuals in such companies can be expected to have an increasing advantage of significant proportions over other chemists in sophistication, power, and completeness of chemical information storage, searching, and retrieval, except in highly specialized fields of chemistry where personal contacts are sufficient and, in fact, may predominate.

19.7. SEARCHING OF PATENT PUBLICATIONS

Patent searching has been greatly improved by introduction of important new online patent databases (some think almost too many) and by improvement of existing services. New services introduced include: (a) Mead *LEXPAT* (see Section 13.14); (b) Pergamon *PATSEARCH®* (see Section 13.15); (c) Pergamon *INPADOC* File (see Section 13.17); and (d) U.S. Patent Office "front page" files (see Table 13.1). Patent services that have been improved and expanded include those of CAS, Derwent, and IFI.

The *LEXPAT* service provides full-text search capabilities for recent U.S. patents and is a particularly important advance.

Availability of *INPADOC* files is noteworthy because this is a very comprehensive concordance to equivalents in many countries.

U.S. Patent Office files (front page information from about 1970 or later to date) are available online through no less than four competing files, for example, those of IFI/Plenum, Derwent, *BRS®*, and Pergamon. These files, along with a special file offered by IFI only, also facilitate searching for patent citations.

Two new multinational patent systems, those of the Patent Cooperation Treaty and the European Patent Office, are now well established, and their patent publications are well indexed and abstracted in both printed and online sources (see Section 13.20).

Despite all of this, there is continued and growing concern about keeping up with new patent information and more specifically, with keeping up with a flood of unexamined patent applications from Japan. Meanwhile, the U.S. Patent Office is still struggling to keep up with new U.S. applications and is attempting to speed its actions, with only very modest success so far. However, it now has well-defined plans to automate fully by about 1990, and a team consisting of Planning Research Corporation and Chemical Abstracts Service has a contract to help implement the plans.

19.8. *KIRK–OTHMER*

The third edition of the highly important *Kirk–Othmer Encyclopedia of Chemical Technology* has been completed. The total set comprises 26 volumes. There is also a *concise* version in one volume, and there are one–volume groupings of *Kirk–Othmer (KO)* articles on special topics. Availability of all volumes online should prove useful to many. It will be interesting to watch whether future revisions of *KO* (or of other major treatises) appear first, or even only, online. *Kirk–Othmer* is recommended as the preferred starting point for research workers beginning new projects or for others dealing with unfamiliar technologies.

19.9. LANGUAGES

Publishing of chemical and other scientific literature in the English language has maintained and strengthened its surprisingly strong position over the years, just when it almost appeared as if the linguistic abilities of English reading chemists would be seriously challenged by literature in Russian, and when courses of study on the reading of scientific Russian appeared to be essential for the future. Even some of the classical sources traditionally in German, for example, *Beilstein* and *Gmelin*, are now published in English, and the new edition of *Ullmanns* is in English (see Chapter 12). The dominant language of the book and journal chemical literature is English, and it is the language of choice for those who want to be recognized and read by their peers.

The ability to handle literature in one or more major foreign languages is still an asset for any chemist, but the relative importance of this skill has declined, except for those connected with specialized assignments such as in defense efforts or in abstracting and indexing. However, in addition to Russian, one non-English language that may still be "troublesome" is Japanese because of the large number of unexamined patent applications being issued in that technologically important country. More than one half of all patents covered by *CA* originate in Japan.

19.10.   COSTS AND REPACKAGING

The subscription and purchase costs of most major printed chemical information sources have been increasing rapidly, in some cases well beyond *normal* inflation. The result has been that more expensive sources have become increasingly available primarily (and perhaps only) to chemists in large, well-financed organizations, and less affordable to individual chemists and those in smaller or less well financed organizations. Even some larger organizations have been hard pressed to keep up with purchasing habits and policies of previous years. Thus, what was once a full set of *Beilstein* or *Gmelin* may, unfortunately, need to be allowed to *temporarily* lag in acquisition of newer volumes, or even lapse. However, both *Beilstein* and *Gmelin* may be available online during the next few years. Discounts are frequently provided to colleges and universities, but this may not always be enough.

Another significant trend, counter to the one just described, is increasing availability of relatively low-priced or medium-priced repackaged subsets, for example, the extremely popular *CA SELECTS* and *ASCATOPICS*® series (see Section 4.4.C), the separately available special interest volume groupings and one volume version of the *Kirk–Othmer Encyclopedia of Chemical Technology* (see Section 12.2.A), and the separately available categories of the *Dictionary of Organic Compounds* (see Section 12.4.B). Related to this, is increasing availability of full-text online databases (see Section 10.13) as, for example, those covering ACS primary journals and U.S. patents, which is another form of repackaging. Chemists who have online access to these databases need not concern themselves with whether or not original printed versions are immediately available, except for graphics or other material not in the full-text databases. Costs of searching online databases have increased somewhat over the last few years but at a rate that seems to be acceptable. Reduced rates for use during off-peak hours on special systems may become increasingly popular. One unresolved question is long-term impact, if any, on costs of online databases of increases in costs of printed full-text originals and vice versa.

19.11. SAFETY–ENVIRONMENT

Newer sources of safety information emphasize safe handling and disposal of hazardous materials, especially wastes. The U.S. Defense Department has contracted to establish a Hazardous Materials Technical Center (HMTC) to make this type of data more readily available, but existence of HMTC has, until recently, been little known.

The National Library of Medicine (NLM) seems to be clearly expanding its role from coverage of medical and toxicity data, to include also other pertinent safety data such as that processed by HMTC and the International Labor Office. NLM files, which now also include data from the National Referral Center, can be readily accessed online, and there is also a Toxicology Information Response Center supported by NLM.

Certain U.S. government rules regarding Material Safety Data Sheets (required in connection with shipment of many chemicals) went into full effect in 1985. These sheets can be expected to be an increasingly important source of property and safe handling information.

Newer toxicity prediction tools, for example, those based on structure-activity relationships, are available, but observers regard many of these as relatively embryonic and just one of several potential aids to judgments of toxicologists.

The area of safety–environment is expected to be of continuing and increasing importance from the perspective of chemical information. In particular, more data will be needed on the effects of potentially hazardous wastes and on the most appropriate methods for handling, storage, and disposal of such wastes.

19.12. PHYSICAL AND OTHER PROPERTIES

Considerably more attention is being paid to issues of data quality, although much progress is still needed. However, improvements in availability of property data are here or on the near horizon.

A surprising number of new computer-based and other systems for calculating or otherwise determining properties are now readily available for use. One of the most important of the newer systems is that of Design Institute For Physical Property Data (DIPPR), which operates under the aegis of the American Institute of Chemical Engineers, with cooperation of the chemical industry and the U.S. National Bureau of Standards (Section 15.15).

More new online databases containing hard data (numerical values) are becoming available. Leaders in this regard include the online database and systems of Fein–Marquart (*CIS*) as described in Chapter 10 and Technical Database Services as described in Chapter 15. Additional important efforts relate to corrosion, electronics materials, and plastics (see Section 15.16).

An increasing number of databases with hard data is anticipated, a trend that is expected to be driven by manufacturers trying to inform potential customers about their products more quickly and efficiently. Finally, coordinated systems for material properties are being actively studied (Section 15.17), but so far there has been little progress of interest to chemists.

19.13.   SPECIALISTS

Effective use of chemical information increasingly is (1) being made easier for all individual chemists by availability of such tools as *CA SELECTS* and *ASCATOPICS®*, and (2) being made more complex, and more the domain of the chemical information specialist, by continuing development of sophisticated online databases. These are conflicting, contradictory trends. However, a third trend, increasing availability and power of personal computers and associated software, should provide the potential of much greater capabilities for almost all chemists. This may tend to significantly slow the rate of growth of the influence of the specialist over the long term, even though the specialist will probably occupy a role of somewhat more importance than at present. Regardless of other developments, personal familiarity with original (primary) chemical information sources, as well as with principal chemical information tools (secondary sources), will continue to be essential for all chemists who want to develop and maintain a competitive edge.

19.14.   PRODUCTIVITY, KNOWLEDGE, INNOVATION, AND CREATIVITY

Will all of the changes in chemical information described in this book affect the quality of chemical research and development? It is safe to say that chemists should now potentially be able to locate and access most of the newer, and some of the older, chemical information more fully and precisely than was ever before thought possible. They should also have the capability of organizing findings, both their own and those of others, more efficiently than ever before, and of communicating these results in more rapid ways directly to one another. If findings can be organized more readily for ease of understanding, this should mean improved opportunity to build on a foundation of previous work. Further, the chances of missing something important due to lack of search tools would appear to be getting considerably less remote. Thus, increases in productivity can be logically expected. What is most important, however, is impact on the processes of knowledge acquisition, innovation, and creativity. One thing is certain: All of the new chemical information tools are simply a means to an end. Except for their developers, employees, and some chemical information specialists, and except as they may require new chemicals (materials) for their operation, they are not an end unto themselves. The true goal is enhancement of human knowledge, innovation, and creativity.

NAME INDEX

Please note: This index does not include (1) the names of the project leaders of the National Bureau of Standards Standard Reference Data System Data Centers and Projects beginning on page 309, or (2) the names of the Patent Depository Managers in Table 5.1.

SUBJECT INDEX